Lecture Notes in Artificial Intelligence   11100

Subseries of Lecture Notes in Computer Science

More information about this series at http://www.springer.com/series/1244

Lev Rozonoer · Boris Mirkin
Ilya Muchnik (Eds.)

Braverman Readings in Machine Learning

Key Ideas from Inception to Current State

International Conference Commemorating
the 40th Anniversary of Emmanuil Braverman's Decease
Boston, MA, USA, April 28–30, 2017
Invited Talks

 Springer

Editors
Lev Rozonoer
West Newton, MA
USA

Boris Mirkin [iD]
National Research University Higher School
  of Economics
Moscow
Russia

Ilya Muchnik
Rutgers University
Piscataway, NJ
USA

ISSN 0302-9743   ISSN 1611-3349 (electronic)
Lecture Notes in Artificial Intelligence
ISBN 978-3-319-99491-8   ISBN 978-3-319-99492-5 (eBook)
https://doi.org/10.1007/978-3-319-99492-5

Library of Congress Control Number: 2018952242

LNCS Sublibrary: SL7 – Artificial Intelligence

Cover illustration: Example detector data from the Daya Bay Reactor Neutrino experiment (p. 287, Fig. 14)

This Springer imprint is published by the registered company Springer Nature Switzerland AG
The registered company address is: Gewerbestrasse 11, 6330 Cham, Switzerland

Preface

The monograph is dedicated to the memory of Emmanuil Markovich Braverman (1931–1977), a pioneer in developing the machine learning theory. The very term "machine learning" was introduced by him (see, for example, Braverman and Arkadiev, Machine Learning for Classification of Objects, Moscow, published in 1971 by Nauka Publishers, (1st Edition 1964, in Russian). E. M. Braverman was the first to propose a geometrical interpretation of the problem of pattern recognition as a problem of finding compact classes of points in the feature space. He, with co-authors, proposed the method of potential functions, later dubbed "kernels," to perform what is called "the kernel trick," so that the complex problem of separation of patterns is converted into a linear hyperplane separation problem. Main theorems stating convergence of the method of potential functions for pattern recognition, both supervised and unsupervised, have been proved by Braverman and co-authors. Overall, Braverman's work featured an emphasis on the geometric nature, as well as interpretational aspects, of the proposed approaches and methods. E. M. Braverman's work was halted by his untimely death at the age of 46, at the very dawn of data science.

This volume is a tribute to Braverman's memory and an overview of some of his ideas and approaches today. These papers were presented at the conference "Braverman Readings in Machine Learning: Key Ideas from Inception to Current State" held during April 28–30, 2017, at the Northeastern University in Boston. This conference marked the 40th anniversary of the death of E. M. Braverman. The conference program and related materials can be found on the conference website at https://yandexdataschool. com/conference/2017/about. Both the conference and this publication were supported by Yandex, a pioneering search technology company and provider of intelligent products and services powered by machine learning. From its very inception more than 20 years ago, Yandex has been honoring the legacy of E. M. Braverman.

The collection is divided in three parts.

The first part bridges the past and present. Its main content relates to the concept of kernel function and its application to signal and image analysis (Sulimova and Mottl; Mottl, Seredin and Krasotkina; Vovk, Nouretdinov, Manokhin and Gammerman; and Aizerman, Braverman and Rozonoer), as well as clustering (Mirkin), bringing together the first, naive, attempts with matured mathematics-loaded approaches. Rozonoer describes an approach to the issue of data complexity. Mandel develops an interesting computational approach to causality modeling by using "immediate" causes as regressors.

The second part presents a set of extensions of Braverman's work to issues of current interest both in theory and applications of machine learning. Applications range from natural sciences (Bottou, Arjovsky, Lopez-Paz, and Oquab) to drug design (Borisov, Tkachev, Buzdin, and Muchnik) to robot motion planning (Lumelsky). Among machine learning theory issues are one-class learning (see E. Bauman and K. Bauman), various distances between distributions (Bottou, Arjovsky, Lopez-Paz,

Oquab), and deep learning mechanisms (Sadowski and Baldi, as well as Agostinelli, Hocquet, Singh and Baldi).

The third part, on E. M. Braverman's personality and his circle, is intentionally made brief. It includes short essays by a friend, L. Rozonoer, a student, M. Levin, and a colleague, B. Mirkin. These also shed light on another aspect of E. M. Braverman's research — his deep insights into modeling of an unbalanced economy, like that of the Soviet Union back in the 1950s to 1970s, in which no free price adjustment was permitted, and the only method of balancing was by using production and consumption quotas (see Levin). The final material is a list of refereed papers published by E. M. Braverman in the *Automation and Remote Control* journal and available in English.

The material in this volume is instructive for several segments of the audience. Most of all, it is oriented at students, developers, and practitioners in machine learning and data analysis. They will find in the volume a number of constructive ideas regarding issues of current interest for the analysis of data of complex structure. The presented material provides useful insights into the role of parameters, such as the number of clusters or a threshold, which are usually considered sealed in the algorithms, but in fact should be open to user–machine interaction. The book will be interesting to historians too.

Reference

A.G. Arkadiev, E.M. Braverman (1971) *Machine Learning for Classification of Objects*, Moscow, The Main Editorial of Physics and Mathematics Literature, Nauka Publishers, 192 p. (In Russian, 1st Edition 1964).

May 2018

Lev Rozonoer
Boris Mirkin
Ilya Muchnik

Organization

Program Committee

Boris Mirkin (Chair)	National Research University Higher School of Economics in Moscow, Russia, and Birkbeck, University of London, UK
Pierre Baldi	University of California, Irvine, USA
Leon Bottou	Facebook, USA
Alexander Gammerman	Royal Holloway, University of London, USA
Ilya Muchnik	Rutgers University, NJ, USA and Yandex School of Data Analysis, Moscow, Russia
Vladimir Vapnik	Facebook, USA
Vladimir Vovk	Royal Holloway, University of London, UK

Organizing Committee

Elena Bunina	Yandex School of Data Analysis, Moscow, Russia
Ilya Muchnik	Yandex School of Data Analysis, Moscow, Russia
Evgenia Kulikova	Yandex School of Data Analysis, Moscow, Russia
Elena Schiryaeva	Yandex School of Data Analysis, Moscow, Russia

Acknowledgments

The Editors express their sincere gratitude to the organizations and individuals whose help was crucial in bringing the Conference and the volume into reality.

Our biggest thanks go to

Yandex

a pioneering search technology company and provider of intelligent products and services, and Yandex' representative, Evgenia Kulikova, Head of Academic Programs Services, Yandex, Russia.

We are also grateful to: Prof. Maxim Braverman, Northeastern University, USA, Prof. Alexander Gammerman, Royal Holloway, University of London, UK, and Benjamin Rozonoyer, Brandeis University, USA.

Contents

Personal and Beyond

Bridging Past and Future

Potential Functions for Signals and Symbolic Sequences

Valentina Sulimova[1(✉)] and Vadim Mottl[2]

[1] Tula State University, Tula 300012 Lenin Ave. 92, Russia
vsulimova@yandex.ru
[2] Computing Center of the Russian Academy of Sciences,
Moscow 119333 Vavilov St. 40, Russia
vmottl@yandex.ru

Abstract. This paper contains a comprehensive survey of possible ways for potential functions design on sets of signals and symbolic sequences. Significant emphasis is placed on a generalized probabilistic approach to construction of potential functions. This approach covers both vector signals and symbolic sequences at once and leads to a large family of potential functions based on the notion of a random transformation of signals and sequences, which can underlie, in particular, probabilistic models of evolution of biomolecular sequences. We show that some specific choice of the sequence random transformation allows to obtain such important particular cases as Global Alignment Kernel and Local Alignment Kernel. The second part of the paper addresses the multi-kernel situation, which is extremely actual, in particular, due to the necessity to combine information from different sources. A generalized probabilistic featureless SVM-based approach to combining different data sources via supervised selective kernel fusion was proposed in our previous papers. In this paper we demonstrate significant qualitative advantages of the proposed approach over other methods of kernel fusion on example of membrane protein prediction.

Keywords: Potential (Kernel) functions · Featureless approach
Probabilistic models · Sequence alignment · SVM · Multi-kernel learning
Membrane protein prediction

1 Introduction

Sequences of different length are typical objects of data analysis. There are a variety of well-known applications of sequence analysis, such as on-line signature verification, speech recognition, medical signal analysis, protein annotation from amino acid sequences forming them, and so on.

Elementary primitives constituting a sequence may be real vectors or symbols from some finite alphabet. In the former case sequences are vector signals, and in the latter one they are traditionally named symbolic sequences.

Historically, methods of signal and symbolic sequence analysis were developing independently from each other. But by virtue of the fact that mathematical structures of

© Springer Nature Switzerland AG 2018
L. Rozonoer et al. (Eds.): Braverman Readings in Machine Learning, LNAI 11100, pp. 3–31, 2018.
https://doi.org/10.1007/978-3-319-99492-5_1

such objects are very similar, methods of their analysis are based on a number of common principles. Therefore it would be useful to consider them together.

Here and later, we will use the term 'sequence' instead of terms 'signal' and 'symbolic sequence' when a statement will hold true for both of them.

It should be noticed, that all practical applications of sequence analysis have an important common characteristic. It is very hard for all of them to choose, in advance, some finite set of features, forming a space, which would be convenient for further sequence analysis, ensuring it's high quality.

In this connection, the most appropriate way for sequence analysis is the featureless approach [1, 2]. According to this approach, each sequence is represented via pairwise comparison of it with other sequences, eliminating, thereby the need of introducing feature description for individual sequences.

But it is evident that not any comparison measure can be useful for applied sequence analysis.

First of all, a sequence comparison measure must be appropriate for the specificity of the respective application. This means that it must form a space, which would satisfy the compactness hypothesis [3], i.e. such a space, in which sequences belonging to the same class (for example signatures of the same person or proteins, performing the same function) should form a compact set. At the same time, sets corresponding to different classes should be as distant from each other as possible.

Second, we would like to have the possibility of using effective and convenient linear methods, such as Support Vector Machines (SVM) [4], which had been initially designed for feature spaces.

The problem of using linear methods in the featureless situation can be solved by constructing appropriate kernel functions (or potential functions, in the original terminology proposed by Braverman [5]).

Any two-argument function can be used as potential function if it satisfies Mercer's conditions [6], namely, for any finite set of sequences the matrix of its values must be positive definite.

Any potential function embeds the respective set of sequences into a hypothetical linear space, in which it plays the role of inner product [7]. Besides, any potential function is bilinear relative to summation and multiplication by coefficient, so, its values can be easily used instead of inner products of feature vectors in the classical SVM, proposed by Vladimir Vapnik [4].

This paper contains a comprehensive survey of possible ways of design of potential functions on sets of signals and symbolic sequences of different length. Significant emphasis is placed on a generalized probabilistic approach. This approach covers both vector signals and symbolic sequences at once, and leads to a large family of potential functions based on the notion of a random transformation, which can define, in particular, probabilistic models of evolution in biomolecular sequences.

We show that some specific choice of the sequence random transformation allows to obtain such important particular cases as Global Alignment Kernels for vector signals [9] and biological sequences [8], as well as Local Alignment Kernels [11].

The second part of the paper addresses the multi-kernel situation, which is extremely actual because of the possibility to introduce a number of different potential functions and, in particular, due to necessity of combining information from different data sources.

In our previous papers, we proposed a generalized probabilistic featureless SVM-based approach to combining several potential functions via supervised selective kernel fusion. In this paper, we demonstrate significant qualitative advantages of the proposed approach over the other methods of kernel fusion on example of membrane protein prediction.

2 Design of Potential Functions (Kernels)

2.1 Potential Functions and Their Properties

Let Ω be a set of objects, in particular, a set of signals or symbolic sequences, maybe, of different length.

Function $K(\omega', \omega''), \omega', \omega'' \in \Omega$ is potential (kernel) function on a set Ω if it satisfies for Mercer's condition [6], which boils down to the requirement to be considered as inner product of some hidden hypothetical features of objects $\varphi_\omega(\vartheta)$ in a hypothetical linear space of all functions $\varphi(\vartheta)$ of some variable $\vartheta \in \Theta$:

$$K(\omega',\omega'') = \int_{\vartheta \in \Theta} \varphi(\omega', \vartheta)\varphi(\omega'', \vartheta)d\vartheta. \tag{1}$$

Thus, any potential function is symmetric $K(\omega',\omega'') = K(\omega'',\omega')$.

In practice, it was determined that the condition (1) holds true if for any finite set of objects $\{\omega_j, j = 1, \ldots M\} \subset \Omega$ the matrix $\{K(\omega_i, \omega_j), i,j = 1, \ldots, M\}$ is positive definite.

Any potential function embeds the set of sequences Ω into some hypothetical linear space $\tilde{\Omega} \supset \Omega$ and plays the role of inner product in it [7].

Sum and product of any two potential functions $K_1(\omega',\omega'')$ and $K_2(\omega\prime,\omega'')$ its are also potential functions:

$$K_3(\omega',\omega'') = K_1(\omega',\omega'') + K_2(\omega',\omega''), \tag{2}$$

$$K_4(\omega',\omega'') = K_1(\omega',\omega'')K_2(\omega',\omega''). \tag{3}$$

Besides, any potential function is bilinear relative to summation

$$K(\omega_1, \omega_2 + \omega_3) = K(\omega_1, \omega_2) + K(\omega_1, \omega_3)$$

and multiplication by coefficient

$$K(c\omega_1, \omega_2) = cK(\omega_1, \omega_2)$$

Finally, it follows from (1) that $K(\omega,\omega) \geq 0$.

So, the values of a potential function can be easy used instead of inner products of feature vectors, in particular, in the Vladimir Vapnik's classical SVM [4]. The respective kernel-based formulation of the two-class SVM is described in Sect. 3.

2.2   Feature-Based Potential Functions

It is evident, that potential functions can be easily constructed on the basis of feature vector representations.

Indeed, if $\mathbf{x}(\omega_i) = [x_{i,1}, \ldots, x_{i,n}]^T = \mathbf{x}_i \in R^n$ and $\mathbf{x}(\omega_j) = [x_{j,1}, \ldots, x_{j,n}]^T = \mathbf{x}_j \in R^n$ are vector representations of sequences ω_i and ω_j, then function

$$K(\omega_i, \omega_j) = K(\mathbf{x}_i, \mathbf{x}_j) = \mathbf{x}_i^T \mathbf{C} \mathbf{x}_j, \tag{4}$$

is potential function, if $\mathbf{C}$ is some positive definite matrix $[n \times n]$. Another popular feature-based potential function is the radial basis potential function (RBF) [5, 12], which can be defined as

$$K(\omega_i, \omega_j) = \exp[-\alpha \rho^2(\omega_i, \omega_j)], \tag{5}$$

where $\rho(\omega_i, \omega_j)$ is the Euclidian distance between ω_i, ω_j, for instance,

$$\rho(\omega_i, \omega_j) = \sqrt{\sum_{i=1}^{n} \left(\mathbf{x}(\omega_i) - \mathbf{x}(\omega_j) \right)^2}. \tag{6}$$

The trivial potential function (4) immediately defines a linear scalar function $z(\mathbf{x}) = K(\mathbf{a}, \mathbf{x}) = \mathbf{a}^T \mathbf{x}$ in R^n. But function of the same structure with respect to potential function (5)–(6) $z(\mathbf{x}) = K(\mathbf{a}, \mathbf{x}) = \exp[-\alpha \rho^2 (\mathbf{a}-\mathbf{x})^T (\mathbf{a}-\mathbf{x})]$, which is linear in R^∞, will be nonlinear in R^n. Thus, potential functions incorporate nonlinearity into linear methods.

However, in the case of sequence analysis, it is hard to form a feature space, which would satisfy the compactness hypothesis, i.e., adaptability of all these functions is very restricted.

2.3   Similarity or Dissimilarity-Based Potential Functions

Similarity and Dissimilarity Measures for Sequences. Traditional sequence comparison measures, such as Needleman-Wunsch [14] and Smith-Waterman [15] similarity measures, as well as a number of other ones [16–18] for biological sequences, and extremely popular Dynamic Time Warping (DTW) algorithm [19] with its modifications [20–22] for vector signals, are based on finding an optimal pairwise alignment of sequences being compared.

An alignment $\mathbf{w}$ of two sequences is understood as bringing them to a common length by inserting gaps ("-") into some positions, if symbolic sequences are compared (Fig. 1a), or by local stretching the axis, when vector signals are compared (Fig. 1b).

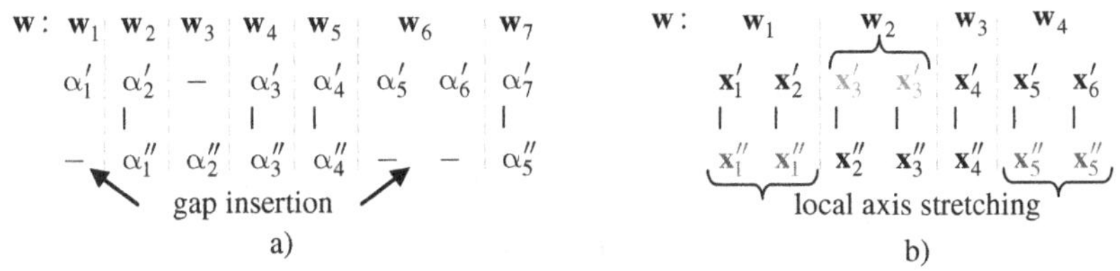

Fig. 1. Examples of sequence alignments for (a) symbolic sequences and (b) vector signals.

Generally speaking, any two sequences can be aligned by a number of different ways. But traditionally some criterion function $S(\omega', \omega''|\mathbf{w})$ is defined, which determines quality of any alignment $\mathbf{w}$ and allows to choose the optimal alignment:

$$\hat{\mathbf{w}} = \arg\max_{\mathbf{w}} S(\omega', \omega''|\mathbf{w}) \text{ or } \hat{\mathbf{w}} = \arg\min_{\mathbf{w}} S(\omega', \omega''|\mathbf{w}).$$

The value $S(\omega', \omega'') = S(\omega', \omega''|\hat{\mathbf{w}})$ can be considered as similarity or dissimilarity measure of two sequences ω', ω'' in accordance with the accepted criterion.

But the resulting two-argument functions may be not positive definite. And so, they could not be strictly used as potential functions in SVM [23]. But at the same time, there are, at least, three known ways to use them for constructing potential functions.

Potential Functions on the Basis of Secondary Features. This way to construct a potential function is based on the fact that each sequence ω can be represented as a vector of so-called secondary features - pairwise similarities (or dissimilarities) of ω to other sequences of some set $\{\omega_i, i = 1, \ldots, M\}$:

$$\mathbf{x}(\omega) = [S(\omega, \omega_1) \ldots S(\omega, \omega_M)]^T \quad (7)$$

The obtained feature representation (7) can be used to construct a potential function in accordance with Sect. 2.2.

Such an approach is applied in a number of papers [24–27], but, it should be noticed, that it leads to the necessity to save the full training set instead of an only small part of it, which is called the subset of support vectors. This approach requires to compare each new sequence to all the sequences of the basic set $\{\omega_i, i = 1, \ldots, M\}$.

Correction of Sequence Similarity Matrix. One more way to obtain potential function on the basis of existing sequence comparison measure is to provide it the required properties of inner product [26]. It can be easily done for any similarity measure $S(\omega', \omega'')$ and some finite set of objects $\{\omega_i, i = 1, \ldots, M\}$:

$$K(\omega_i, \omega_j) = S(\omega_i, \omega_j) - \lambda_{\min}(S)\mathbf{I} \quad (8)$$

where $\lambda_{\min}(S)$ is the minimal eigenvalue of the similarity matrix $[S(\omega_i, \omega_j), i, j = 1, \ldots, M]$.

For any finite set $\{\omega_i, i = 1, \ldots, M\}$ matrix $\{K(\omega_i, \omega_j), i, j, = 1, \ldots, M\}$ is positive definite and so, it can be used as potential function matrix.

The essential disadvantage of this approach consists in the necessity to have the whole set of sequences at once to have the possibility to correct the full matrix. In addition, it is evident, that such correction.

Heuristic Transformation of the Sequence Dissimilarity Measure to a Radial Basis Potential Function. In a number of papers it is supposed, that sequence dissimilarity measure is similar to the Euclidean metric [28]. It can be proved that for any dissimilarity measure potential function (5) is positive definite kernel for sufficiently large value of the coefficient α [29]. This fact allows to obtain positive definite matrix and to use it in SVM. However, it may happen that this property will be achieved only with too large α, which would be of no practical sense.

Such a way does not require to memorize the full set of basic sequences and compare new objects to all of them. But, in general, if useful value of α is not sufficiently large, such transformation can lead to the presence of negative eigenvalues and, of course, it is more preferable to construct potential function strictly, without heuristic transformation and the necessity of having the whole set of sequences at once.

2.4   Alignment-Based Potential Functions (Convolution Kernels)

The disadvantages of potential function construction ways described in the Sects. 2.2 and 2.3, make the more desirable direct construction of potential functions, without introducing interim notions, such as feature vectors and similarity or dissimilarity measures.

Generally speaking, there are known a big number of approaches to construction of potential functions for sequences [26, 30–34, 43, etc.]. The respective survey can be found, for example, in [43].

Most well-performing and, so, most interesting potential functions are the convolution ones [8].

The Conception of Convolution. The term "convolution" denotes a special algebraic operation on sequences, which is based on their decomposition into subsequences by all possible ways and multiplication of some potential functions for the respective subsequences.

Let $\omega', \omega'' \in \Omega$ be two sequences, $\omega' = \omega'_1 \omega'_2$ - decomposition of sequence ω' into two subsequences $\omega'_1, \omega''_2 \in \Omega$.

Then convolution potential function, convolving two potential functions K_1 and K_2, is the function

$$K(\omega', \omega'') = \sum_{\omega' = \omega'_1 \omega'_2, \ \omega'' = \omega''_1 \omega''_2} K_1(\omega'_1, \omega''_1) K_2(\omega'_2, \omega''_2), \qquad (9)$$

where the sum is to be computed over all possible decompositions of sequences ω', ω''.

Convolution potential functions are especially popular in bioinformatics. There exist a number of convolution potential functions, such as Global Alignment Kernel for vector signals [9], Global Alignment Kernel [8] (in some papers it is also named Sequence Alignment Kernel [10]) and Local Alignment Kernel for biological sequences [11].

All of them are introduced in convolution terms, but it is more convenient to express convolution potential functions in terms of sequence alignment.

Convolution potential functions in terms of sequence alignment. The main principle. Let $\boldsymbol{\omega}' = (\omega_1', \omega_2', \ldots, \omega_{N'}')$ and $\boldsymbol{\omega}'' = (\omega_1'', \omega_2'', \ldots, \omega_{N''}'')$ be two sequences.

Any alignment $\mathbf{w} = \mathbf{w}(\boldsymbol{\omega}', \boldsymbol{\omega}'')$ of length $N_{\mathbf{w}}$

$$\mathbf{w} = [\mathbf{w}_i, i = 1, \ldots, N_{\mathbf{w}}], \mathbf{w}_i = \begin{bmatrix} w_i' \\ w_i'' \end{bmatrix}. \tag{10}$$

splits each of aligned sequences $\boldsymbol{\omega}'$ and $\boldsymbol{\omega}''$ into $N_{\mathbf{w}}$ parts: $\boldsymbol{\omega}' = (\boldsymbol{\omega}_{w_1'}' \boldsymbol{\omega}_{w_2'}' \ldots \boldsymbol{\omega}_{w_{N\mathbf{w}}'}')$ and $\boldsymbol{\omega}'' = (\boldsymbol{\omega}_{w_1''}'' \boldsymbol{\omega}_{w_2''}'' \ldots \boldsymbol{\omega}_{w_{N\mathbf{w}}''}'')$.

Alignment $\mathbf{w}$ for symbolic sequences and vector signals has slightly different splitting properties because of different nature of connection between elements.

So, for symbolic sequences, each part $\mathbf{w}_i, i = 1, \ldots, N_{\mathbf{w}}$ of alignment $\mathbf{w}$ can indicates either two one-length subsequences $|\boldsymbol{\omega}_{w_i'}'| = |\boldsymbol{\omega}_{w_i''}''| = 1$, either zero-length and nonzero-length subsequences: $(|\boldsymbol{\omega}_{w_i'}'| \neq 0; |\boldsymbol{\omega}_{w_i''}''| = 0)$ or $(|\boldsymbol{\omega}_{w_i'}'| = 0; |\boldsymbol{\omega}_{w_i''}''| \neq 0)$. At that zero-length subsequence $|\boldsymbol{\omega}_{w_i'}'| = 0$ corresponds to insertion a $|\boldsymbol{\omega}_{w_i''}''|$-length gap series in the respective place of $\boldsymbol{\omega}'$.

For vector signals each part $\mathbf{w}_i, i = 1, \ldots, N_{\mathbf{w}}$ of alignment $\mathbf{w}$ indicates subsequences with length $(|\boldsymbol{\omega}_{w_i'}'| \geq 1; |\boldsymbol{\omega}_{w_i''}''| \geq 1)$, at that at least one of them must be one-length: $(|\boldsymbol{\omega}_{w_i'}'| = 1; |\boldsymbol{\omega}_{w_i''}''| \geq 1)$ or $(|\boldsymbol{\omega}_{w_i'}'| \geq 1; |\boldsymbol{\omega}_{w_i''}''| = 1)$. If some subsequence length is greater than one, for example, $|\boldsymbol{\omega}_{w_i'}'| \geq 1$, it corresponds to local axis stretching of the respective part of $\boldsymbol{\omega}''$.

Examples of alignments of both vector signals and symbolic sequences are presented in Fig. 1.

Each alignment $\mathbf{w} = \mathbf{w}(\boldsymbol{\omega}', \boldsymbol{\omega}'')$ can be characterized by alignment's quality $K(\boldsymbol{\omega}', \boldsymbol{\omega}''|\mathbf{w})$.

In contrast to traditional similarity and dissimilarity measures (as described in Sect. 2.3), any convolution potential function is based on all possible alignments, but not only on the optimal one:

$$K(\boldsymbol{\omega}', \boldsymbol{\omega}'') = \sum_{\mathbf{w}} K(\boldsymbol{\omega}', \boldsymbol{\omega}''|\mathbf{w}). \tag{11}$$

This is an entire family of potential functions, which are different from each other by the type of alignment and by the way of evaluating the alignment's quality.

Global Alignment Kernel for Vector Signals. To compare two vector signals $\boldsymbol{\omega}' = (\omega_1', \omega_2', \ldots, \omega_{N'}')$ and $\boldsymbol{\omega}'' = (\omega_1'', \omega_2'', \ldots, \omega_{N''}'')$, where $\omega_i' = \mathbf{x}_i' \in \mathbb{R}^m, i = 1, \ldots, N'$ and $\omega_i'' = \mathbf{x}_i'' \in \mathbb{R}^m, i = 1, \ldots, N''$ are real-valued vectors, the alignment's quality is usually defined as product of radial basis potential functions (5) for each of N_w pairs of aligned vector elements:

$$K(\omega', \omega'' | \mathbf{w}) = \prod_{i=1}^{N_w} \kappa(\omega'_{w'_i}, \omega''_{w''_i}) = \prod_{i=1}^{N_w} \kappa(\mathbf{x}'_{w'_i}, \mathbf{x}''_{w''_i}) = \prod_{i=1}^{N_w} \exp\left(-\alpha \left\| \mathbf{x}'_{w'_i} - \mathbf{x}''_{w''_i} \right\|^2\right). \quad (12)$$

The respective potential function is named Global Alignment Kernel [9] because the alignment's quality (12) is responsive to all elements of the signals.

Global Alignment Kernel for Symbolic Sequences. The respective Global Alignment Kernel for symbolic sequences [8] $\omega' = (\omega'_1, \omega'_2, \ldots, \omega'_{N'})$ and $\omega'' = (\omega''_1, \omega''_2, \ldots, \omega''_{N''})$, which, as distinct from signals, consist of elements of some finite alphabet $\omega'_i = \alpha'_i \in A = \{\alpha^1, \ldots, \alpha^m\}, i = 1, \ldots, N'$ and $\omega''_i = \alpha''_i \in A, i = 1, \ldots, N''$, is different from the previous case by the fact, that each alignment $\mathbf{w} = \mathbf{w}(\omega', \omega'')$ may contain zero positions corresponding to gaps.

The alignment quality in this case is defined as product of potential functions, which are different for parts of alignment, containing gaps and parts without gaps (matches):

$$K(\omega', \omega'' | \mathbf{w}) = \prod_{\substack{i=1, \\ i \, : \, gaps}}^{N_w} \kappa_1(\omega'_{\mathbf{w}_i}, \omega''_{\mathbf{w}_i}) \prod_{\substack{i=1, \\ i \, : \, matches}}^{N_w} \kappa_2(\omega'_{\mathbf{w}_i}, \omega''_{\mathbf{w}_i}), \quad (13)$$

where the potential function for gaps

$$\kappa_1(\omega'_{\mathbf{w}_i}, \omega''_{\mathbf{w}_i}) = \exp\left(\beta\left[g(|\omega'_{\mathbf{w}_i}|) + g(|\omega''_{\mathbf{w}_i}|)\right]\right), g(n) = \begin{cases} 0, & \text{if } n = 0, \\ d + e(n-1), & \text{if } n \geq 1 \end{cases} \quad (14)$$

has the meaning of gap penalties and is very similar by its structure to gap penalties for optimal alignment-based similarity measures [14–17], and the potential function for matches is some potential function on the set of sequences elements:

$$\kappa_2(\omega'_{\mathbf{w}_i}, \omega''_{\mathbf{w}_i}) = \begin{cases} \kappa(\alpha', \alpha''), & \alpha', \alpha'' \in A, \quad \text{if} \quad |\omega'_{\mathbf{w}_i}| = |\omega''_{\mathbf{w}_i}| = 1, \\ 0, & \text{otherwise.} \end{cases} \quad (15)$$

Local Alignment Kernel for Symbolic Sequences. The Local Alignment Kernel for symbolic sequences [11] is based on local sequence alignments.

In this case, the alignment quality function does not include penalties for starting and finishing gaps:

$$K(\omega', \omega'' | \mathbf{w}) = \prod_{\substack{i=2, \\ i \, : \, gaps}}^{N_w-1} \kappa_1(\omega'_{\mathbf{w}_i}, \omega''_{\mathbf{w}_i}) \prod_{\substack{i=1, \\ i \, : \, matches}}^{N_w} \kappa_2(\omega'_{\mathbf{w}_i}, \omega''_{\mathbf{w}_i}). \quad (16)$$

So, the Local Alignment Kernel is more appropriate for situations, when we need to estimate sequence similarity by presence of similar common subsequences, in contrast to the Global Alignment Kernel, which compares whole sequences.

Some Problems of Practical Application of Alignment-based Potential Functions. It should be noticed there are some problems, which arise when one tries to apply alignment-based potential functions to practice.

The Diagonal Dominance Problem. This is the problem of extremely expressed diagonal dominance in the alignment-based potential function: $K(\omega', \omega') > > K(\omega', \omega'')$. In practice, it has been observed that SVM does not perform well in this situation.

In order to decrease the effect of diagonal dominance, J.-P. Vert and his colleagues [11] proposed to consider the logarithmic function of the initial potential function $S(\omega', \omega'') = \ln K(\omega', \omega'')$.

But it is evident, that this operation leads to depriving the potential function of its mathematical properties. And so, the authors make correction (8) of the obtained similarity matrix.

A more soft way of solving this problem was proposed by V. Mottl and his coauthors in [35]. It is based on the observation that the normalized alignment-based potential function

$$K_n(\omega', \omega'') = \frac{K(\omega', \omega'')}{\sqrt{K(\omega', \omega')K(\omega'', \omega'')}} \tag{17}$$

is very similar by its behavior to the radial basis potential function (5).

So, assuming that the normalized potential function is a radial one, i.e. $K_n(\omega', \omega'') = \exp[-\beta\rho^2(\omega'', \omega'')]$, it is proposed to decrease a degree of its slope:

$$\tilde{K}(\omega', \omega'') = [K_n(\omega', \omega'')]^\gamma, \ 0 < \gamma < 1. \tag{18}$$

In many practical applications, for sufficiently large values of $\gamma < 1$ such heuristic allows to solve the problem of diagonal dominance without losing mathematical properties of the potential function.

The Problem of Potential Functions on the Finite Alphabet of Primitives. This problem may arise when we would like to use a Global or Local Alignment Kernel to compare biological sequences.

To introduce a potential function $\kappa(\alpha', \alpha'')$ on the set of biological sequences A (amino acids or nucleotides), it is proposed in [11] to use an exponential transformation $\kappa(\alpha', \alpha'') = \exp(\beta s(\alpha', \alpha''))$ of the traditional similarity measure of amino acids or nucleotides $s(\alpha', \alpha'')$, such as PAM [36] and BLOSUM [37] substitution matrixes and others. However, in the general case the resulting function may be not positive definite, and so it is no potential function.

But at the same time, it is easy to obtain a valid potential function on sets of amino acids on the basis of the notion of the traditional Point Accepted Mutation (PAM) evolution model [36]. The respective approach was proposed in our previous paper [38], and is shortly described here in Sect. 2.5.

2.5   Potential Functions on Elements of Biological Sequences

Potential Functions on the Set of Amino Acids on the basis of the PAM Model.
There are known 20 different amino acids $A = \{\alpha^1, \ldots, \alpha^m\}, m = 20$. The main theoretical concept underlying the proposed method of comparing amino acids is the well-known probabilistic model of evolution of amino acids by Margaret Dayhoff, called PAM (Point Accepted Mutation) [36]. Its main instrument is the notion of Markov chain of evolution of amino acids in some separate point of protein's chain, which is determinate by the matrix of transitional probabilities $\Psi = \left(\psi_{[1]}(\alpha^j | \alpha^i)\right)$ of changing an amino acid α^i into amino acid α^j at the next step of evolution. The index [1] means that the initial one-step Markov chain is considered.

At that, it is supposed that this Markov chain is an ergodic and reversible random process, i.e. process, which is characterized by a final probability distribution $\xi(\alpha^j)$:

$$\sum_{\alpha^i \in A} \xi(\alpha^i) \psi_{[1]}(\alpha^j | \alpha^i) = \xi(\alpha^j),$$

and satisfies the reversibility condition:

$$\xi(\alpha^i) \psi_{[1]}(\alpha^j | \alpha^i) = \xi(\alpha^j) \psi_{[1]}(\alpha^i | \alpha^j).$$

We have proved [38], that for any matrix of transitional probabilities $\Psi_{[s]} = [\underbrace{\Psi_{[1]} \times \cdots \times \Psi_{[1]}}_{s}]$, which corresponds to a sparse Markov chain (i.e. probabilistic process of evolution with a bigger evolutionary step), normalized similarity measures

$$\pi_s(\alpha^i, \alpha^j) = \frac{\Psi_{[s]}(\alpha^i | \alpha^j) \xi(\alpha^j)}{\xi(\alpha^i) \xi(\alpha^j)} = \frac{\Psi_{[s]}(\alpha^i | \alpha^j)}{\xi(\alpha^i)} \tag{19}$$

form nonnegative definite matrixes of pairwise similarity of amino acids for any even step $s = 2, 4, \ldots, 250, \ldots$ So, they are potential functions.

Traditional PAM substitution matrixes with s-PAM evolutionary distance are defined as

$$D_{[s]} = [d_{[s]}(\alpha^i, \alpha^j), i, j = 1, \ldots, 20], \quad d_{[s]}(\alpha^i, \alpha^j) = [10 \log_{10} \hat{\pi}_{[s]}(\alpha^i, \alpha^j)], \tag{20}$$

where $\hat{\pi}_{[s]}(\alpha^i, \alpha^j)$, $s = 1, 2, \ldots, 250, \ldots$ are computed in accordance with (19) as estimates of probabilities $\hat{\psi}_{[s]}(\alpha^i | \alpha^j)$ and $\hat{\xi}(\alpha^i)$ obtained from some set of protein data.

It should be noticed that $d_{[s]}(\alpha^i, \alpha^j)$ is never a potential function because of taking logarithm and rounding to the nearest integer. But the similarity measure (19) can be used as $\kappa(\alpha', \alpha'')$ for comparing amino acids in (15).

Potential Functions on the Set of Amino Acids on the basis of BLOSUM. One more traditional way to compare amino acids had been proposed by S. and J. Henikoff in the form of BLOSUM substitution matrixes [37]. It was initially meant as obtained from some statistics without any reference to a model of amino acids evolution, in contrast to PAM.

But we have proved [38], that BLOSUM substitution matrixes B_s, $s = 1, 2$, ..., 62, ..., which are constructed on the basis of different sets of protein data Ω_s, $s = 1, 2, \ldots, 62, \ldots$, may be strictly explained in the same evolutionary terms as PAM:

$$B_s = [b_s(\alpha^i, \alpha^j), i, j = 1, \ldots, 20], \ b_s(\alpha^i, \alpha^j) = [2 \log_2 \hat{\pi}^s_{[2]}(\alpha^i, \alpha^j)].$$

The difference between PAM and BLOSUM lies only in the different initial data for estimating the parameters $\hat{\pi}^s_{[2]}(\alpha^i, \alpha^j)$ of the same model.

So, BLOSUM, absolutely like PAM, also produces a series of potential functions $\hat{\pi}^s_{[2]}(\alpha^i, \alpha^j)$, $s = 1, 2, \ldots, 62, \ldots$, on the set of amino acids. They lose their initial positive definiteness only because of inappropriate final representation.

Potential Functions on the set of Nucleotides. From the mathematical point of view, the only distinction of nucleotides from amino acids is that the size of the respective alphabet consist of 4 symbols instead of 20. No mathematical problems prevent construction of a potential function on this alphabet. In particular, it can be organized like potential functions on the set of amino acids.

2.6   The Generalized Probabilistic Approach to Constructing Potential Functions for Sequences

The Main Principle. It appears to be a natural idea, especially for biological sequences [39], to evaluate similarity of two sequences $\omega', \omega'' \in \Omega$ by evaluating the probability of the hypothesis that they originate from the same random ancestor $\vartheta \in \Omega$ as results of two independent branches of evolution, both of which are defined by the same known random transformation $(\varphi(\omega|\vartheta), \ \omega \in \Omega, \vartheta \in \Omega)$:

$$\kappa(\omega', \omega'') = \sum_{\vartheta \in \Omega} p(\vartheta) \varphi(\omega'|\vartheta) \varphi(\omega''|\vartheta) d\vartheta. \tag{21}$$

A particular interest to use the function (21) is determined by the fact, that it is a valid potential function by its structure for any distributions $p(\vartheta)$ and $\varphi(\omega|\vartheta)$.

A number of variants of random transformation of amino acid and nucleotide sequences have been proposed [40, 41], but they are oriented only on biological sequences and, besides, practically all of them are incomputable because of impossibility to compute the sum over the entire infinite set of ancestors for the proposed type of random transformation.

In this section. We:

(1) on the one hand, describe a more generalized approach, the main principle of which was initially proposed by us in [43] and partially in [42]. This approach covers at once both vector signals and symbolic sequences (what is possible due to taking into account that elements of symbolic sequences are embedded into some hypothetical linear space [7] via a valid potential function, described in Sect. 2.5);

(2) on the other hand, present a specific random transformation of sequences, which allows for numerical computation of potential functions. Moreover, it will be shown that this approach leads to a large family of probabilistically justified potential functions (which could be obtained by a special choice of distributions) and, particularly, to potential functions, having the same structure, as alignment-based ones, considered in Sect. 2.4.

Let Ω be a set of all sequences $\omega = (\omega_t \in \tilde{A}, \ t = 1, \ldots, N_\omega) \in \Omega$, where $\tilde{A}$ is a linear space. In particular, $\tilde{A} = \mathbb{R}^m$ if vector signals are considered. In the case of symbolic sequences, $\tilde{A}$ is a hypothetical linear space, in which some potential function (Sect. 2.5) embeds the finite set of sequence elements in accordance with the principle, proposed in [7].

Let also $\Omega_n = \{\omega : N_\omega = n\} \subseteq \Omega$ be a subset of n-length sequences and $\vartheta = \{\vartheta_i \in \tilde{A}, i = 1, \ldots, n\} \in \Omega_n$ be an n-length sequence of this set considered as some unknown ancestor. At that, we will suppose that there exists a probability distribution on the set of n-length ancestors $p_n(\vartheta) = p(\vartheta|n), \vartheta \in \Omega_n$, and, in addition, a distribution on the set of all possible lengths $r(n)$.

We shall consider a the family of random transformations $\varphi_n(\omega|\vartheta)$, which is meant to transform the ancestor sequence $\vartheta \in \Omega_n$ into any sequence $\omega \in \Omega$. It is not hard to see that the function

$$\kappa(\omega', \omega'') = \sum_{n=0}^{\infty} r(n) \int_{\vartheta \in \Omega_n} p_n(\vartheta) \varphi_n(\omega'|\vartheta) \varphi_n(\omega''|\vartheta) d\vartheta \tag{22}$$

is a potential function on the set of sequences of arbitrary kind $\omega \in \Omega$ for any $r(n)$, $p_n(\vartheta)$, $\varphi_n(\omega|\vartheta)$ [43].

We shall consider two-step random transformations $\varphi_n(\omega|\vartheta)$ of the ancestor sequence into a resulting sequence.

At the first step, an n-length structure of transformation is randomly chosen. This structure $\mathbf{v} = (v_1, \ldots, v_n)$, named one-side alignment. For each ancestor's positions $(1, \ldots, n)$, this structure defines positions of the resulting sequence $v_i = \{\check{v}_i \leq t \leq \hat{v}_i\}$, into which the ancestor elements will be transform. An example of such structure is presented in Fig. 2. The transformation structure $\mathbf{v}$ is chosen from a set of n-length structures V_n in accordance with some distribution $\sum_{\mathbf{v} \in V_n} q_n(\mathbf{v}) = 1$.

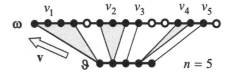

- ● - key subsequence
- ○ - additional subsequence

Fig. 2. An example n-length sequence transformation structure.

At the second step, elements, forming the key subsequence and the additional one, are randomly chosen in accordance with a structure-dependent random transformation

$$\eta_n(\omega|\vartheta, \mathbf{v}) \ge 0, \eta_n(\omega|\vartheta, \mathbf{v}) = 0 \text{ if } \omega \notin \Omega_k, \ k \ge \ddot{v}_n.$$

As a result, the random sequence transformation $\varphi_n(\omega|\vartheta)$ is defined as

$$\varphi_n(\omega|\vartheta) = \sum_{\mathbf{v} \in V_n} q_n(\mathbf{v})\eta_n(\omega|\vartheta, \mathbf{v}).$$

Besides we assume that:

(1) the elements of the ancestor sequence are formed independently in accordance with some distribution $\xi(\vartheta)$, $\vartheta \in \tilde{A}$;

(2) the elements of the key subsequence are independently randomly generated from those of the original sequence in accordance with some random transformation $\psi(\omega|\vartheta)$, $\omega, \vartheta \in \tilde{A}$;

(3) distributions $\xi(\vartheta)$, $\vartheta \in \tilde{A}$ and $\psi(\omega|\vartheta)$, $\omega, \vartheta \in \tilde{A}$, allow for easy computation of the integral

$$\int_{\vartheta \in \tilde{A}} \xi(\vartheta)\psi(\omega'|\vartheta)\psi(\omega''|\vartheta)d\vartheta.$$

It should be noticed that the assumptions 1 and 2 are the same as the assumptions used in the traditional PAM model of amino acids evolution [25]. So, they are natural for biological sequences, and, in addition, can be also applied to vector signals.

Besides, it is not hard to fulfil assumption 3. It holds for any finite set of symbolic sequence elements A if some ergodic and reversible Markov chain is defined on it [43] with final distribution $\xi(\vartheta)$, $\vartheta \in A$ and transitional probabilities $\psi(\omega|\vartheta)$, $\omega, \vartheta \in A$.

We proved in [43] that in this case there exists a potential function for sequence elements

$$\kappa(\omega', \omega'') = \int_{\vartheta \in A} \xi(\vartheta)\psi(\omega'|\vartheta)\psi(\omega''|\vartheta)d\vartheta = \psi_{[2]}(\omega''|\omega')\xi(\omega') = \psi_{[2]}(\omega'|\omega'')\xi(\omega''),$$

$$(23)$$

which can be directly computed without necessity to compute the integral. The normalized version of this potential function has the form:

$$\bar{\kappa}(\omega', \omega'') = \frac{\psi_{[2]}(\omega''|\omega')}{\xi(\omega'')} = \frac{\psi_{[2]}(\omega'|\omega'')}{\xi(\omega')}. \tag{24}$$

As to vector signals, it is natural to model their random transformations by normal distributions, for which computation of the integral is also no problem.

Coming back to the sequences transformation structures, it should be noticed, that two one-side alignments $\mathbf{v}' = (v_i',\ i = 1, \ldots, n) \in V_n$ and $\mathbf{v}'' = (v_i'',\ i = 1, \ldots, n) \in V_n$ define one two-side (that is pairwise) alignment

$$\mathbf{w} = (\mathbf{v}', \mathbf{v}'') = \left[\begin{pmatrix} v_1' \\ v_1'' \end{pmatrix}, \ldots, \begin{pmatrix} v_n' \\ v_n'' \end{pmatrix} \right] \in W_n,$$

which directly connects the elements of two sequences mentally formed by such transformation. An example of two one-side alignments and the respective pair-wise alignment is presented in Fig. 3.

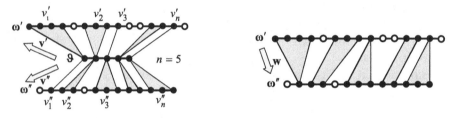

Fig. 3. An example of two one-side sequence alignments (left) and the respective pair-wise alignment (right).

It should be noticed that properties of this model allow to express the initial potential function (22) in an absolutely equivalent, but essentially more simple form, which does not contain the integral over all possible ancestor sequences:

$$\mathcal{K}(\omega', \omega'') = \sum_{n=0}^{\infty} r(n) \sum_{\mathbf{w} \in W_{nN'N''}} q_n(\mathbf{v_w'}) q_n(\mathbf{v_w''}) K(\omega', \omega''|\mathbf{w}), \tag{25}$$

where the inner sum is taken over all admissible pairwise alignments $W_{nN'N''} = \{\mathbf{w} : v_{n,\mathbf{w}}' \leq N',\ v_{n,\mathbf{w}}'' \leq N''\}$, and

$$K(\omega', \omega''|\mathbf{w}) = \left(\prod_{t=1}^{\left|\bar{\bar{\omega}}_{\mathbf{w}}'\right|} \xi \bar{\bar{\omega}}_{\mathbf{v_w'},t}' \right) \left(\prod_{t=1}^{\left|\bar{\bar{\omega}}_{\mathbf{w}}''\right|} \xi \bar{\bar{\omega}}_{\mathbf{v_w''},t}'' \right) \prod_{i=1}^{n} \kappa\left(\omega_{v_{i,\mathbf{w}}'}', \omega_{v_{i,\mathbf{w}}''}'' \right)$$

is a joint conditional probability density that randomly forms the two sequences $\omega', \omega'' \in \Omega$.

Particular forms of the probabilistic potential function for sequences.

Potential Functions of Fixed and Absolutely Unfixed Order. In addition to the main form of the potential function for sequences (25), we can consider a number of particular forms. So, depending on the a priori distribution on a set of ancestor lengths $r(n)$, we can obtain the potential functions of fixed order n, if we will put $r(n) = 1$ and $r(k) = 0 \; \forall k \neq n$

$$\mathcal{K}_n(\omega', \omega'') = \sum_{\mathbf{w} \in W_{nN'N''}} q_n(\mathbf{v}'_{\mathbf{w}}) q_n(\mathbf{v}''_{\mathbf{w}}) K(\omega', \omega'' | \mathbf{w}),$$

or the potential function of so-called absolutely unfixed order for an improper "almost uniform" distribution $r(n) = const$:

$$\mathcal{K}(\omega', \omega'') \propto \sum_{n=0}^{\infty} \sum_{\mathbf{w} \in W_{nN'N''}} q_n(\mathbf{v}'_{\mathbf{w}}) q_n(\mathbf{v}''_{\mathbf{w}}) K(\omega', \omega'' | \mathbf{w}).$$

Potential functions for symbolic sequences and vector signals. Aspecial choice of the distribution on the set of transformation structures $q_n(\mathbf{v})$ and the distributions on the set of sequence elements allows taking into account the specificity of symbolic sequences and vector signals.

So, to compare amino acid sequences we should take $q_n(\mathbf{v}) \neq 0$ only if $\dot{v}_i = \ddot{v}_i = v_i$ for all $i = 1, \ldots, n$. An example of the respective sequence transformation structure is presented in Fig. 4. Also, we should define $\psi(\omega'|\omega'')$, $\omega', \omega'' \in \tilde{A}$ and $\xi(\omega)$, $\omega \in \tilde{A}$, as the final and transitional probabilities of the PAM model.

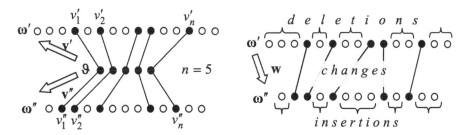

Fig. 4. An example of transformation structures for symbolic sequences.

At the same time, to compare vector signals, we should define $q_n(\mathbf{v})$ in such a way that the full segmentation of positions of the signal being formed $\dot{v}_1 = 1$, $\dot{v}_i = \ddot{v}_{i-1} + 1$, $i = 1, \ldots, n$ would be defined with the probability equal to 1. An example of the respective sequence transformation structure is presented in Fig. 5.

The distributions for signal elements can be chosen as follows:

$$\xi(\vartheta) = \xi(\mathbf{y}) = \mathcal{N}(\mathbf{y}|\mathbf{0}, \sigma^2 \mathbf{I}), \psi(\omega|\vartheta) = \psi(\mathbf{x}|\mathbf{y}) = \mathcal{N}(\mathbf{x}|\mathbf{y}, \delta^2 \mathbf{I}).$$

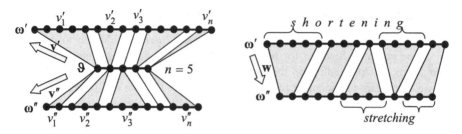

Fig. 5. An example of transformation structures for vector signals.

Potential Functions on the Basis of Global and Local Alignments of Symbolic Sequences. The local and global alignments can also be easily modeled.

To model global alignments, we should make a special choice of the distribution of sequence structure transformation:

$$q_n(\mathbf{v}) = g(v_1|a,b)\left(\prod_{i=2}^{n} g(v_i - v_{i-1}|a,b)\right)g(N_\omega - v_n). \qquad (26)$$

Here $g(d|a,b)$ is some function, which has the meaning of penalty for series of d gaps and can be defined, for example, as

$$g(d|a,b)\propto\begin{cases} 1, & d = 1, \\ \exp[-\beta(a + bd)], & d > 1, \end{cases}$$

where $a \geq 0, b \geq 0, \beta \geq 0$ are numerical parameters of the gap penalty function. In case $a > 0$ one long gap will be more preferable than a number of short gaps making the same summary length.

To model local alignments, the distribution $q_n(\mathbf{v})$ can be taken in the following way:

$$q_n(\mathbf{v}) \propto \prod_{i=2}^{n} g(v_i - v_{i-1}|a,b). \qquad (27)$$

Examples of Specific Potential Functions for Sequences. Each specific choice of distributions leads to some kind of the potential function (25). Some most popular examples are presented below. In particular, we can obtain the convolution potential functions, considered in Sect. 2.4.

So, the normalized global potential function of absolutely unfixed order for symbolic sequences with the distribution $q_n(\mathbf{v})$ defined in accordance with (26)

$$\mathcal{K}^{[s]}(\omega', \omega'') = \sum_{n=0}^{\infty} \sum_{\mathbf{w}\in W_{nN'N''}} q_n(\mathbf{v}'_\mathbf{w})q_n(\mathbf{v}''_\mathbf{w})$$
$$\times \left(\prod_{t=1}^{\left|\overline{\omega}'_\mathbf{w}\right|} \xi(\overline{\omega}'_{v'_{t,\mathbf{w}}})\right)\left(\prod_{t=1}^{\left|\overline{\omega}''_\mathbf{w}\right|} \xi(\overline{\omega}''_{v''_{t,\mathbf{w}}})\right)\prod_{i=1}^{n} \kappa_{[s]}\left(\omega'_{v'_{i,\mathbf{w}}}, \omega''_{v''_{i,\mathbf{w}}}\right) \qquad (28)$$

is nothing else, then Global Alignment Kernel for symbolic sequences (13).

Two more examples are the Local Alignment Kernel (16) for symbolic sequences

$$\bar{\mathcal{K}}_{LA}^{[s]}(\omega', \omega'') = \sum_{n=0}^{\infty} \sum_{\mathbf{w} \in W_{nN'N''}} q_n(\mathbf{v}'\mathbf{w}) q_n(\mathbf{v}''\mathbf{w}) \prod_{i=1}^{n} \bar{\kappa}_{[s]}\left(\omega'_{v'_{i,\mathbf{w}}}, \omega''_{v''_{i,\mathbf{w}}}\right), \quad (29)$$

where $q_n(\mathbf{v})$ is defined in accordance with (27), and the Global Alignment Kernel (12) for vector signals:

$$\tilde{\mathcal{K}}(\omega', \omega'') = \sum_{n=0}^{\infty} \sum_{\mathbf{w} \in W_{nN'N''}} \left\{ q_n(\mathbf{v}'_\mathbf{w}) q_n(\mathbf{v}'_\mathbf{w}) \times \right.$$
$$\left. \left[\prod_{i=1}^{n} \left(1/\mathcal{N}\left(\bar{\mathbf{x}}''_{v''_{i,\mathbf{w}}} | \mathbf{0}, (\sigma^2 + \delta^2)\mathbf{I}\right)\right) \mathcal{N}\left(\bar{\mathbf{x}}''_{v''_{i,\mathbf{w}}} | \bar{\mathbf{m}}_{\bar{\mathbf{x}}''_{v''_{i,\mathbf{w}}}} | \bar{\mathbf{x}}'_{v'_{i,\mathbf{w}}}, \gamma^2_{|v'_{i,\mathbf{w}}|} \mathbf{1} + \delta^2 \mathbf{I}\right) \right] \right\}, \quad (30)$$

where $q_n(\mathbf{v})$ is defined as in (26).

So, algebraically motivated alignment-based potential functions have a probabilistic justification, what explains the empirical fact that they perform well.

But at the same time, it should be noticed that the presented generalized approach produces much more particular potential functions, for example, such interesting ones as the potential functions of fixed order

$$\mathcal{K}_n^{[s]}(\omega', \omega'') = \sum_{\mathbf{w} \in W_{nN'N''}} q_n(\mathbf{v}'_\mathbf{w}) q_n(\mathbf{v}''_\mathbf{w})$$
$$\times \left(\prod_{t=1}^{\left|\overline{\overline{\omega}}'_\mathbf{w}\right|} \xi(\overline{\overline{\omega}}'_{v'_{t,\mathbf{w}}}) \right) \left(\prod_{t=1}^{\left|\overline{\overline{\omega}}''_\mathbf{w}\right|} \xi(\overline{\overline{\omega}}''_{v''_{t,\mathbf{w}}}) \right) \prod_{i=1}^{n} \kappa_{[s]}\left(\omega'_{v'_{i,\mathbf{w}}}, \omega''_{v''_{i,\mathbf{w}}}\right),$$

which can be successfully used for multiple alignments of biological sequences [44], for finding the most probable common ancestor sequence [43], and so on.

3   Kernel-Based SVM for Two-Class Pattern Recognition

In kernel terms, the SVM decision function [4] for classification of sequences into two classes $y = 1$ and $y = -1$ can be represented as a discriminant hyperplane

$$\hat{y}(\omega) = K(\mathbf{a}, \omega) + b > 0 \rightarrow y = 1, y(\omega) \leq 0 \rightarrow y = -1,$$

formed by the direction vector $\mathbf{a}$, which is element of a hypothetical linear space $\mathbf{a} \in \tilde{\Omega}$, into which the potential function $K(\omega', \omega'')$ embeds the set of sequences $\tilde{\Omega} \supset \Omega$. It can be found as linear combination of elements of the training set $\{\omega_j, j = 1, \ldots, N\}$

$$\mathbf{a} = \sum_{j:\lambda_j > 0} y_j \lambda_j \omega_j$$

with coefficients $\lambda_j \geq 0$, which are solutions of the dual formulation of the SVM training problem

$$\begin{cases} \sum_{j=1}^{N} \lambda_j - (1/2) \sum_{j=1}^{N} \sum_{l=1}^{N} \left[y_j y_l K(\omega_j, \omega_l) \right] \lambda_j \lambda_l \rightarrow \max, \\ \sum_{j=1}^{N} y_j \lambda_j = 0, 0 \leq \lambda_j \leq C/2, j = 1, \ldots, N. \end{cases}$$

4 Multi-kernel SVM-Based Learning

4.1 Short Review of Multi-kernel Learning

The multi-kernel situation is extremely actual because of the possibility to combine (fuse) a number of different potential (kernel) functions.

There have thus been a number of attempts to combine potential functions for sequence analysis. The simplest approach is unweighted sum of kernels. Different linear (or even non-linear) combinations with fixed or heuristically chosen weights have also been considered; however, overall performance is generally poor.

The most general method of kernel fusion is the approach of Lanckriet et al. [45], which seeks to directly find the optimal linear combination of kernels, and gives rise to a quadratically-constrained algorithm for determining the nonnegative adaptive weights of kernel matrices. The respective kernel combination is incorporated into a decision rule, in which each particular kernel influences on the decision proportionally to its weight.

A number of authors have carried this work further in various ways, generalizing the approach onto problems other than classification [46, 47], working on algorithmic improvements [48, 49], or deriving theoretical variations, applying different restrictions for weights [50], and making certain theoretical extensions, e.g. weighting not only potential functions but also features [51, 52]. These variants typically perform well in constrained scenarios, and where the data are initially represented by feature vectors. However, they tend not to outperform [45] on real data.

Furthermore, most of existing multiple kernel learning methods share a common disadvantage – the absence of a mechanism for supervising the so-called "sparseness" of the obtained vector of kernel weights. In practice, the obtained vector of weights is frequently too sparse, with many informative kernels excluded from the decision rule, with the resulting loss of decision quality.

Only few methods are explicitly oriented towards elimination of this disadvantage and obtaining non-sparse decisions [53, 54] (more advanced versions utilize a *supervised* sparseness parameter [55–58]). We refer to this property as "*selectivity*", because it provides the algorithm with the ability to select kernels most useful for the classification task at hand. A generalized probabilistic approach for supervised selective kernel fusion was proposed by us in [57, 58] and includes, as particular cases, such well-known approaches as the classical SVM [4], Lasso SVM [59], Elastic Net SVM [60] and others.

In this paper, we apply a further particular case of this approach, called Supervised Selective Support Kernel SVM (SKSVM), initially proposed in [58] for the membrane protein prediction problem.

The proposed approach has the very significant qualitative advantage over the other methods of explicitly indicating a discrete subset of support kernels within the combination, in contrast to other methods that assign some positive (even if small) weight to *each* kernel, requiring significantly greater memory overhead.

4.2 Supervised Selective Support Kernel SVM (SSKSVM)

Let $\{(\omega_j, y_j), j = 1, \ldots, N\}$ be the training set of sequences $\omega_j \in \Omega$ (for example, proteins) and $y_j = y(\omega_j) \in \{-1, 1\}$ defines their class-membership. Let also n competitive potential functions $K_i(\omega', \omega''), \omega', \omega'' \in \Omega, i = 1, \ldots, n$ be initially considered.

Each potential function $K_i(\omega', \omega''), i = 1, \ldots, n$, embeds the set of objects Ω into some hypothetical linear space $\mathbb{X}_i$ by a hypothetical mapping $x_i = x_i(\omega) \in \mathbb{X}_i, \omega \in \Omega$, and plays the role of inner product within it $K_i(\omega', \omega'') = <x_i(\omega'), x_i(\omega'')> : \quad \mathbb{X}_i \times \mathbb{X}_i \to \mathbb{R}$.

For combination using several potential functions we here utilize the generalized probabilistic formulation of the SVM, which was proposed in [54, 56, 57] as an instrument for making Bayesian decisions on the discriminant hyperplane $\sum_{i=1}^{n} K_i(a_i, \omega) + b \gtrless 0$ within the Cartesian product of the kernel-induced hypothetical linear spaces $\mathbf{a} = (a_1, \ldots, a_n) \in \mathbb{X} = \mathbb{X}_1 \times \ldots \times \mathbb{X}_n, b \in \mathbb{R}$.

The main idea of the proposed probabilistic formulation consists in assuming a specific system of probabilistic assumptions regarding the two distribution densities of hypothetical feature vectors for the two classes $\varphi(\mathbf{x}|y = +1)$ and $\varphi(\mathbf{x}|y = -1)$, defined by the (as yet) undetermined hyperplane in the combined linear space $\mathbf{x} = (x_1, \ldots, x_n) \in \mathbb{X} = \mathbb{X}_1 \times \ldots \times \mathbb{X}_n$ under certain *a priori* probabilistic assumptions.

The generalized training criterion proposed in [54, 56, 57] can be formulated in terms of potential functions:

$$\begin{cases} -\ln\Psi(a_1, \ldots, a_n) + c \to \min(a_i \in \mathbb{X}_i, b \in \mathbb{R}, \delta_j \in \mathbb{R}), \\ y_j\left(\sum_{i=1}^{n} K_i(a_i, x_j) + b\right) \geq 1 - \delta_j, \delta_j \geq 0, j = 1, \ldots, N. \end{cases} \tag{31}$$

Each specific choice of *a priori* distribution density $\Psi(a_1, \ldots, a_n)$ expresses a specific *a priori* preference about the hyperplane orientation, and endows the training criterion (31) with the ability to select informative kernel representations and suppress redundant ones.

In particular, a number of well-known SVM-based training criteria can be obtained from the proposed probabilistic approach, for example, the traditional SVM, Lasso SVM and Elastic Net SVM, differing from one another in the regularization function, which has the form, respectively, $\sum_{i=1}^{n} K_i(a_i, a_i)$, $\sum_{i=1}^{n} \sqrt{K_i(a_i, a_i)}$ and $\sum_{i=1}^{n} K_i(a_i, a_i) + \mu \sum_{i=1}^{n} \sqrt{K_i(a_i, a_i)}$.

We apply here a very specific case of the general problem formulation (31), which was initially proposed in [58]. The *a priori* density of orientation distributions is represented here as a composite of the Laplace and Gaussian distributions in connection with the given threshold μ:

$$
\psi(a_i|\mu) \propto \exp(-q(a_i|\mu)),
$$
$$
q(a_i|\mu) = \begin{cases} 2\mu\sqrt{K_i(a_i,a_i)}, & \sqrt{K_i(a_i,a_i)} \leq \mu, \\ \mu^2 + \sqrt{K_i(a_i,a_i)}, & \sqrt{K_i(a_i,a_i)} > \mu. \end{cases} \tag{32}
$$

The *a priori* assumption (32) along with the generalized training criterion (31) define together a training optimization problem in the form:

$$
\begin{cases}
J_{SKSVM}(a_1,\ldots,a_n,b,\delta_1,\ldots,\delta_N,c,\mu) = \\
\sum_{i=1}^{n} q(a_i|\mu) + c\sum_{j=1}^{N} \delta_j \rightarrow \min(a_i \in \mathbb{X}_i, b \in \mathbb{R}, \delta_j \in \mathbb{R}), \\
q(a_i|\mu) = \begin{cases} 2\mu\sqrt{K_i(a_i,a_i)} & \text{if } \sqrt{K_i(a_i,a_i)} \leq \mu, \\ \mu^2 + K_i(a_i,a_i) & \text{if } \sqrt{K_i(a_i,a_i)} > \mu, \end{cases} \\
y_j\left(\sum_{i=1}^{n} K_i(a_i,x_j) + b\right) \geq 1 - \delta_j, \delta_j \geq 0, j = 1,\ldots,N
\end{cases} \tag{33}
$$

The proposed training criterion is thus a generalized version of the classical SVM that implements the principle of *kernel selection*.

The threshold μ is named here "selectivity" parameter because it regulates the ability of the criterion to enact selection of potential functions. When μ is equal to 0, the criterion is equivalent to the kernel-based SVM with the minimal ability to select potential functions. At the same time, values much greater than zero are equivalent to the Lasso SVM with increasing selectivity as μ grows up to full suppression of all potential functions.

The solution of the problem (33) is equivalent to the solution ($\hat{\xi}_i \geq 0, i \in I = \{1,\ldots, n\}, \hat{\lambda}_j \geq 0, j = 1,\ldots,N$) of the dual problem

$$
\begin{cases}
L(\lambda_1,\ldots,\lambda_N|c,\mu) = \sum_{j=1}^{N} \lambda_j - \sum_{i\in I}(1/2)\xi_i \rightarrow \max(\lambda_1,\ldots,\lambda_N), \\
\xi_i \geq 0, \quad \xi_i \geq \sum_{j=1}^{N}\sum_{l=1}^{N} y_j y_l K_i(\omega_j,\omega_l)\lambda_j\lambda_l - \mu^2, \quad i\in I = \{1,\ldots,n\}, \\
\sum_{j=1}^{N} y_j\lambda_j = 0, \quad 0 \leq \lambda_j \leq (c/2), \quad j = 1,\ldots,N,
\end{cases}
$$

and can be expressed in the form

$$
\begin{cases}
\hat{a}_i = \sum_{j:\hat{\lambda}_j > 0} y_j\hat{\lambda}_j x_i(\omega_j), & i \in I^+ = \left\{i \in I : \sum_{j=1}^{N}\sum_{l=1}^{N} y_j y_l K_i(\omega_j,\omega_l)\hat{\lambda}_j\hat{\lambda}_l > \mu^2\right\} \\
\hat{a}_i = \hat{\eta}_i \sum_{j:\hat{\lambda}_j > 0} y_j\hat{\lambda}_j x_i(\omega_j), & i \in I^0 = \left\{i \in I : \sum_{j=1}^{N}\sum_{l=1}^{N} y_j y_l K_i(\omega_j,\omega_l)\hat{\lambda}_j\hat{\lambda}_l = \mu^2\right\} \\
\hat{a}_i = 0, & i \in I^- = \left\{i \in I : \sum_{j=1}^{N}\sum_{l=1}^{N} y_j y_l K_i(\omega_j,\omega_l)\hat{\lambda}_j\hat{\lambda}_l < \mu^2\right\}
\end{cases}
$$

where $\{0 \geq \eta_i \geq 1, \ i \in I^0\}$ are additionally computed coefficients.

It should be noticed, that the criterion (33), in contrast to other criteria of kernel fusion, explicitly splits the entire set of potential functions into two subsets: "support" potential functions $I^+ \cup I^0$ (which will participate in the resulting discriminant hyperplane) and excluded ones I^-.

As a result, the optimal discriminant hyperplane, which is defined by the solution of the SKSVM training problem (33), can be expressed as

$$\sum_{j:\lambda_j > 0} y_j\lambda_j \left(\sum_{i \in I^+} K_i(\omega_j, \omega) + \sum_{i \in I^0} \eta_i K_i(\omega_j, \omega)\right) + b \gtrless 0,$$

where the numerical parameters $\{0 \leq \eta_i \leq 1, i \in I^0; b\}$ are solutions of the linear programming problem:

$$\begin{cases} 2\mu^2 \sum_{i \in I^0} \eta_i + c \sum_{j=1}^{N} \delta_j \rightarrow \min(\eta_i, i \in I^0; b; \delta_1, \ldots, \delta_N), \\ \sum_{i \in I^0} \left(\sum_{l=1}^{N} y_j y_l K_i(\omega_j, \omega_l)\lambda_l\right)\eta_i + y_j b + \delta_j \geq 1 - \sum_{i \in I^+} \sum_{l=1}^{N} y_j y_l K_i(\omega_j, \omega_l)\lambda_l, \\ \delta_j \geq 0, j = 1, \ldots, N, 0 \leq \eta_i \leq 1, i \in I^0. \end{cases}$$

So, the proposed approach has a very significant qualitative advantage over the other methods – it explicitly indicates a discrete subset of support kernels within the combination, in contrast to other methods that assign some positive (even if small) weight to *each* kernel, requiring significantly greater memory.

5   Experimental Design

5.1   Multi-Kernel Membrane Protein Prediction

Membrane Protein Prediction Problem. Membrane proteins comprise 20–30% of all proteins encoded by a genome and perform a variety of functions vital to the survival of organisms. Membrane proteins serve as receptors (i.e. sensors of the cells), transport molecules across the membrane, participate in energy production (ATP biosynthesis), in cell-cell interaction (cell adhesion), etc. [61]. They are targets of over 50% of all modern medicinal drugs [62]. Consequently, membrane protein prediction, i.e. the classification of proteins as either a membrane or non-membrane is a biomedically important problem, and the subject of much research [63, 65, 66].

Data Description. In this paper we use the same data set as Lanckriet et al. (described in [67]). We thus use as a gold standard of annotations provided by the Munich Information Center for Protein Sequences Comprehensive Yeast Genome Database (CYGD) [68]. The CYGD assigns subcellular locations to 2318 yeast proteins, of which 497 belong to various membrane protein classes. The remaining approximately 4000 yeast proteins have uncertain location and are therefore not used in these experiments.

Potential Functions for Membrane Protein Prediction. For membrane protein prediction, we use 14 potential functions, derived from 3 different data types: protein sequences, hydrophaty profile, protein-protein interactions data and gene expression information. Short description of these potential functions is presented in Table 1.

Table 1. Potential functions for membrane protein prediction.

N	Potential function	Data type	Description
1	K_B	protein sequences	BLAST similarity measure + Linear kernel in secondary feature space
2	K_{SW}	protein sequences	Smith-Waterman similarity measure + Linear kernel in secondary feature space
3	K_{LA}	protein sequences	Local Alignment Kernel
4	K_{GA}	protein sequences	Global Alignment Kernel
5	K_{Pfam}	protein sequences	HMM model, trained on Pfam database
6	K_{FFT}	hydropathy profile	Radial basis kernel for vectors of FFT values
7	K_{Li}	protein interactions	Linear kernel
8	K_D	protein interactions	Diffusion kernel
9	K_E	gene expression	Radial basis kernel
10–14	K_{Rnd1}–K_{Rnd5}	random numbers	Radial basis kernel for random features

Potential functions $K_B, K_{SW}, K_{Pfam}, K_{FFT}, K_{Li}, K_D$ and K_E have been collected by Lanckriet et al. [67]. The respective kernel matrices, along with the data from which they were generated are available at https://noble.gs.washington.edu/proj/sdp-svm.

Additionally, two alignment-based potential functions were computed – Local Alignment Kernel K_{LA} (29) and Global Alignment Kernel K_{GA} (28), and also five random kernels $K_{Rnd1}, \ldots, K_{Rnd5}$ on the basis of 100-length feature vectors were randomly generated without taking into account labeling information about the classes of proteins. These non-informative kernels were introduced in order to check the ability of the proposed procedure to eliminate non-useful information.

Matrices of potential functions K_{LA} and K_{GA} were computed with using the equipment of the shared research facilities of HPC computing resources at Lomonosov Moscow State University [63].

For each of K_{LA} and K_{GA} matrices normalization was made in accordance with (17) and a degree of its slope was decreased (18).

Experimental Setup.
The full set of 2318 proteins (497 membrane proteins and 1821 non-membrane proteins) was randomly split 30 times into training and test sets in the proportion 80:20. As a result, each training set contained 397 membrane proteins and 1456 non-membrane proteins. Each of the test sets contain, respectively, 100 membrane proteins and 365 non-membrane proteins.

For each of 30 obtained training sets, we derived 21 different decision rules for membrane protein prediction:

(1) For each of 9 informative and 5 random potential functions the traditional SVM training procedure was performed separately (Sect. 3);
(2) SVM classification on the unweighted sum of all 14 potential functions was also applied;
(3) For all 14 potential functions, the proposed Selective Supervised Selective Kernel SVM was performed 6 times with 6 different values of the selectivity-parameter (Sect. 4.2);
(4) The optimal decision rule was selected for the proposed method via 5-fold cross-validation.

As a pre-processing step, each kernel matrix was centered and normalized to be an identity diagonal matrix.

The quality of each decision was estimated via the ROC-score using the hyperplane bias b to vary sensitivity.

Results of Membrane Protein Prediction. The averages and standard deviations of ROC-scores, computed across 30 randomly generated 80:20 splits for each of 20 training conditions listed in the previous section, are presented in Table 2.

Table 2. Results of membrane protein prediction.

Kernels	Algorithm	μ	ROC-score	Kernels	Algorithm	μ	ROC-score
K_B	SVM	–	0.8302±0.031	K_{Rnd3}	SVM	–	0.515±0.030
K_{SW}	SVM	–	0.854±0.021	K_{Rnd4}	SVM	–	0.521±0.029
K_{LA}	SVM	–	0.873±0.022	K_{Rnd5}	SVM	–	0.509±0.029
K_{GA}	SVM	–	0.554±0.033	All 14	SVM	–	0.877±0.015
K_{Pfam}	SVM	–	0.859±0.022	All 14	SKSVM	0	0.877±0.015
K_{FFT}	SVM	–	0.776±0.014	All 14	SKSVM	5	0.915±0.015
K_{Li}	SVM	–	0.635±0.044	All 14	SKSVM	7	0.917±0.015
K_D	SVM	–	0.640±0.041	All 14	SKSVM	8	0.917±0.015
K_E	SVM	–	0.752±0.023	All 14	SKSVM	10	0.916±0.016
K_{Rnd1}	SVM	–	0.510±0.029	All 14	SKSVM	15	0.911±0.016
K_{Rnd2}	SVM	–	0.517±0.028	All 14	SKSVM	optimal	**0.918±0.015**

As we can see from Table 2, the results of the proposed supervised selective support kernel SVM outperform those obtained for each of 14 potential functions individually, and also those of the unweighted kernel sum with SVM training. The result obtained at the zero-selectivity level is exactly equal to the result obtained for the unweighted sum of potential functions, what supports the theoretical results above.

Moreover, it may be seen that practically all reasonable values of the selectivity parameter provide good results. The performance obtained using the optimal selectivity value selected via 5-fold cross-validation for each of 30 training sets individually only slightly outperforms the best result obtained using fixed selectivity levels.

This implies that the same selectivity-level is near optimal across the range of training sets (though of course a fixed selectivity level may be not appropriate for different tasks, for example, for recognition of different classes of proteins).

The reported results of membrane protein prediction obtained by another multiple kernel learning techniques [25, 51, 67] for the same data set lie in the range [0.87–0.917]. The proposed approach therefore achieves the top-ranked position of the methods reported in the literature. It should be noted, furthermore, that the proposed approach has the unique qualitative advantage of clearly delineating the subset of support kernels that participate in the decision rule, being thereby directly scientifically interpretable, and potentially assisting with further generation of experimental hypotheses.

To demonstrate this delineation of support kernels for one of 30 training sets, Table 3 contains the results of the partitionings of the full set of 14 kernels into three subsets: (1) the subset of kernels I^-, which were classified by the algorithm as non-support ones, and which are not weighted or included in the decision rule; (2) the subset of kernels I^+ having unit weight, and (3) the subset of kernels I^0 having a weight between 0 and 1. Only the kernels of subsets I^+ and I^0 are support kernels and participate in the decision rule.

As we can see from Table 3, the highest selectivity value excludes the random kernels from the set of support kernels entirely, as well as Global Alignment Kernel K_{GA}, what is expectable because of the presence of sequences of essentially different

Table 3. Kernel fusion results: support kernel weights for different selectivity values μ (non-support kernels are indicated by '$\times$').

Kernel	$\mu = 0$	$\mu = 5$	$\mu = 7$	$\mu = 8$	$\mu = 10$	$\mu = 15$
K_B	1	1	0,195	$\times$	$\times$	$\times$
K_{SW}	1	1	1	1	0,643	$\times$
K_{LA}	1	1	1	1	1	1
K_{GA}	1	$\times$	$\times$	$\times$	$\times$	$\times$
K_{Pfam}	1	1	1	0,776	0,513	0,400
K_{FFT}	1	1	1	1	0,859	0,533
K_{Li}	1	0,547	$\times$	$\times$	$\times$	$\times$
K_D	1	1	1	1	0,894	0,462
K_E	1	1	1	0,794	0,558	0,223
K_{Rnd1}	1	0,214	$\times$	$\times$	$\times$	$\times$
K_{Rnd2}	1	0,135	$\times$	$\times$	$\times$	$\times$
K_{Rnd3}	1	0,073	$\times$	$\times$	$\times$	$\times$
K_{Rnd4}	1	0,147	$\times$	$\times$	$\times$	$\times$
K_{Rnd5}	1	0,242	$\times$	$\times$	$\times$	$\times$
ROC	0,875	0,917	0,920	0,919	0,918	0,909

lengths within one class of proteins. Local Alignment Kernel K_{LA} is more useful in this situation, what is confirmed by its best ROC-score among all potential functions and unit weight for all selectivity values.

Also, interaction-based linear kernel K_{Li} was excluded in most cases, while another interaction-based kernel KD was always excluded.

This result can be explained as follows. The information about protein-protein interactions can be useful for membrane protein prediction for two reasons. First, hydrophobic molecules or regions of molecules tend to interact with each other. Second, transmembrane proteins are often involved in signaling pathways, and therefore different membrane proteins are likely to interact with a similar class of molecules upstream and downstream in these pathways. At the same time, the diffusion kernel involves the information about interactions more carefully and provides essentially more accurate way for comparing protein sequences in contrast to the linear kernel.

Finally, we can see that only half (6 of 12) of the full kernel set are support kernels in this example, saving on memory requirements. Thus, in sum, this particular feature of the proposed method makes it preferable to other multi-kernel methods within the literature, which generally assign positive weights to all kernels.

6 Conclusion

This paper contains a comprehensive survey of possible ways for potential functions design on sets of signals and symbolic sequences. Significant emphasis is placed on a generalized probabilistic approach to construction of potential functions. This approach covers both vector signals and symbolic sequences at once and leads to a large family of potential functions based on the notion of a random transformation of signals and sequences, which can underlie, in particular, probabilistic models of evolution of biomolecular sequences. We show that some specific choice of the sequence random transformation allows to obtain such important particular cases as Global Alignment Kernel and Local Alignment Kernel. The second part of the paper addresses the multi-kernel situation, which is extremely actual, in particular, due to the necessity to combine information from different sources. A generalized probabilistic featureless SVM-based approach to combining different data sources via supervised selective kernel fusion was proposed in our previous papers. In this paper we demonstrate significant qualitative advantages of the proposed approach over other methods of kernel fusion on example of membrane protein prediction.

Acknowledgements. The research is carried out using the equipment of the shared research facilities of HPC computing resources at Lomonosov Moscow State University.

The results of the research project are published with the financial support of Tula State University within the framework of the scientific project № 2017-63PUBL.

References

1. Duin, R.P.W., De Ridder, D., Tax, D.M.J.: Experiments with a featureless approach to pattern recognition. Pattern Recogn. Lett. **18**(11–13), 1159–1166 (1997)
2. Mottl, V.V., Dvoenko, S.D., Seredin, O.S., Kulikowski, C.A., Muchnik, I.B.: Featureless pattern recognition in an imaginary hilbert space and its application to protein fold classification. In: Proceedings of the II-th International Workshop on MLDM in Pattern Recognition, 2001, pp. 322–336 (2001)
3. Braverman, E.M.: Experiments on training a machine for pattern recognition. Ph.D. Thesis. Moscow (1961)
4. Vapnik, V.N.: Statistical Learning Theory. Wiley-Interscience (1998). 768 p
5. Aizerman, M.A., et al.: Potential functions method in machine learning theory (in Russian). M.: Nauka (1970). 384 p
6. Mercer, T.: Functions of positive and negative type and their connection with the theory of integral equations. Trans. London. Philos. Soc. **A 209**, 415–416 (1999)
7. Mottl, V.V.: Metric spaces, assuming introducing linear operations and inner products. Reports of the RAS. — 2003. 388(3):1–4 (2003). (In Russian)
8. Haussler, D.: Convolution kernels on discrete structures. Technical report. University of California (1999)
9. Cuturi, M., Vert, J.-P., Birkenes, Ø., Matsui, T.: A kernel for time series based on global alignments. In: Proceedings of ICASSP, vol. II, pp. 413–416 (2007)
10. Gordon, L., Chervonenkis, A., Gammerman, A., Shahmuradov, I., Solovyev, V.: Sequence alignment kernel for recognition of promoter regions. Bioinformatics **19**(15), 1964–1971 (2003). https://doi.org/10.1093/bioinformatics/btg265
11. Saigo, H., Vert, J.-P., Ueda, N., Akutsu, T.: Protein homology detection using string alignment kernels. Bioinformatics **20**, 1682–1689 (2004)
12. Rogen P., Fain B.: Automatic classification of protein structure by using Gauss integrals. Proc. Natl. Acad. Sci. USA. 200;100(1), 119–24. https://doi.org/10.1073/pnas.2636460100
13. Genton, M.G.: Classes of kernels for machine learning: a statistics perspective. J. Mach. Learn. Res. **2**, 299–312 (2001)
14. Needleman, S.B., Wunsch, C.D.: A general method applicable to the search for similarities in the amino acid sequence of two proteins. J. Mol. Biol. **48**(3), 443–453 (1970). https://doi.org/10.1016/0022-2836(70)90057-4
15. Smith, T.F., Waterman, M.S.: Identification of common molecular subsequences. J. Mol. Biol. **147**(1), 195–197 (1981). https://doi.org/10.1016/3960022-2836(81)90087-5
16. Zhang, Z., Schwartz, S., Wagner, L., Miller, W.: A greedy algorithm for aligning DNA sequences. J. Comput. Biol. **7**, 203–214 (2000). https://doi.org/10.1089/3991066527005 0081478
17. Durbin, R., Eddy, S., Krogh, A., Mitchison, G.: Biological Sequences Analysis: Probabilistic Models of Proteins and Nucleic Acid. Cambridge University Press, Cambridge (1998)
18. Sulimova, V.V., Seredin, O.S., Mottl, V.V.: Metrics on the basis of optimal alignment of biological sequences (In Russian). J. Mach. Learn. Data Min. **2**(3), 286–304 (2016)
19. Sakoe, H., Chiba, S.: Dynamic programming algorithm optimization for spoken word recognition. IEEE Trans. Acoust. Speech Signal Process. **26**(1), 43–49 (1978)
20. Malenichev, A., Sulimova, V., Krasotkina, O., Mottl, V., Markov, A.: An automatic matching procedure of ultrasonic railway defectograms. In: Perner, P. (ed.) MLDM 2014. LNCS (LNAI), vol. 8556, pp. 315–327. Springer, Cham (2014). https://doi.org/10.1007/978-3-319-08979-9_24

21. Salvador, S., Chan, P.: Toward accurate dynamic time wrapping in linear time and space. Intell. Data Anal. **11**(5), 561–580 (2007)
22. Al-Naymat, G., Chawla, S., Taheri, J.: SparseDTW: A Novel Approach to Speed up Dynamic Time Warping (2012)
23. Lei, H., Sun, B.: A study on the dynamic time warping in kernel machines. In: Proceedings of the 2007 Third International IEEE Conference on Signal-Image Technologies and Internet-Based System, pp. 839–845 (2007)
24. Pekalska, E., Paclic, P., Duin, R.: A generalized kernel approach to dissimilarity-based classification. J. Mach. Learn. Res. **2001**(2), 175–211 (2001)
25. Liao, L., Noble, W.S.: Combining pairwise sequence similarity and support vector machines for remote protein homology detection. In: Proceedings of the Sixth Annual International Conference on Computational Molecular Biology, pp. 225–232 (2002)
26. Schölkopf, B., Tsuda, K., Vert, J.-P.: Kernel Methods in Computational Biology. MIT Press, Cambridge (2004). 410 p
27. Ben-Hur, A., Ong, C.S., Sonnenburg, S., Schölkopf, B., Rätsch, G.: Support vector machines and kernels for computational biology. PLoS Comput. Biol. **4**(10), 1–10 (2008)
28. Mottl, V., Lange, M., Sulimova, V., Yermakov, A.: Signature verification based on fusion of on-line and off-line kernels. In: 19-th International Conference on Pattern Recognition. Florida, Tampa (2008)
29. Mottl, V., Seredin, O., Krasotkina, O.: Compactness hypothesis, potential functions, and rectifying linear space in machine learning. In: Key Ideas in Learning Theory from Inception to Current State: Emmanuel Braverman's Legacy. Springer (2017)
30. Vert, J.-P., Saigo, H., Akutsu, T.: Local alignment kernels for biological sequences. In: Schölkopf, B., Tsuda, K., Vert, J. (eds.) Kernel Methods in Computational Biology, pp. 131–154. MIT Press (2004)
31. Qiu, J., Hue, M., Ben-Hur, A., Vert, J.-P., Noble, W.S.: A structural alignment kernel for protein structures. Bioinformatics **23**(9), 1090–1098 (2007)
32. Sun, L., Ji, S., Ye, J.: Adaptive diffusion kernel learning from biological networks for protein function prediction. BMC Bioinf. **9**, 162 (2008)
33. Cuturi, M., Vert, J.-P.: The context-tree kernel for strings. Neural Network (2005)
34. Jaakkola, T.S., Diekhans, M., Haussler, D.: Using the Fisher kernel method to detect remote protein homologies. In: Proceedings of the Seventh International Conference on Intelligent Systems for Molecular Biology, pp. 149–158 (1999)
35. Mottl, V.V., Muchnik, I.B., Sulimova, V.V.: Kernel functions for signals and symbolic sequences of different length. In: International Conference on Pattern Recognition and Image Analysis: New Information technologies, pp. 155–158 (2007)
36. Dayhoff, M., Schwarts, R., Orcutt, B.: A model of evolutionary change in proteins Atlas of prot seq and structures. Nat. Biometr. Res. Found. **5**(3), 345–352 (1978)
37. And, H.S., Henikoff, J.: Amino acid substitution matrices from protein blocks. Proc. Nat. Acad. Sci. **1992**, 10915–10919 (1992)
38. Sulimova, V., Mottl, V., Kulikowski, C., Muchnik, I.: Probabilistic evolutionary model for substitution matrices of PAM and BLOSUM families. DIMACS Technical Report 2008-16. DIMACS, Center for Discrete Mathematics and Theoretical Computer Science, Rutgers University, New Jersey, USA (2008). 17 p., ftp://dimacs.rutgers.edu/pub/dimacs/TechnicalReports/TechReports/2008/2008-16.pdf
39. Watkins, C.: Dynamic alignment kernels. Technical Report (1999)
40. Seeger, M.: Covariance kernels from bayesian generative models. Adv. Neural Inform. Process. Syst. **14**, 905–912 (2002)

41. Miklos, I., Novak, A., Satija, R., Lyngso, R., Hein, J.: Stochastic models of sequence evolution including insertion-deletion events. Statistical Methods in Medical Research: 29 (2008)

42. Mottl, V.V., Muchnik, I.B., Sulimova, V.V.: Kernel functions for signals and symbolic sequences of different length. In: International Conference on Pattern Recognition and Image Analysis: New Information Technologies. Yoshkar-Ola, pp. 155–158 (2007)

43. Sulimova, V.V.: Potential functions for analysis of signals and symbolic sequences of different length. Tula. Ph.D. Thesis (2009). 122 p

44. Sulimova, V., Razin, N., Mottl, V., Muchnik, I., Kulikowski, C.: A maximum-likelihood formulation and EM algorithm for the protein multiple alignment problem. In: Dijkstra, Tjeerd M.H., Tsivtsivadze, E., Marchiori, E., Heskes, T. (eds.) PRIB 2010. LNCS, vol. 6282, pp. 171–182. Springer, Heidelberg (2010). https://doi.org/10.1007/978-3-642-16001-1_15

45. Lanckriet, G., et al.: A statistical framework for genomic data fusion. Bioinformatics **20**, 2626–2635 (2004)

46. Ong, C.S., et al.: Learning the kernel with hyperkernels. J. Mach. Learn. Res. **6**, 1043–1071 (2005)

47. Bie, T., et al.: Kernel-based data fusion for gene prioritization. Bioinformatics **23**, 125–132 (2007)

48. Bach, F.R., et al.: Multiple kernel learning, conic duality, and the SMO algorithm. In: Proceedings of the Twenty-first International Conference on Machine Learning (ICML04). Omnipress, Banff, Canada (2004)

49. Sonnenburg, S., Röatsch, G., Schöafer, C., Schölkopf, B.: Large scale multiple kernel learning. J. Mach. Learn. Res. **7**, 1531–1565 (2006)

50. Hu, M., Chen, Y., Kwok, J.T.-Y.: Building sparse multiple-kernel SVM classifiers. IEEE Trans. Neural Networks **20**(5), 827–839 (2009)

51. Gönen, M., Alpayd, E.: Multiple kernel machines using localized kernels. In: Proceedings of PRIB (2009)

52. Gönen, M., Alpayd, E.: Localized algorithms for multiple kernel learning. Pattern Recogn. **46**, 795–807 (2013)

53. Cortes, C., Mohri, M., Rostamizadeh, A.: Learning non-linear combinations of kernels. In: Bengio, Y. et al. (eds.) Advances in Neural Information Processing Systems, vol. 22, pp. 396–404 (2009)

54. Mottl, V., Tatarchuk, A., Sulimova, V., Krasotkina, O., Seredin, O.: Combining pattern recognition modalities at the sensor level via kernel fusion. In: Proceedings of the IW on MCS (2007)

55. Kloft, M., Brefeld, U., Sonnenburg, S., et al.: Efficient and accurate lp-norm multiple kernel learning. In: Bengio, Y., et al. (eds.) Advances in Neural Information Processing Systems, vol. 22, pp. 997–1005. MIT Press (2009)

56. Tatarchuk, A., Mottl, V., Eliseyev, A., Windridge, D.: Selectivity supervision in combining pattern-recognition modalities by feature- and kernel-selective Support Vector Machines. In: Proceedings of the ICPR (2008)

57. Tatarchuk, A., Sulimova, V., Windridge, D., Mottl, V., Lange, M.: Supervised selective combining pattern recognition modalities and its application to signature verification by fusing on-line and off-line kernels. In: Proceedings of the IW on MCS (2009)

58. Tatarchuk, A., Urlov, E., Mottl, V., Windridge, D.: A support kernel machine for supervised selective combining of diverse pattern-recognition modalities. In: El Gayar, N., Kittler, J., Roli, F. (eds.) MCS (2010)

59. Bradley P., Mangasarian O.: Feature selection via concave minimization and support vector machines. In: International Conference on Machine Learning (1998)

60. Wang, L., Zhu, J., Zou, H.: The doubly regularized support vector machine. Stat. Sinica **16**, 589–615 (2006)
61. Alberts, B., Bray, D., Lewis, J., et al.: Molecular Biology of the Cell, 3rd edn, p. 1361. Garland Publishing, New York and London (1994)
62. Overington, J.P., Al-Lazikani, B., Hopkins, A.L.: How many drug targets are there? Nat. Rev. Drug. Discov. **5**(12), 993–996 (2006)
63. Voevodin, V.V., Zhumatiy, S.A., Sobolev, S.I., Antonov, A.S., Bryzgalov, P.A., Nikitenko, D.A., Stefanov, K.S., Voevodin, V.V.: Practice of 'Lomonosov' supercomputer. Open Syst. **7**, 36–39 (2012). Moscow: "Open Systems" Publishing house, (in Russian)
64. Krogh, A., Larsson, B., von Heijne, G., Sonnhammer, E.L.L.: Predicting transmembrane protein topology with a hidden markov model: application to complete genomes. J. Mol. Biol. **305**, 567–580 (2001)
65. Chen, C.P., Rost, B.: State-of-the-art in membrane protein prediction. Appl. Bioinf. **1**, 2135 (2002)
66. Gao, F.P., Cross, T.A.: Recent developments in membrane-protein structural genomics. Genome Biol. **6**, 244 (2005)
67. Lanckriet, G., et al.: A statistical framework for genomic data fusion. Bioinformatics **20**, 2626–2635 (2004)
68. Mewes, H.W., et al.: MIPS: a database for genomes and protein sequences. Nucleic Acids Res. **28**, 37–40 (2000)

Braverman's Spectrum and Matrix Diagonalization Versus iK-Means: A Unified Framework for Clustering

Boris Mirkin[1,2(✉)]

[1] Department of Data Analysis and Machine Intelligence,
National Research University Higher School of Economics,
Moscow, Russian Federation
bmirkin@hse.ru
[2] Department of Computer Science, Birkbeck University of London, London, UK

Abstract. In this paper, I discuss current developments in cluster analysis to bring forth earlier developments by E. Braverman and his team. Specifically, I begin by recalling their Spectrum clustering method and Matrix diagonalization criterion. These two include a number of user-specified parameters such as the number of clusters and similarity threshold, which corresponds to the state of affairs as it was at early stages of data science developments; it remains so currently, too. Meanwhile, a data-recovery view of the Principal Component Analysis method admits a natural extension to clustering which embraces two of the most popular clustering methods, K-Means partitioning and Ward agglomerative clustering. To see that, one needs just adjusting the point of view and recognising an equivalent complementary criterion demanding the clusters to be simultaneously "large-sized" and "anomalous". Moreover, this paradigm shows that the complementary criterion can be reformulated in terms of object-to-object similarities. This criterion appears to be equivalent to the heuristic Matrix diagonalization criterion by Dorofeyuk-Braverman. Moreover, a greedy one-by-one cluster extraction algorithm for this criterion appears to be a version of the Braverman's Spectrum algorithm – but with automated adjustment of parameters. An illustrative example with mixed scale data completes the presentation.

1 Two Early Approaches by Braverman and His Team

1.1 Braverman's Algorithm Spectrum

The problem of clustering has been formulated by Misha Braverman as related to a set of objects $I = \{i_1, i_2, ..., i_N\}$ in the so-called potential field which is specified by a similarity function between objects $A(i, j)$, $i, j = 1, 2, ..., N$ [1,4,5]. A preferred potential function is defined by equation

$$\phi(d) = \frac{1}{1 + \alpha d^2} \tag{1}$$

© Springer Nature Switzerland AG 2018
L. Rozonoer et al. (Eds.): Braverman Readings in Machine Learning, LNAI 11100, pp. 32–51, 2018.
https://doi.org/10.1007/978-3-319-99492-5_2

where d is Euclidean distance between feature vectors (see, [4]). As is currently well recognised, this is what is referred to as a kernel, one of the most important concepts in machine learning theory [12]. That is a similarity function which forms a positive semidefinite function at every finite set of objects. Moreover, when depending on the Euclidean distance between objects as elements of a Euclidean space of a finite dimension, any kernel function admits a finite set of "eigen-functions" such that $\phi(d(i,j))$ can be expressed as a linear combination of products of values of the eigen-functions on objects $x_i, x_j, i, j \in I$.

To define his early batch, or parallel, clustering heuristic, Braverman introduces the concept of average similarity between a point x_i and subset of points S referred to as the potential of x_i inflicted by S,

$$A(x_i, S) = \sum_{x_j \in S} A(i, j)/|S| \tag{2}$$

where $|S|$ is the cardinality of S, that is, the number of elements in S. His algorithm *Spectrum* begins at arbitrary point x_1 to build a sequence $x_1, x_2, ..., x_N$ over I so that each next point x_{k+1} maximizes the similarity $A(x_{k+1}, S_k)$ (2) where $S_k = \{x_1, x_2, ..., x_k\}$, $k = 1, 3, ..., N-1$. The sequence of points is accompanied by the sequence of the average similarity values $A(x_2, S_1), A(x_3, S_2), ...,$ $A(x_N, S_{N-1})$. These two sequences form a spectrum, in Braverman's terminology, which can be illustrated with Fig. 1 replicating an image from Arkadiev and Braverman's book [4], p. 107. On this Figure, x-axis represents the sequence of objects, hand-written images of digits 1, 2, 3, 4, 5, and y-axis shows the levels of the average similarity $A(x_{k+1}, S_k)$. Normally, when a set of homogeneous clusters is present in the data, the graph of the spectrum should look like that on Fig. 1.

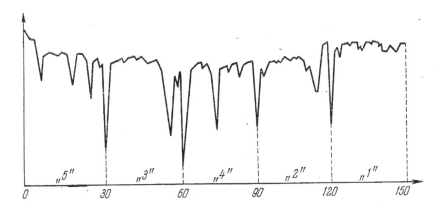

Fig. 1. A spectrum of hand-written images of digits 1, 2, 3, 4, 5; 30 copies of each. The x-axis represents the sequence of objects after application of algorithm Spectrum, and y-axis shows the levels of the average similarity $A(x_{k+1}, S_k)$.

In the current author's view, the graph on Fig. 1 looks somewhat unlikely. Consider, say, the deep drop between images for "5" and "3". Indeed, the image number 31, of "3", would look much different from the 30 images of "5". But the image number 32 is also much different from the previous 30 images, which makes the return of the curve to high levels immediately highly unlikely. A simple modification, though, can save the picture. Assume that the spectrum sequence breaks immediately just before the drop, and a new ordering procedure starts again. Then the averaging of the similarity with the previous cluster discontinues, and new averages are computed starting from the object 31. This is, I think, how a practical version of Spectrum algorithm was working. A threshold value for the drop of similarity has to be pre-chosen, so that a drop of the similarity value below that level would stop the algorithm's run at a found cluster. After this, the found cluster is removed, and another run of the algorithm is performed at the remaining objects, possibly with a different threshold value. This goes on till the set of remaining objects gets empty. In the follow-up this version of Spectrum will be referred to as Spectrum-B. It should be mentioned that there is a degree of similarity between an earlier attempt at clustering by Holzinger and Harman [11] and Spectrum-B.

1.2   Diagonalization of Similarity Matrices

This is another idea of Braverman's team, probably generated by the work over the PhD thesis by Alex Dorofeyuk [6,8]. Given a similarity matrix $A = (a_{ij})$, consider criterion of goodness of a partition $S = \{S_1, S_2, ..., S_K\}$ with a prespecified number of clusters K by scoring it according to formula

$$f(S) = \sum_{k=1}^{K} \frac{1}{N_k} \sum_{i,j \in S_k} a_{ij} \tag{3}$$

where N_k is the cardinality of cluster S_k $(k = 1, 2, ..., K)$.

Criterion (3) has been selected by the authors as the best performer out of a family of criteria

$$f(S) = \sum_{k=1}^{K} \frac{1}{\phi(N_k)} \sum_{i,j \in S_k} a_{ij},$$

where $\phi(S)$ is either $\phi(S) = 1$ or $\phi(S) = 1/|S|$ or $\phi(S) = 1/|S|(|S|-1)$. Of course, the experimental base was much limited at the time. And there are obvious drawbacks of the other two criteria in the family. Indeed, at $\phi(S) = 1$ the criterion is just the sum of within-cluster similarities. At non-negative similarity matrix $A = (a_{ij})$ this criterion would lead to a trivial optimal partition S at which all elements gather into the same "big" cluster, whereas all other clusters would be singletons consisting of the weakest links. In contrast, at $\phi(S) = 1/|S|(|S| - 1)$, the criterion would be proportional to the average within cluster similarity. Maximization of this criterion normally would prohibit large-sized clusters because the average similarity may only decrease when a cluster size grows. This leaves criterion (3) as the only option remaining for getting normal-size clusters. In the follow-up we will see that this is not just a lucky occurrence but rather a model-based property.

2   K-Means Clustering as a Data Recovery Method

2.1   K-Means Algorithm and Criterion

K-Means is arguably the most popular clustering algorithm. For an empirical proof of this statement, one may wish to consult [3] and references therein. Specifically, the following Table 1 from [3] clearly demonstrates the prevalence of K-means over other clustering techniques.

Table 1. Numbers of relevant web pages returned by the most popular search engines with respect to queries of the named methods at a computer in Birkbeck University of London (15 November 2015).

Search engine	Google	Bing	Yahoo
K-means	2,070,000	481,000	537,000
Hierarchical clustering	677,000	251,000	268,000
Neighbor-joining	591,000	146,000	148,000
Spectral clustering	202,000	71,500	78,100
Single linkage	140,000	30,900	32,800
Agglomerative clustering	130,000	33,100	33,000

Another, less controversial, statement would be that K-means has nothing to do with Braverman's team developments described above. Here is a conventional formulation of K-Means as a method for the analysis of an object-to-feature dataset.

Batch K-Means

0. *Data pre-processing.* Transform data into a standardized quantitative $N \times V$ matrix Y where N is the number of objects and V, the number of quantified features.
1. *Initial setting.* Choose the number of clusters, K, and tentative centers $c_1, c_2, ..., c_K$, frequently referred to as seeds. Assume initial cluster lists S_k empty.
2. *Clusters update.* Given K centers, determine clusters S'_k ($k = 1, ..., K$) with the Minimum distance rule assigning any object to its nearest center.
3. *Stop-condition.* Check whether $S' = S$. If yes, end with clustering $S = \{S_k\}$, $c = \{c_k\}$. Otherwise, change S for S'.
4. *Centers update.* Given clusters S_k, calculate within cluster means c_k ($k = 1, ..., K$) and go to Step 2.

This algorithm usually converges fast, depending on the initial setting. Location of the initial seeds may affect not only the speed of convergence but, more importantly, the final results as well.

As is well known, there is a scoring function, which is minimized by K-Means. To formulate the function, let us define the within cluster error. For a cluster S_k with center $c_k = (c_{kv})$, $v \in V$, its square error is defined as the summary distance from its elements to c_k:

$$W(S_k, c_k) = \sum_{i \in S_k} d(y_i, c_k) = \sum_{i \in S_k} \sum_{v \in V} (y_{iv} - c_{kv})^2. \tag{4}$$

The square error criterion is the sum of these values over all clusters:

$$W(S, c) = \sum_{k=1}^{K} W(S_k, c_k) = \sum_{k=1}^{K} \sum_{i \in S_k} d(y_i, c_k) \tag{5}$$

Criterion $W(S, c)$ (5) depends on two groups of arguments: cluster lists S_k and centers c_k. An alternating minimization algorithm for this criterion would proceed in a series of iterations. At each of the iterations, $W(S, c)$ is, first, minimized over S, given c, and, second, minimized over c, given the resulting S. It is not difficult to see that the batch K-Means above is such an alternating minimization algorithm This warrants that K-Means converges in a finite number of steps because the set of all partitions S over a finite I is finite and $W(S, c)$ is decreased at each change of c or S. Moreover, as experiments show, K-Means typically does not move far away from the initial setting of c. Considered from the perspective of minimization of criterion (5), this leads to the conventional strategy of repeatedly applying the algorithm starting from various randomly generated sets of seeds to reach as deep a minimum of (5) as possible. This strategy may fail especially if the feature set is large because in this case random settings cannot cover the space of solutions in a reasonable time.

Yet, there is a different perspective, of typology making, in which the criterion is considered not as something that must be minimized at any cost but rather a beacon for direction. In this perspective, the algorithm is a model for developing a typology represented by the centers. The centers should come from an external source such as advice by experts, leaving to data analysis only their adjustment to real data. In this perspective, the property that the final centers are not far away from the original ones, is more of an advantage than not. What is important in this perspective, though, is defining an appropriate, rather than random, initial setting.

2.2   Data Recovery Equation: Encoder and Decoder

According to conventional wisdom, the data recovery approach is a cornerstone of contemporary thinking in statistics and data analysis. It is based on the assumption that the observed data reflect a regular structure in the phenomenon of which they inform. The regular structure A, if known, would produce data $F(A)$ that should coincide with the observed data Y up to small residuals which are due to possible flaws in any or all of the following three aspects: (a) sampling entities, (b) selecting features and tools for their measurements, and (c) modeling the phenomenon in question. Each of these can drastically affect results.

However, so far only the simplest of the aspects, (a), has been addressed by introduction of probabilities to study the reliability of statistical inference in data analysis. In this text we are not concerned with these issues. We are concerned with the underlying equation:

Observed data Y = Recovered data $F(A)$ + Residuals E $(*)$

In this equation, the following terminology applied in the context of unsupervised learning is getting popular [15]. Data model A such as, for example, partition, is referred to as "encoded data" produced with an encoder, whereas the recovered data, $F(A)$, are those decoded with a decoder. The quality of the encoded data A is assessed according to the level of residuals E: the smaller the residuals, the better the model. Since both encoder and decoder methods involve unknown coefficients and parameters, this naturally leads to the idea of fitting these parameters to data in such a way that the residuals become as small as possible, which can be captured by the least squares criterion.

Data analysis involves two major activities: summarization and correlation [15]. In machine learning, their counterparts are unsupervised learning and supervised learning, respectively. In a correlation problem, there is a target feature or a set of target features that are to be related to other features in such a way that the target feature can be predicted from values of the other features. Such is the linear regression problem. In a summarization problem, such as the Principal component analysis or clustering, all the features available are considered target features so that those to be constructed as a summary can be considered as "hidden input features" (see Fig. 2).

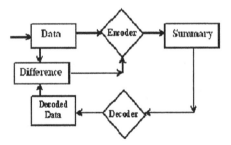

Fig. 2. A diagram for data recovery summarization. Rectangles are for data, both observed and computed, rhombs are for computational constructions. A double feedback shows the ways for adjustment of the encoder according to both the data and decoder.

Considering the structure of a summarization problem as a data recovery problem, one should rely on existence of a rule, decoder, providing a feedback from the summary back to the data. This makes it possible to use the criterion of minimization of the difference between the original data and the output so that the less the difference, the better. In supervised learning, this criterion works only for the feature(s) which is being predicted. Here all the data is to be approximated by the encoded model. This leads to a number of difficulties such as, for example, the issue of data standardization, which, in the supervised context, can be addressed by using mainly the feature being predicted.

The least squares perspective gives us a framework in which K-Means, Spectrum and Matrix diagonalization become comparable, as will be shown further on.

2.3 Principal Component Analysis Extended to Clustering

Principal Component Analysis is a major tool for approximating observed data with model data formed by a few 'hidden' factors. Observed data such as marks of students $i \in I$ at subjects labeled by $v = 1, ..., V$ constitute a data matrix $X = (x_{iv})$. Assume that each mark x_{iv} reflects student i abilities over a set of K hidden talent factors z_{ik}, c_{vk} up to coefficients $z_{ik}, c_{vk}, (i \in I, k = 1, 2, ..., K)$. The principal component analysis model [15], suggests that the student i's marks x_{iv} reflect the inner product of the hidden talent factor scores of the student i, $z_i = (z_{ik})$ and subject v loadings, $c_v = (c_{vk})$. Then equation (∗) can be examplified as

$$x_{iv} = <c_v, z_i> + e_{iv}, \tag{6}$$

where the inner product $<c_v, z_i>$ is a specific decoder leading to a rather remarkable, spectral, solution. The least squares criterion is $L^2 = \sum_{i \in I} \sum_{v \in V} (x_{iv} - \sum_{k=1}^{K} c_{vk} z_{ik})^2$.

In matrix terms Eq. 6 can be rewritten as

$$X = ZC^T + E, \tag{7}$$

where Z is $N \times K$ matrix of hidden factor scores $Z = (z_{ik})$, C is $V \times K$ matrix of subject loadings $C = (c_{vk})$, and $E = (e_{iv})$ is the matrix of residuals. The least squares criterion can be put as $L^2 = ||E||^2 = Tr(E^T E)$ to minimize. Of course, the solution Z, C to this problem is specified up to any unitary $K \times K$ matrix U so that pair ZU, CU leads to the same product $ZUU^T C^T = ZC^T$ since UU^T is an identity matrix. That means that the solution to the problem is not unique but rather specifies a subspace of rank K. Let us denote non-trivial singular triplets of matrix X as (μ_k, z_k, c_k), $k = 1, ..., r$, where r is the rank of X, $\mu_k > 0$ is a singular value, z_k is a normed N-dimensional singular vector $(z_k = (z_{ik})$, c_k is V-dimensional normed vector $c_k = (c_{vk})$ such that $Xc_k = \mu_k z_k$ and $X^T z_k = \mu_k c_k$.

It is not difficult to prove that (μ_k, z_k, c_k) is a singular triplet for matrix X if and only if μ_k^2 is eigen-value of the square matrix $X^T X$ corresponding to eigen-vector c_k, and $z_k = Xc_k/\mu$. This implies that the singular vectors $c_1, c_2, \ldots, c_K$ are mutually orthogonal as well as vectors $z_1, z_2, ..., z_K$.

Provided that $K < r$, the first-order optimality conditions for the least-squares criterion lead to the first K singular triplets forming its minimizer as the matrix $Z_K M_K C_k^T$ where Z_K and C_K are matrices whose columns are the first K singular vectors, z_k and c_k, respectively, and M_K is the diagonal matrix of the first K singular values μ_k, $k = 1, 2, ..., K$. This implies that $Z = Z_K(M_K)^{1/2}$ and $C = C_K(M_K)^{1/2}$ form solution to (7). Moreover, this solution provides for a Pythagorean decomposition of the data scatter:

$$||X||^2 = Tr(X^T X) = \mu_1^2 + \mu_2^2 + ... + \mu_K^2 + L^2 \tag{8}$$

This implies a one-by-one method for low-rank approximation of the data. You want a one-dimensional approximation? Take the first, the maximum, singular value. You want a visualization of the data on a plane? Take two. The descending order of the singular values implies the order of extraction of the principal components, one-by-one.

This one-by-one extraction approach of PCA was extended by the author to cluster analysis [14]. Specifically, Eq. (7) with criterion (8) was extended to the constraint that the unknown Z must be zero-one binary N-dimensional vectors to represent K clusters to be found. More precisely, any binary z_k one-to-one corresponds to subset of I, $S_k = \{i : z_{ik} = 1\}$.

The requirements that clusters do not overlap is translated as the constraint that binary vectors z_k are to be mutually orthogonal. Because of this, the square error criterion can be reformulated by putting the sum over k in the beginning while using the binary membership values to limit summation over i by the cluster S_k only:

$$L^2 = \sum_{i \in I} \sum_{v \in V} \left(x_{iv} - \sum_{k=1}^{K} c_{vk} z_{ik} \right)^2 = \sum_{k=1}^{K} \sum_{i \in S_k} \sum_{v \in V} (x_{iv} - c_{vk})^2.$$

This proves that the least-squares criterion for the model (7) in the clustering context, that is, under the orthogonal binarity of Z constraint, is but the square-error K-Means clustering criterion.

Thus, a proven fact is that K-Means is a clustering analogue to the Principal Component Analysis. A few conclusions from that:

i. A data scatter decomposition should hold.
ii. One-by-one strategy for extracting clusters should be valid.
iii. Reformulation in terms of feature-to-feature covariance and object-to-object similarity should be tried.

We do not mention some other analogies such as, for instance, a possibility of trying the spectral relaxation of the clustering model for obtaining clusters. These three will be covered in brief in the follow-up Sects. 2.4, 2.5 and 2.6.

2.4 Data Scatter Clustering Decomposition

Although the data scatter decomposition involving the square error clustering criterion (4) can be derived by using matrix algebra [14], we, however, do this here from the criterion itself:

$$W(S,c) = \sum_{k=1}^{K} \sum_{i \in S_k} \sum_{v=1}^{V} (y_{iv} - c_{kv})^2$$

$$= \sum_{k=1}^{K} \sum_{i \in S_k} \sum_{v=1}^{V} (y_{ik}^2 - 2y_{iv}c_{kv} + c_{kv}^2)$$

$$= \sum_{k=1}^{K} \sum_{i \in S_k} \sum_{v=1}^{V} y_{iv}^2 - \sum_{k=1}^{K} N_k <c_k, c_k>$$

where N_k is the number of elements in cluster S_k. The last equation is derived from the definition that $\sum_{k=1}^{K} \sum_{i \in S_k} \sum_{v=1}^{V} y_{iv}c_{kv} = \sum_{k=1}^{K} N_k \sum_{v=1}^{V} c_{kv}c_{kv}$ because $c_{kv} = \sum_{i \in S_k} y_{iv}/N_k$.

As $\sum_{k=1}^{K} \sum_{i \in S_k} \sum_{v=1}^{V} y_{iv}^2$ is but the data scatter by $T(Y) = \sum_{i=1}^{N} \sum_{v=1}^{V} y_{iv}^2$, denoting the right-hand term in the equation above, by

$$F(S,c) = \sum_{k=1}^{K} N_k <c_k, c_k>, \tag{9}$$

the equation above can be expressed as

$$T(Y) = F(S,c) + W(S,c) \tag{10}$$

which is the Pythagorean decomposition. It should be mentioned that this decomposition is well known in the analysis of variance, a classical part of mathematical statistics: at a centered matrix Y, the data scatter $T(Y)$ is proportional to the summary feature variance, whereas $F(S,c)$ and $W(S,c)$, respectively, to the inter-group and within-group summary variances.

In clustering, however, it is important on it own, because of the complementary criterion in (9) which is to be maximized to make $W(S,c)$ minimum. The value $F(S,c)$ is the part of the data scatter taken into account, that is, the contribution of clustering (S,c) to the data scatter.

The complementary criterion in (9) is the sum of contributions by individual clusters, $f(S_k, c_k) = N_k<c_k, c_k>$; each is the product of the cluster's cardinality and the squared distance from the cluster's center to the origin, 0. Provided that the origin preliminarily has been shifted into the point of 'norm' such as the gravity center, the problem of maximization of $F(S,c)$ is of finding as large-sized and as anomalous clusters as possible, to maximize the sum of cluster contributions $f(S_k, c_k) = N_k<c_k, c_k>$, $k = 1, 2, ..., K$.

2.5   One-By-One Strategy for Extracting Clusters

A procedure proposed by Mirkin [13] is an extension of the one-by-one principal component analysis method to the case at which the scoring components are constrained to represent clusters by having only 1/0 values. This procedure was later renamed as the method of anomalous clusters because of the criterion minimized by the algorithm. Denoting an anomalous cluster sought by S and its center by c, the criterion can be put as

$$D(S, c) = \sum_{i \in S} d(y_i, c) + \sum_{i \notin S} d(y_i, 0) \tag{11}$$

where 0 is the space origin.

   This criterion is akin to that of k-means. Two points of departure from the batch K-Means formulation are: (i) the number of clusters K = 2 and (ii) one of the centers is at 0 and never changes.

Anomalous cluster algorithm

   Input: $N \times V$ data matrix X.
   Output: A single cluster S and its center c as far away from 0, as possible.
   Step 1. **Initialization.** Choose an object as far away from 0 as possible and put c at this location.
   Step 2. **Cluster update.** Assign to S all those i for which $d(y_i, c) < d(y_i, 0)$.
   Step 3. **Center update.** Recompute c as the mean of the newly found S.
   Step 4. **Test.** If the new c coincides with the previous one, halt and output the current S and c. Otherwise, take the newly computed c and go to Step 2.

   After S is found, its elements can be removed from the entity set, so that next anomalous cluster could be found with the very same process, while the origin remains unchanged. Continuing the process of one-by-one extraction of anomalous clusters, one arrives at the situation when no non-clustered objects remain. Then all the small anomalous clusters are to be removed and centers of the remaining clusters are to be taken to initialise the K-means clustering process – this whole procedure is referred to as the intelligent K-Means (iK-Means) in [2,7,16]. The "small" clusters are defined as those containing a predefined number t or less elements.

   By default, t is usually taken as unity, $t = 1$. At synthetic data with Gaussian clusters generated, the iK-Means at $t = 1$ tends to produce about twice the number K^* of generated clusters. Indeed, in our computations, it never ever led to a smaller than K^* number of clusters [7]. This is why, in a recent paper, Amorim et al. [2] proposed further agglomeration of iK-Means clusters with the same criterion. It appears, such a hybrid method is much faster than the classic Ward agglomeration method while maintaining similar cluster recovery capabilities [2].

   Another comment that should be made is that the criterion in (11) is not exactly equivalent to the criterion of maximization of cluster's contribution $N_k < c_k, c_k >$, see [18] where a different algorithm is proposed. Here is an example

Table 2. Illustrative example of the difference between two single cluster criteria.

Object	x	y
A	−1	3
B	0	3
C	2	2
D	1	1
E	−1	1
F	0	1
G	−2	−1
H	0	−1

of a dataset at which the two criteria lead to different solutions from [18]. Consider a set of eight two-dimensional observations A to H in Table 2. Assume that the "norm" here is specified as the origin, point $0 = (0,0)$, and no normalization is required. Then the criterion would lead to two non-trivial clusters S1 = {A, B, C} and S2 = {D, E, F}, leaving G and H singletons. Anomalous Cluster method outputs only one nontrivial cluster, S1 = {A, B, C} here.

2.6 Reformulation of the Complementary Criterion in Terms of Object-To-Object Similarity

Consider the complementary clustering criterion

$$F(S,c) = \sum_{k=1}^{K} \sum_{v \in V} c_{kv}^2 N_k = \sum_{k=1}^{K} N_k d(0, c_k) \tag{12}$$

To maximize this criterion, the clusters should be as far away from 0 as possible. This idea is partly implemented in the Anomalous clustering algorithm above. A batch version is developed in [11] by using an accordingly modified version of the Affinity Propagation algorithm developed by Frey and Dueck [10] (see also URL http://scikit-learn.org/stable/modules/clustering.html\/# affinity-propagation).

However, in this account, I am not going to concentrate on this, but rather on reformulation of the criterion by using row-to-row inner products. Indeed, let us substitute one within cluster average c_k by its definition in (12): $<c_k, c_k> = <c_k, \sum_{i \in S_k} y_{kv}/N_k> = \sum_{i \in S_k} <c_k, y_i>/N_k$. This implies that

$$F(S,c) = \sum_{k=1}^{K} \sum_{i \in S_k} <y_i, c_k> \tag{13}$$

This expression shows that the K-Means criterion that minimizes the within-cluster distances to centroids is equivalent to the criterion (13) for maximization of the within-cluster inner products with centroids. By further substituting the same formula into (13), we arrive at equation

$$F(S,c) = \sum_{k=1}^{K} \sum_{i,j \in S_k} <y_i, y_j>/N_k \qquad (14)$$

expressing the criterion in terms of entity-to-entity similarities $a_{ij} = <y_i, y_j>$, with centroids c_k present implicitly. Criterion (14) is the total within-cluster semi-averaged similarity that should be maximized to minimize the least-squares clustering criterion.

2.7   Returning to Matrix Diagonalization

Obviously, by denoting $a_{ij} = \sum_v y_{iv} y_{jv}$, criterion (14) gets the form of Braverman-Dorofeyuk criterion in (3), that is, the semi-average within cluster similarity to maximize. Indeed, $a_{ij} = \sum_v a_{v,ij}$ is the sum of $a_{v,ij} = y_{iv} y_{jv}$, that are scores of the level of similarity between entities due to single features:

$$F(S,c) = \sum_{k=1}^{K} \frac{1}{N_k} \sum_{i,j \in S_k} a_{ij} \qquad (15)$$

Assuming that y_{iv} has been pre-processed by subtracting a reference value such as the grand mean $m_v = \sum_i^N y_{iv}/N$, one can see that $a_{v,ij}$ is negative if i and j differ over m_v so that y_{iv} and y_{jv} lie on v-axis at the different sides of zero. In contrast, $a_{v,ij}$ is positive if y_{iv} and y_{jv} lie on v-axis at the same side, either positive or negative. The greater the distance from 0 to y_{iv} and y_{jv}, the greater the value of $a_{v,ij}$. If the distribution is similar to a Gaussian one, most entities fall near grand mean, which is 0 after the normalization, so that most similarities are quite small.

Moreover, there is a claim in the literature that the inner product is much beneficial at larger sizes of the feature set if the data is pre-normalized in such a way that the rows, corresponding to entities, are normed so that each has its Euclidean norm equal to unity, so that the inner product becomes the cosine of the angle between the vectors.

This criterion has something to do with the spectral approach. Indeed, matrix $A = (a_{ij})$ can be expressed through the data matrix Y as $A = YY^T$ so that criterion (14), in a matrix form, is:

$$F(S,c) = \sum_{k=1}^{K} \frac{s_k^T A s_k}{s_k^T s_k} \qquad (16)$$

where $s_k = (s_{ik})$ is the binary membership vector for cluster S_k so that $s_{ik} = 1$ if $i \in S_k$ and $s_{ik} = 0$, otherwise. In mathematics, s_k is referred to as indicator of S_k. Proof of (16) follows from the fact that $s_k^T s_k = N_k$ and $\sum_{i,j \in S_k} a_{ij} = s_k^T A s_k$. This is the sum of Rayleigh quotients whose extremal values are eigenvalues of

A. Of course, this is similar to Rayleigh quotients in the Principal Component Analysis, except that solutions here must be binary zero-one vectors. Therefore, the contributions of individual clusters, $\sum_v c_{kv}^2 N_k$, are akin to the contributions of individual principal components, μ_k^2, both expressing the maximized values of the corresponding Rayleigh quotients, albeit over different domains.

One more comment should be about potentially using kernels, that is, functions $K(y_i, y_j)$ forming positive semi-definit matrices, to imitate the inner products $a_{ij} = <y_i, y_j>$. These functions were introduced by E. Braverman and his team to be used as a computationally feasible possibility of non-linearly transforming the feature space. Kernel "trick" is quite popular in clustering. However, it applies to the K-Means criterion itself leading to a much more complex and less interpretable formula than the expression (15) – see, for example, review in [9].

There can be several approaches to optimization of the criterion (15). Of course one of them is the agglomerative approach by Braverman and Dorofeyuk. However, I concentrate on an extension of the Anomalous clustering one-by-one approach to the case of similarity matrices.

2.8   Returning to Spectrum

Consider any item in the criterion (15) as a maximized criterion for finding an individual cluster. One may drop the index k in this section, since one cluster S is sought here. (Note that S here is not a partition but just a subset.)

Given a similarity matrix $A = (a_{ij})$, $i, j \in I$, let us define within-S average similarity as $a(S) = \sum_{i,j \in S} a_{ij}/N_S^2$ where N_S is the cardinality of S. Then the semi-average clustering criterion in (15) can be converted to

$$g(S) = \sum_{i,j \in S} a_{ij}/N_S = N_S a(S). \qquad (17)$$

This criterion combines two:

(i) maximize the cluster's tightness as measured by the within-cluster mean $a(S)$,
(ii) maximize the cluster size, measured by N_S.

These two goals are not quite compatible: the greater the cluster, the weaker the average within-cluster similarity. The product in (17), thus, balances them and, in this way, lead to a relatively tight cluster of a reasonable size.

Indeed, the tightness of S can be mathematically described by using the following concept [15]. For any entity $i \in I$, define its attraction to subset S as the difference between its average similarity to S, $a(i, S) = \sum_{j \in S} a_{ij}/N_S$, and half the average within-cluster similarity:

$$\alpha(i, S) = a(i, S) - a(S)/2. \qquad (18)$$

The fact that $a(S)$ is equal to the average $a(i, S)$ over all $i \in S$, leads us to expect that normally $a(i, S) \geq a(S)/2$ for the majority of elements $i \in S$, that is, normally $\beta(i, S) \geq 0$ for $i \in S$. It appears, a cluster S maximizing the criterion $A(S)$ is much more than that: S is cohesive internally and separated externally because all its members are positively attracted to S whereas non-members are negatively attracted to S [15].

Maximizing criterion $g(S)$ in (17) is akin to a combinatorially feasible problem of finding a subgraph of maximum density when the similarity matrix is non-negative [14]. Otherwise, it is non-polynomial. A local search algorithm would use a pre-defined neighborhood system to locally maximize the criterion at the neighborhood of a pre-specified subset S. In [15], the neighborhood system was considered to include $S \pm k$ for any k. Here, we take on a simpler neighbourhood system to include only those subsets $S + k$ obtained from S by adding to S any $k \notin S$.

Specifically, a version of algorithm ADDI [14] can be defined as follows.

Start from a singleton $S = \{i\}$ consisting of any object $i = 1, 2, ..., N$. Then proceed iteratively as follows. Given S, take an object $k \notin S$ maximizing the difference $\Delta(S, k) = g(S + k) - g(S)$. Check whether $\Delta(S, k) > 0$. If Yes, make $S = S + k$ and go to the beginning of the iteration. If not, stop and output S and its contribution to the data scatter $g(S)$ as well as the within-cluster average similarity $a(S)$.

This algorithm is a model-based version of Spectrum algorithm. Indeed, it is not difficult to prove that

$$\Delta(S, k) = \frac{N_S}{N_S + 1} \left(a(k, S) - \frac{a(S)}{2} + \frac{a_{kk}}{2N_S} \right) \tag{19}$$

Consider a simplifying assumption that $a_{kk} = 0$, so that clusters are defined by similarities between different objects only. Computationally, this assumption is easy to maintain by zeroing the main diagonal immediately before the start of a run of the ADDI algorithm. Then the maximum of $\Delta(S, k)$ corresponds to the maximum of the average similarity $a(k, S)$ between k and S, exactly as in Spectrum algorithm. But in ADDI, there is a natural stopping condition according to the sign of (19). Adding of elements stops whenever condition $a(k, S) < \frac{a(S)}{2}$ holds.

In this way, ADDI algorithm may be considered a theory based version of Spectrum. Moreover, ADDI not only defines a cluster-specific stopping condition, but also, the starting object: that must be that i maximizing $g(S)$ over S found at different i.

2.9   Illustrative Example

Consider an illustrative dataset in Table 3 after [16]. It relates to eight fictitious colleges in the UK described by five features.

Two features are of college sizes:

(1) Stud - The number of full time students;
(2) Acad - The number of full time teaching staff.

   Three features of teaching environment:

(3) NS - The number of different schools in a college;

(4) DL - Yes or No depending on whether the college provides distant e-learning courses or not;

(5) Course type - The course type is a categorical feature with three categories: (a) Certificate, (b) Master, (c) Bachelor, depending on the mainstream degree provided by the college.

The data in Table 3 can be utilized to cluster the set of colleges and describe clusters in terms of the features. One would ask whether the clusters are in line with the three main areas: Science, Engineering, Arts.

To analyze the data, one should first quantify the table by enveloping all the qualitative categories into binary zero-one features. Therefore, the Type of course will be represented by three yes/no features: is it MSc; is it Bsc; is it Certificate. Then Yes answer is coded by 1 and No answer by 0 (see Table 4).

Table 3. Colleges: Eight colleges: the first three mostly in Science, the next three in Engineering, and the last two in Arts. These categories are not supposed to be part of the data. They can be seen via first letters of the names, S, E and A, respectively.

College	Stud	Acad	NS	DL	Course type
Soli	3800	437	2	No	MSc
Semb	5880	360	3	No	MSc
Sixpe	4780	380	3	No	BSc
Etom	3680	279	2	Yes	MSc
Efin	5140	223	3	Yes	BSc
Enkee	2420	169	2	Yes	BSc
Ayw	4780	302	4	Yes	Certif.
Ann	5440	580	5	Yes	Certif.

Table 4. Quantitative representation of the Colleges data as an 8 × 7 entity-to-attribute matrix.

Entity	Stud	Acad	NS	DL	MSc	BSc	Certif.
1	3800	437	2	0	1	0	0
2	5880	360	3	0	1	0	0
3	4780	380	3	0	0	1	0
4	3680	279	2	1	1	0	0
5	5140	223	3	1	0	1	0
6	2420	169	2	1	0	1	0
7	4780	302	4	1	0	0	1
8	5440	580	5	1	0	0	1

Table 5. Range standardized Colleges matrix with the additionally rescaled nominal feature attributes; Mean is grand mean, Range the range and Cntr the relative contribution of a feature to the data scatter.

Item	Stud	Acad	NS	DL	MSc	BSc	Cer.
1	−0.20	0.23	−0.33	−0.63	0.36	−0.22	−0.14
2	0.40	0.05	0.00	−0.63	0.36	−0.22	−0.14
3	0.08	0.09	0.00	−0.63	−0.22	0.36	−0.14
4	−0.23	−0.15	−0.33	0.38	0.36	−0.22	−0.14
5	0.19	−0.29	0.00	0.38	−0.22	0.36	−0.14
6	−0.60	−0.42	−0.33	0.38	−0.22	0.36	−0.14
7	0.08	−0.10	0.33	0.38	−0.22	−0.22	0.43
8	0.27	0.58	0.67	0.38	−0.22	−0.22	0.43
Mean	4490	341.3	3.0	0.6	0.4	0.4	0.3
Range	3460	411	3.00	1.00	1.73	1.73	1.73
Cntr, %	12.42	11.66	14.95	31.54	10.51	10.51	8.41

Now this table can be standardized by subtracting from each column its average and dividing it by its range. To further balance the total contribution of the three categorical features on the right so that it is equal to the contribution of one feature represented by them, we divide them by the square root of 3. Then their part in the data scatter will be divided by 3 and the effect of tripling the original feature, Type of course, will be reversed [15]. The resulting data table is in Table 5.

Example 1. **Centers of subject clusters in Colleges data**

Let us consider the subject-based clusters in the Colleges data. The cluster structure is presented in Table 6 in such a way that the centers are calculated twice, once for the raw data in Table 4 and the second time, for the standardized data in Table 5.

Table 6. Means of the variables in Table 5 within K = 3 subject-based clusters, real (upper row) and standardized (lower row).

Cl.	List	Mean						
		St (f1)	Ac (f2)	NS (f3)	DL (f4)	B (f5)	M (f6)	C (f7)
1	1, 2, 3	4820	392	2.67	0	0.67	0.33	0
		0.095	0.124	−0.111	−0.625	0.168	−0.024	−0.144
2	4, 5, 6	3740	224	2.33	1	0.33	0.67	0
		−0.215	−0.286	−0.222	0.375	−0.024	0.168	−0.144
3	7, 8	5110	441	4.50	1	0.00	0.00	1
		0.179	0.243	0.500	0.375	−0.216	−0.216	0.433

Example 2. **Minimum   distance   rule   at   subject   cluster   centers in Colleges data**

Let us apply the Minimum distance rule to entities in Table 5, given the standardized centers c_k in Table 6. The matrix of distances between the standardized eight row points in Table 5 and three centers from Table 6 is in Table 7.

The table, as expected, shows that points 1, 2, 3 are nearest to centers c_1, 4, 5, 6 to c_2, and 7, 8 to c_3. This means that the rule does not change clusters. These clusters will have the same centers. Thus, no further calculations can change the clusters: the subject-based partition is to be accepted as the result.

But of course the algorithm may bring wrong results if started with a wrong set of centers, even if the initial setting fits well into clustering by subject.

Table 7. Distances between the eight standardized College entities and centers; within column minima are highlighted.

Centers	Entity, row point from Table 5							
	1	2	3	4	5	6	7	8
c_1	**0.22**	**0.19**	**0.31**	1.31	1.49	2.12	1.76	2.36
c_2	1.58	1.84	1.36	**0.33**	**0.29**	**0.25**	0.95	2.30
c_3	2.50	2.01	1.95	1.69	1.20	2.40	**0.15**	**0.15**

Table 8. Distances between the standardized Colleges entities and entities 1, 4, 7 as tentative centers.

Centers	Row-point							
	1	2	3	4	5	6	7	8
1	**0.00**	**0.51**	**0.88**	1.15	2.20	2.25	2.30	3.01
4	1.15	1.55	1.94	**0.00**	0.97	**0.87**	1.22	2.46
7	2.30	1.90	1.81	1.22	**0.83**	1.68	**0.00**	**0.61**

Example 3. **Unsuccessful K-Means run with subject-based initial seeds**

With the initial centers at rows 1, 4, and 7, all of different subjects, the entity-to-center matrix in Table 8 leads to cluster lists $S_1 = \{1, 2, 3\}$, $S_2 = \{4, 6\}$ and $S_3 = \{5, 7, 8\}$ which do not change in the follow-up operations. These results put an Engineering college among the Arts colleges. Not a good outcome.

Example 4. **Explained part of the data scatter**

The explained part of the data scatter, $F(S, c)$, is equal to 43.7% of the data scatter $T(Y)$ for partition $\{\{1, 4, 6\}, \{2\}, \{3, 5, 7, 8\}\}$, found with entities 1,2,3 as initial centroids. The score is 58.9% for partition $\{\{1, 2, 3\}, \{4, 6\}, \{5, 7, 8\}\}$, found with entities 1,4,7 as initial centroids. The score is maximum, 64.0%, for the subject based partition $\{\{1, 2, 3\}, \{4, 5, 6\}, \{7, 8\}\}$, which is thus superior.

Example 5. **Similarity matrix and clusters using ADDI algorithm**

The similarity matrix $A = YY^T$ in Table 9 is obtained from the standardized matrix Y in Table 5. Its structure pretty much corresponds to that underlying the Spectrum algorithm: there are three groups, S, E, and A, so that all the within-group similarities are positive, whereas almost all the inter-group similarities are negative, or quite small when positive, see a(Soli, Etom) = 0.086 and a(Efin, Aiw) = 0.09.

Let us apply algorithm ADDI to it starting from the most anomalous object, which is Ann with its squared Euclidean distance to 0 equal to 1.279. Then Aiw joins in, with a(Ann, Aiw) = 0.612. All the other objects have negative similarities with this cluster (assuming zeroing of all the diagonal elements while computing or not), so that the computation stops here: cluster A is complete.

After removal of the two A-colleges, college Enkee becomes the most anomalous, with the diagonal element 0.983. Its nearest is Efin, a(Efin, Enkee) = 0.347. Merge them into one cluster. The only positive average similarity to this is frim Etom, $(0.05 + 0.320)/2 = 0.162$. Now the results split. Assuming the diagonals zeroed, this value is less than half of the within-cluster similarity, $0.347/2 = 0.174$, so that Etom should not be added to the cluster. By sticking to the inequality (19), one can see that $a(k, S) - a(S)/2 + a_{kk}/(2N_S) = 0.162 - 0.368/2 + 0.527/4 = 0.1098 > 0$, that is, adding Etom to the cluster will increase the criterion $g(S)$ and, thus, must be done. This would complete E-cluster.

This leaves three unclustered S-colleges. Of them, Soli is the most anomalous; it goes into the cluster first. Its nearest is Semb with a(Soli, Semb) = 0.519. The average similarity of Sixpe to the current cluster {Soli, Semi} is $(0.260 + 0.293)/2 = 0.276$. This is greater than half the within-cluster similarity (with the diagonal zeroed) $0.519/2 = 0.260$; the more so at the diagonal taken into account.

Therefore, ADDI leads to the course based clusters only, unlike K-Means itself, which may fail on this.

Table 9. The matrix of similarity between objects obtained as the result of multiplication of the standardized matrix in Table 5 by its transpose.

College	1	2	3	4	5	6	7	8
Soli	0.794	0.519	0.260	0.086	−0.474	−0.237	−0.478	−0.470
Semb	0.519	0.752	0.293	−0.137	−0.307	−0.629	−0.299	−0.191
Sixpe	0.260	0.293	0.604	−0.404	−0.048	−0.126	−0.330	−0.250
Etom	0.086	−0.137	−0.404	0.527	0.005	0.320	−0.069	−0.328
Efin	−0.474	−0.307	−0.048	0.005	0.457	0.347	0.090	−0.069
Enkee	−0.237	−0.629	−0.126	0.320	0.347	0.983	−0.074	−0.583
Aiw	−0.478	−0.299	−0.330	−0.069	0.090	−0.074	0.549	0.612
Ann	−0.470	−0.191	−0.250	−0.328	−0.069	−0.583	0.612	1.279

3   Conclusion

This paper gives a review of author's contribution to K-Means clustering, which includes the two seemingly unrelated E. Braverman's approaches described in the beginning of the presentation. Main concepts and results reported are:

1. K-Means can be considered as a procedure emerging within the data-recovery approach. Specifically, this is a method for fitting an extension of the SVD-like Principal Component Analysis data model towards binary hidden factor scores.
2. This opens up a bunch of equivalent reformulations of K-Means criterion including:
   (a) Maximum of partition's contribution to the data scatter, that is the sum of squared Euclidean distances between centers and the origin weighted by cluster cardinalities;
   (b) Maximum of the summary inner product between object points and corresponding centers;
   (c) Spectral reformulation;
   (d) Maximum of Dorofeyuk-Braverman's semi-average within-cluster similarity.
3. One-by-one Principal Component Analysis strategy applies to clustering. This leads to:
   (a) One-by-one extraction of anomalous clusters leading to a natural initialization of K-Means. The initialization has proven competitive experimentally in application to such issues as speeding-up agglomerative Ward clustering and determining the number of clusters.
   (b) One-by-one extraction of clusters over a similarity matrix. This approach appears to be much similar to Braverman's Spectrum algorithm, leading additionally to automation of starting and stopping conditions.

Other extensions emerging within the data-recovery approach, such as clustering over mixed data, feature weighting, Minkowski metric clustering, consensus clustering, and hierarchical clustering can be found in the author's monograph [16] and papers [2,3].

References

1. Aiserman, M.A., Braverman, E.M., Rosonoer, L.I.: Method of Potential Functions in the Theory of Machine Learning. Nauka Publishers: Main Editorial for Physics and Mathematics, Moscow (1970). (in Russian)
2. de Amorim, R., Makarenkov, V., Mirkin, B.: A-Ward$_{p\beta}$: effective hierarchical clustering using the Minkowski metric and a fast k-means initialisation. Inf. Sci. **370**, 343–354 (2016)
3. de Amorim, R.C., Shestakov, A., Mirkin, B., Makarenkov, V.: The Minkowski central partition as a pointer to a suitable distance exponent and consensus partitioning. Patt. Recogn. **67**, 62–72 (2017)

4. Arkadiev, A.G., Braverman, E.M.: Machine Learning for Classification of Objects. Nauka Publishers: Main Editorial for Physics and Mathematics, Moscow (1971). (in Russian)

5. Bashkirov, O.A., Braverman, E.M., Muchnik, I.B.: Algorithms for machine learning of visual patterns using potential functions. Autom. Remote Control **5**, 25 (1964). (in Russian)

6. Braverman, E., Dorofeyuk, A., Lumelsky, V., Muchnik, I.: Diagonalization of similarity matrices and measuring of hidden factors. In: Issues of extension of capabilities of automata, pp. 42–79. Institute of Control Problems Press, Moscow (1971). (in Russian)

7. Chiang, M., Mirkin, B.: Intelligent choice of the number of clusters in k-means clustering: an experimental study with different cluster spreads. J. Classif. **27**(1), 3–40 (2010)

8. Dorofeyuk, A.A.: Machine learning algorithm for unsupervised pattern recognition based on the method of potential functions. Autom. Remote Control (USSR) **27**, 1728–1737 (1966)

9. Filippone, M., Camastra, F., Masulli, F., Rovetta, S.: A survey of kernel and spectral methods for clustering. Patt. Recogn. **41**(1), 176–190 (2008)

10. Frey, B., Dueck, D.: Clustering by passing messages between data points. Science **315**(5814), 972–976 (2007)

11. Holzinger, K.J., Harman, H.H.: Factor Analysis. University of Chicago Press, Chicago (1941)

12. Kung, S.Y.: Kernel Methods and Machine Learning. Cambridge University Press, Cambridge (2014)

13. Mirkin, B.G.: The method of principal clusters. Autom. Remote Control **48**(10), 1379–1388 (1987)

14. Mirkin, B.: Sequential fitting procedures for linear data aggregation model. J. Classif. **7**, 167–195 (1990)

15. Mirkin, B.: Core Concepts in Data Analysis: Summarization, Correlation, Visualization. Springer, London (2011)

16. Mirkin, B.: Clustering: A Data Recovery Approach. Chapman and Hall/CRC Press (2012)

17. Mirkin, B., Tokmakov, M., de Amorim, R., Makarenkov, V.: Capturing the number of clusters with K-Means using a complementary criterion, affinity propagation, and Ward agglomeration (2017). (Submitted)

18. Taran, Z., Mirkin, B.: Exploring patterns of corporate social responsibility using a complementary k-means clustering criterion (2017). (Submitted)

Compactness Hypothesis, Potential Functions, and Rectifying Linear Space in Machine Learning

Vadim Mottl[1]($\boxtimes$), Oleg Seredin[1], and Olga Krasotkina[2]

[1] Tula State University, Tula 300012 Lenin Ave. 92, Russia
{vmottl, oseredin}@yandex.ru
[2] Moscow State University, Moscow 119991 Lenin Hills 1, Russia
o.v.krasotkina@yandex.ru

Abstract. Emmanuel Braverman was one of the very few thinkers who, during his extremely short life, managed to inseminate several seemingly completely different areas of science. This paper overviews one of the knowledge areas he essentially affected in the sixties years of the last century, namely, the area of Machine Learning. Later, Vladimir Vapnik proposed a more engineering-oriented name of this knowledge area – Estimation of Dependencies Based on Empirical Data. We shall consider these titles as synonyms. The aim of the paper is to briefly trace the way how three notions introduced by Braverman formed the core of the contemporary Machine Learning doctrine. These notions are: (1) compactness hypothesis, (2) potential function, and (3) the rectifying linear space, in which the former two have resulted. There will be little new in this paper. Almost all the constructions we are going to speak about had been published by numerous scientists. The novelty is, perhaps, only in that all these issues will be systematically considered together as immediate consequences of Braveman's basic principles.

Keywords: Set of real-world objects · Pattern recognition
Numerical regression · Ordinal regression · Compactness hypothesis
Precedent-based learning · Distance representation modalities
Pseudo Euclidean liner space · Regularized empirical risk minimization
Potential function · Distance transformation · Selective fusion of distances

1 Introduction

1.1 The Generalized Machine Learning (Dependence Estimation) Problem

First of all, let us accept a common terminology. The most general understanding of the machine learning problem implies a set of objects $\omega \in \Omega$ imagined by the observer to exist in the real world, each of which is assumed to be provided by the Nature with a hidden characteristic $y \in \mathbb{Y}$:

$$y(\omega) \ : \ \Omega \to \mathbb{Y}. \tag{1}$$

© Springer Nature Switzerland AG 2018
L. Rozonoer et al. (Eds.): Braverman Readings in Machine Learning, LNAI 11100, pp. 52–102, 2018.
https://doi.org/10.1007/978-3-319-99492-5_3

The Nature independently draws objects $\omega \in \Omega$ and requires the observer to guess their characteristics $\hat{y}(\omega)$. Errors $\hat{y}(\omega) \neq y(\omega)$ are punished, and denials are forbidden (totally punished). In particular, this problem is said to be that of

$$
\begin{cases}
\text{(a) pattern recognition, when the hidden (goal) characteristic takes values} \\
\qquad \text{from the scale having the structure of a finite unordered set } \mathbb{Y} = \{y_1, \ldots, y_k\}, \\
\qquad \text{in the simplest case } k = 2; \\
\text{(b) regression estimation, when } y \text{ is real number } \mathbb{Y} = \mathbb{R}; \\
\text{(c) ordinal regression, when } \mathbb{Y} \text{ is an arbitrary linearly ordered set} \\
\qquad \mathbb{Y} = \{\ldots \prec y_{i-1} \prec y_i \prec y_{i+1} \prec \ldots\}.
\end{cases}
\tag{2}
$$

The only information available to the observer is a finite set of precedents (training set) $\{(\omega_j, y_j), \ j = 1, \ldots, N\}$. At first sight, this is impossible. Who knows the Nature's design?

Braverman's main supposition stemmed, most likely, from his deep conviction that the Nature is not malevolent. If the Nature has assigned almost the same hidden characteristics to a pair of objects $y(\omega') \approx y(\omega'')$, they must be similar to each other in some sense. It remains only to guess how the Nature understands similarity.

1.2 Compactness Hypothesis

The compactness hypothesis, formulated by Emmanuel Braverman in 1961 as the main point of his PhD Thesis partially published in [1, 2], remains the basic principle that underlies the entire methodology of precedent based Machine Learning [3], or, what is actually the same, Estimation of Dependences Based on Empirical Data [4]. In accordance with Braverman's idea, it is sufficient that the observer, armed with a computer, choses a symmetric real-valued two-argument function on the set of objects, which would measure pair-wise similarity or dissimilarity of objects and appear to him relevant to the respective application area. This choice is meant to be fortunate if the *compactness hypothesis* is met – pairs of objects close to each other in terms of the chosen comparison function $\omega' \approx \omega''$ have also, as a rule, close values of the goal characteristic $y(\omega') \approx y(\omega'')$ [5].

1.3 Potential Function

On the other hand, from the mathematical point of view, it is much easier to build methods of learning if the observer or, more exactly, his computer, are assumed to formally perceive real-world objects as points in some linear space. Braverman proceeded from the simplest assumption that real-world objects $\omega \in \Omega$ allow for measuring a finite number of their assumed elementary properties as real values $\mathbf{x}_\omega = (x_{\omega,i}, i = 1, \ldots, n) \in \mathbb{R}^n$. The Euclidean metric

$$
\rho(\mathbf{x}_{\omega'}, \mathbf{x}_{\omega''}) = \left[\sum\nolimits_{i=1}^{n} (x_{\omega',i} - x_{\omega'',i})^2\right]^{1/2} : \ \Omega \times \Omega \to \mathbb{R}^+,
\tag{3}
$$

is the natural quantitative measure of the pairwise dissimilarity (distance) between objects. Moreover, this is a metric on the set of real-world objects

$$\rho(\omega', \omega'') = \rho(\mathbf{x}_{\omega'}, \mathbf{x}_{\omega''}) : \Omega \times \Omega \to \mathbb{R}^+, \quad \rho(\omega', \omega'') + \rho(\omega'', \omega''') \geq \rho(\omega', \omega'''). \quad (4)$$

The two-argument function

$$K_0(\mathbf{x}', \mathbf{x}'') = \mathbf{x}'^T \mathbf{x}'' = \frac{1}{2} \left[\rho^2(\mathbf{x}', \mathbf{0}) + \rho^2(\mathbf{x}'', \mathbf{0}) - \rho^2(\mathbf{x}', \mathbf{x}'') \right] : \mathbb{R}^n \times \mathbb{R}^n \to \mathbb{R} \quad (5)$$

is natural inner product in the feature space $\mathbb{R}^n$, which is, in the general case, "greater" that the set of images (feature vectors) of real-world objects $\{\mathbf{x}_\omega, \ \omega \in \Omega\} \subset \mathbb{R}^n$. Lower index $K_0(\mathbf{x}', \mathbf{x}'')$ in (5) is a reminder that the linear operations and the inner product in $\mathbb{R}^n$ are understood relative to the zero vector $\mathbf{0} \in \mathbb{R}^n$ chosen as null point.

The restriction of the function (5) from $\mathbb{R}^n$ onto this subset

$$K_\phi(\omega', \omega'') = K_0(\mathbf{x}_{\omega'}, \mathbf{x}_{\omega''}) = \mathbf{x}_{\omega'}^T \mathbf{x}_{\omega''} = \frac{1}{2} \left[\rho^2(\mathbf{x}_{\omega'}, \mathbf{0}) + \rho^2(\mathbf{x}_{\omega''}, \mathbf{0}) - \rho^2(\mathbf{x}_{\omega'}, \mathbf{x}_{\omega''}) \right] :$$
$$\Omega \times \Omega \to \mathbb{R} \quad (6)$$

is a kernel in the terminology introduced by Vladimir Vapnik [3], moreover, this is positive definite kernel since $\rho(\mathbf{x}', \mathbf{x}'')$ is Euclidean metric in $\mathbb{R}^n$ (5).

Actually, this is a linear-space embedding of Ω into $\mathbb{R}^n$. We shall associate the points of the embedding space $\mathbf{x} \in \mathbb{R}^n$, which are no images of any really existing objects $\mathbf{x} \neq \mathbf{x}_\omega \in \mathbb{R}^n$, as images of hypothetical (conceptual) objects $\omega \in \widetilde{\Omega}$, which form an extension $\widetilde{\Omega} \supseteq \Omega$ of the set of real-world objects.

Just like the set of objects Ω, which is isomorphic to the subset of their images $\{\mathbf{x}_\omega, \ \omega \in \Omega\} \subseteq \mathbb{R}^n$, we shall consider the extension $\widetilde{\Omega} \supseteq \Omega$ as a hypothetical n-dimensional linear space isomorphic to the entire embedding space $\mathbb{R}^n$, so that the linear operations in $\widetilde{\Omega}$ are defined by the respective operations on vectors $\mathbf{x} \in \mathbb{R}^n$. The hypothetical point $\phi \in \widetilde{\Omega}$ (6), which is mapped in zero $\mathbf{0} \in \mathbb{R}^n$, plays the role of the null point in the n-dimensional linear space $\widetilde{\Omega}$. The extension of (4) onto the entire hypothetical liner space $\rho(\omega', \omega'') = \rho(\mathbf{x}_{\omega'}, \mathbf{x}_{\omega''}) : \widetilde{\Omega} \times \widetilde{\Omega} \to \mathbb{R}^+$ is Euclidean metric on it $\rho(\omega', \omega'') + \rho(\omega'', \omega''') \geq \rho(\omega', \omega''')$, and the respective extension of (6)

$$K(\omega', \omega'') = \frac{1}{2} \left[\rho^2(\omega', \phi) + \rho^2(\omega'', \phi) - \rho^2(\omega', \omega'') \right] : \widetilde{\Omega} \times \widetilde{\Omega} \to \mathbb{R}$$

is inner product on $\widetilde{\Omega}$. Thus, we have spanned a hypothetical n-dimensional Euclidean linear space over the initial Euclidean metric space of objects (4).

1.4   Rectifying Linear Space

In particular, Braverman considered the exponential function $K(\omega', \omega''|\alpha) = \exp(-\alpha\rho^2(\omega', \omega''))$ on the metric space of real-world objects $\omega \in \Omega$. It will be convenient to consider this function with an additive constant and coefficient (Sect. 5.3.1):

$$K_\phi(\omega', \omega''|\alpha) = \frac{1+\alpha}{\alpha} + \frac{1+\alpha}{2\alpha}\exp(-\alpha\rho^2(\omega', \omega'')), \alpha > 0, \mathbf{x}_\omega \in \mathbb{R}^n,$$

$$\rho(\omega', \omega'') = \sqrt{\sum_{i=1}^n \left(x_{\omega',i} - x_{\omega'',i}\right)^2}. \tag{7}$$

It is a kind of Braverman's famous potential functions [6]. The lower index in $K_\phi(\omega', \omega''|\alpha)$ reflects the fact that this function can be naturally considered as inner product in an infinite-dimensional Hilbert space $\widetilde{\Omega} \sim \mathbb{R}^\infty$ with some conventional null point $\phi \in \widetilde{\Omega}$. This space embeds the initial metric space $\Omega \subset \widetilde{\Omega}$, which, as it will be shown in Sect. 4, maps in $\widetilde{\Omega}$ as a sphere of Euclidean radius $(1+\alpha)/2\alpha$ around the conventional null point.

In Braverman's terminology, this is *the rectifying linear space*. The reasons for such a name, outlined in [6], rest on the fact that, for any coefficient α and any finite collection of objects $\{\omega_1, \ldots, \omega_N\} \in \Omega$, potential function (7) forms positive definite matrices $\mathbf{K}(\alpha) = \left[K(\omega_j, \omega_l|\alpha), j, l = 1, \ldots, N\right]$. Such a function is said to be function of positive type (positive definite kernel [3]). In accordance with the famous Mercer's theorem [7], there always exists an infinite succession of real-valued functions $\tilde{x}_i(\mathbf{x}|\alpha)$, $i = 1, 2, \ldots$, and the respective succession of positive numbers $\tilde{\lambda}_i(\alpha) > 0$, $i = 1, 2, \ldots$, which represent any potential function (7) as an infinite sum (series)

$$K(\mathbf{x}', \mathbf{x}''|\alpha) = \sum_{i=1}^\infty \tilde{\lambda}_i(\alpha)\tilde{x}_i(\mathbf{x}'|\alpha)\tilde{x}_i(\mathbf{x}''|\alpha), \; \widetilde{\mathbf{x}} = (\tilde{x}_1, \tilde{x}_2, \ldots) \in \mathbb{R}^\infty.$$

This means that there exists a mapping $\widetilde{\mathbf{x}}(\mathbf{x}) : \mathbb{R}^n \to \mathbb{R}^\infty$ such that $K(\mathbf{x}', \mathbf{x}''|\alpha)$ is inner product in the Hilbert space $\mathbb{R}^\infty$.

In turn, this means that the set of all linear functions $\psi(\mathbf{z}) : \mathbb{R}^\infty \to \mathbb{R}$ in $\mathbb{R}^\infty$ defines a rich class of non-linear functions $f(\mathbf{x}) : \mathbb{R}^n \to \mathbb{R}$ in $\mathbb{R}^n$, which, for instance, might play the role of nonlinear decision functions of two-class pattern recognition in the initial feature space $f(\omega) = f(\mathbf{x}(\Omega)) \gtrless 0$.

It is just this why the Hilbert space $\mathbb{R}^\infty$ was named by Braverman rectifying linear space. The notion of potential functions actually obliterates the difference between linear and non-linear decision rules in the feature space – a linear function in the rectifying space produced by a pair $(\rho_1(\omega', \omega''), \alpha_1)$ will be nonlinear in terms of another pair $(\rho_2(\omega', \omega''), \alpha_2)$.

1.5   The Aim and the Structure of This Paper

The aim of this paper is to show that Braverman's concepts of (1) compactness hypothesis, (2) potential functions, and (3) embedding the initial feature space into a rectifying linear space practically underlie the entire contemporary framework of Machine Learning. In what follows, we outline our view of the respective conceptual basis as a generalization of the three above-mentioned keystones. As the basis of this generalization, we harness two mathematical ideas originally suggested by Lev Goldfarb [8, 9] and later further refined by Duin, Pękalska and Haasdonk [10–15]. These ideas result in immediate *featureless linear space embedding* of the set of real-world objects and bypass, thereby, the onerous notion of real-valued features[1].

To trace the ways how three Braverman's basic concepts evolved into the contemporary Machine Learning methodology, we introduce the notion of *generalized distance based linear approach to dependence estimation*. The concept of a linear space that embeds the initial set of real-world objects is tacitly exploited by the majority of existing learning methods. We call this concept the generalized linear approach to dependence estimation, because it, first, simplistically retains the main marks of *Generalized Linear Models* (GLM) introduced in mathematical statistics by John Nelder and his colleagues [18, 19] (putting aside, so far, their statistical origin), and, second, embraces all the commonly adopted kinds of dependencies, including, at least, pattern recognition, numerical regression and ordinal regression.

Sections 2 and 3 briefly outline the way how Braverman's compactness hypothesis, potential function and rectifying linear space, considered jointly, lead to the notion of a kernel, indefinite in the general case, which embeds an arbitrary distance space of real-world objects into a pseudo-Euclidean linear space with, generally speaking, indefinite inner product (Krein space).

For the sake of clarity of explanation of the respective conceptual framework, its strict mathematical justification is omitted in Sect. 3. The detailed mathematical argumentation makes the content of the next Sect. 4, which may be skipped in first reading.

The generalized linear approach to dependence estimation, outlined in Sect. 5, is a natural implementation of Braverman's idea of rectifying linear space. In our interpretation, the generalized linear model of the sought-for dependence is completely defined by three heuristics to be made by the observer.

The first heuristic of the observer is distance function (8) chosen in the hope that the compactness hypothesis holds for it.

Further, we slightly crystallize the central concept of the Generalized Linear Model [18, 19], which implies that the observer has to choose the link function $q(y, z)$: $\mathbb{Y} \times \mathbb{R} \to \mathbb{R}^+$ meant to provide the relationship between an appropriate linear predictor applied to a real-world object $z(\omega) : \widetilde{\Omega} \to \mathbb{R}$ and the actual unknown value of its hidden characteristic $y(\omega) : \Omega \to \mathbb{Y}$, which may be not numerical by its nature. We shall call this choice *the second heuristic* of the observer.

Finally, *the third heuristic* of the observer is regularization function meant to quantitatively express his a priori judgement on the complexity of the sought-for dependence.

---

[1] The latter developments are a particular case of a more general relational approach to dependence estimation, which allows for asymmetric comparison functions [16, 17].

Considered jointly, the three heuristics of the observer completely define the unified training criterion which is a generalization of Vladimir Vapnik's principle of regularized empirical risk minimization.

However, the unified training criterion expresses only the concept of empirical risk minimization and cannot be immediately implemented by a computer, because finding the linear predictor $z(\omega|\vartheta, b) = K_\phi(\omega, \vartheta) + b : \widetilde{\Omega} \to \mathbb{R}$ as result of training means finding its direction point (direction vector) in the hypothetical pseudo-Euclidean linear space $\vartheta \in \widetilde{\Omega}$. In contrast to the bias $b \in \mathbb{R}$, which is a real number, this point is an abstract notion that cannot be immediately represented in a computer.

Therefore, in Sect. 5.2 we discuss the ways of parametric representation of the direction vector $\vartheta \in \widetilde{\Omega}$ as a linear combination of a basic set of objects $\{\omega_1, \ldots, \omega_n\} \subset \Omega \subset \widetilde{\Omega}$ mapped into the pseudo-Euclidean rectifying linear space $\vartheta = \sum_{i=1}^{n} c_i \omega_i$.

In Sect. 5.3.1, we return to Braverman's idea of potential function. We consider the exponential potential function (7) as a means of forming a parametric family of distances produced from the initially accepted distance function. This parametric family is conceived to enable constructing a family of dependence models of unrestrictedly growing complexity if $\alpha \to \infty$ in (7).

Further in Sect. 5.3.2 we discuss the ways of selective fusion (combining) of several tentative distance functions (modalities of object representation). The selectivity of fusion, namely, its ability to ignore some irrelevant distances from the initially accepted full set, is controlled by a selectivity parameter $\mu \geq 0$ – the greater μ the less is the number of remaining active distances, and vice versa. This is another axis of controlling the complexity of the family of dependence models, the complexity grows if $\mu \to 0$.

The effect of decision rules of growing complexity is considered in Sect. 5.3.4. The possible principles remain the same as in the case of fusing several kernels in a synthetic training criterion, as, for instance, in [20–22].

Finally, Sect. 6 concludes the paper.

Some mathematical assertions, most of which are well-known, are formulated as Theorems. The proofs are given in Appendix for the readers' convenience.

We concentrate here only on the main concepts of Machine Learning that, in our opinion, issue from Braveman's basic principles. Their probabilistic interpretation and algorithmic implementation remain beyond our attention.

2 Distance Space of Real-World Objects

2.1 The Notion of the Distance Space

The idea of immediate mathematical formalization of Braverman's compactness hypothesis consists in quantitatively evaluating the intuitive perception of pair-wise dissimilarity between real-world objects as a numerical nonnegative symmetric function.

Let $\Omega = \{\omega\}$ be the set of "all" objects considered by the observer. Any two-argument real-valued function is said to be distance function if it takes only non-negative values and is symmetric:

$$\rho(\omega', \omega'') : \Omega \times \Omega \to \mathbb{R}^+, \rho(\omega', \omega'') \geq 0, \quad \rho(\omega', \omega'') = \rho(\omega'', \omega'). \tag{8}$$

The set Ω with a distance function defined on it is called the distance space [8, 24]. If we additionally assume the triangle inequality

$$\rho(\omega', \omega'') + \rho(\omega'', \omega''') \geq \rho(\omega', \omega'''), \tag{9}$$

the distance function will be a metric and, respectively, the set Ω will be metric space [24].

However, there is little essential distinction between distance and metric [24].

Theorem 1. Let the distance space be bounded sup $\rho(\omega', \omega'') < \infty$, $\omega', \omega'' \in \Omega$, and

$$h = \sup_{\omega', \omega'', \omega''' \in \Omega} \{\rho(\omega', \omega''') - [\rho(\omega', \omega'') + \rho(\omega'', \omega''')]\} \geq 0. \tag{10}$$

Then $r(\alpha, \beta) = \rho(\alpha, \beta) + h : \Omega \times \Omega \to \mathbb{R}^+$ is metric.

Proof. In Appendix 1.

In addition, we shall see in Sect. 4 that the extra assumption on the distance to be metric little contributes to the utility of the distance function for mathematical formulation of machine learning problems. But in the particular case of a more specific property, the respective metric, called in Sect. 4 a proto-Euclidean one, will result in a large well-known class of kernel-based machine learning methods [3, 23].

2.2   Distance-Induced Similarity Function

Let us choose an arbitrary element of the set of objects $\phi \in \Omega$ as its conventional "center". We shall see below in Sect. 4 that it will not affect the generalized dependence model we are going to construct. Let us, further, form the two-argument function

$$K_\phi(\omega', \omega'') = \frac{1}{2} [\rho^2(\omega', \phi) + \rho^2(\omega'', \phi) - \rho^2(\omega', \omega'')] : \quad \Omega \times \Omega \xrightarrow{\phi} \mathbb{R}, \tag{11}$$

which will be said the distance-induced center-dependent similarity of a pair of objects in the distance space. This function is an analog of the well-known notion of Gromov product similarity [24]. Its properties:

(a) $K_\phi(\omega', \omega'') = K_\phi(\omega'', \omega') -$ symmetry; $\tag{12}$

(b) $K_\phi(\omega, \omega) = \rho^2(\omega, \phi) \geq 0 -$ non-negativity for $\omega' = \omega''$; $\tag{13}$

(c) $\quad K_{\phi'}(\omega', \omega'') = K_\phi(\omega', \omega'') - K_\phi(\omega', \phi') - K_\phi(\omega'', \phi') + K_\phi(\phi', \phi'),$

$\qquad \phi' \in \Omega$; $-$center translation;

$$\hspace{9cm}(14)$$

(d) $\quad \rho^2(\omega', \omega'') = K_\phi(\omega', \omega') + K_\phi(\omega'', \omega'')$

$\qquad\qquad - 2K_\phi(\omega', \omega'')$; $-$return to the distance function;

$$\hspace{9cm}(15)$$

(e) $\quad \left| K_\phi(\omega', \omega'') \right| \leq \sqrt{K_\phi(\omega', \omega')} \sqrt{K_\phi(\omega'', \omega'')}$ $-$ in the particular case of metric :

$\qquad$ Cauchy-Bunyakowsky inequality.

$$\hspace{9cm}(16)$$

These properties resemble inner product very much! But the set of real-world objects is no linear space, there are no linear operations in Ω as yet.

3 Distance Based Embedding of the Set of Real-World Objects into a Hypothetical Pseudo-Euclidean Linear Space

In this Section, slightly generalizing Lev Goldfarb [8, 9], Robert Duin and Elżbieta Pękalska [10, 11], we consider the center-dependent similarity function on the set of real-world objects (11) as a generalization of Braverman's potential function. The main property of this kind of potential function will be the possibility of a mental extension of the initial distance space Ω to a greater embedding continuum set $\widetilde{\Omega} \supset \Omega$. In this set, the accepted distance function (8) induces linear operations along with an indefinite inner product as a continuation of (11) onto the entire linear space $\widetilde{\Omega}$. We shall call $\widetilde{\Omega}$ the embedding pseudo-Euclidean linear space, as in [8–11]. This will be a usual Euclidean linear space with usual inner product in the particular case when the initial distance is the proto-Euclidean metric (see below Sect. 4.4).

The idea of such an embedding is outlined in this Section without strict mathematical justification. However, this content will be completely sufficient for constructing methods of distance-based dependence estimation below in Sect. 5.

The respective mathematical details are explained in Sects. 4, which may be skipped in first reading.

3.1 Embedding in the General Case of a Distance Space

It will be shown in Sects. 4.1 and 4.2 that, for any distance space Ω, it is possible to imagine a hypothetical greater continuum set $\widetilde{\Omega} \supset \Omega$, in which linear operations are defined with respect to an arbitrary null point $\phi \in \widetilde{\Omega}$ (linear space):

(a) $\quad$ Summation $\omega' + \omega'' = \omega'' + \omega', (\omega' + \omega'') + \omega''' = \omega' + (\omega'' + \omega'''), \omega + \phi = \omega,$

$\qquad$ there exists the additive inverse element $(-\omega) + \omega = \phi$ for each $\omega \in \widetilde{\Omega}.$

$$\hspace{9cm}(17)$$

(b) Multiplication with real-valued scalar
$$c\omega \in \widetilde{\Omega}, \ c \in \mathbb{R}, \ c\phi = \phi, \ 0\omega = \phi, \ c'(c''\omega) = (c'c'')\omega. \tag{18}$$

(c) Distributivity $(c' + c'')\omega = c'\omega + c''\omega, \ c(\omega' + \omega'') = c\omega' + c\omega''. \tag{19}$

Moreover, there exists a two-argument function $K_\phi(\omega', \omega'') : \ \widetilde{\Omega} \times \widetilde{\Omega} \xrightarrow{\phi \in \widetilde{\Omega}} \mathbb{R}$, which is continuation of the similarity function $K_\phi(\omega', \omega'') : \ \Omega \times \Omega \xrightarrow{\phi \in \Omega} \mathbb{R}$ (11) from the initial set of real-world objects Ω onto the entire embedding space $\widetilde{\Omega}$. This function possesses the following properties:

(a) Symmetricity $K_\phi(\omega', \omega'') = K_\phi(\omega'', \omega'), \ \omega', \omega'' \in \widetilde{\Omega}. \tag{20}$

(b) Bilinearity $K_\phi(c'\omega' + c''\omega'', \ \omega''') = c'K_\phi(\omega', \omega''') + c''K_\phi(\omega'', \omega'''). \tag{21}$

(c) Translation of the null point as a generalization of (14)
$$K_{\phi'}(\omega', \omega'') = K_\phi(\omega', \omega'') - K_\phi(\omega', \phi') - K_\phi(\omega'', \phi') + K_\phi(\phi', \phi'), \ \phi' \in \widetilde{\Omega}. \tag{22}$$

(d) In the case if the distance is metric – Cauchy-Bunyakowsky inequality $\left| K_\phi(\omega', \omega'') \right| \leq \sqrt{K_\phi(\omega', \omega')}\sqrt{K_\phi(\omega'', \omega'')}. \tag{23}$

(e) But the property of non − negativity when the arguments coincide $K_\phi(\omega, \omega) \geq 0$ is absent $K_\phi(\omega, \omega) < 0$ for some $\omega \in \widetilde{\Omega}. \tag{24}$

If the property (e) $K_\phi(\omega, \omega) \geq 0$ were met, this would be usual inner product, and $\widetilde{\Omega}$ would be usual Euclidean space. If this inequality is not always valid, the function $K_\phi(\omega', \omega'') : \ \widetilde{\Omega} \times \widetilde{\Omega} \to \mathbb{R}$ and linear space $\widetilde{\Omega}$ are said, respectively, indefinite inner product and pseudo-Euclidean linear space [25, 26].

The choice of the null element is absolutely conventional because the rule of center translation (14)
$$K_{\phi'}(\omega', \omega'') = K_\phi(\omega', \omega'') - K_\phi(\omega', \phi') - K_\phi(\omega'', \phi') + K_\phi(\phi', \phi'), \ \phi' \in \widetilde{\Omega},$$

remains completely valid for (indefinite) inner product in the embedding space just as for similarity function in the initial set of real-world objects.

We shall also see in Sect. 4 that the indefinite inner product, in its turn, defines a continuation of the squared initial distance in Ω (15) onto the embedding space $\widetilde{\Omega}$, and this continuation does not depend on the choice of the null point $\phi \in \Omega$:
$$\rho^2(\omega', \omega'') = K_\phi(\omega', \omega') + K_\phi(\omega'', \omega'') - 2K_\phi(\omega', \omega'') : \widetilde{\Omega} \times \widetilde{\Omega} \to \mathbb{R}^+. \tag{25}$$

In particular,

$$\rho^2(\omega, \phi) = K_\phi(\omega, \omega) + K_\phi(\phi, \phi) - 2K_\phi(\omega, \phi) = K_\phi(\omega, \omega), \text{ since } K_\phi(\phi, \phi)$$
$$= K_\phi(\omega, \phi) = 0. \tag{26}$$

The latter two equalities are analogous to the squared Euclidean metric and squared Euclidean norm in the habitual vector space $\mathbf{x} \in \mathbb{R}^n$, where

$$\phi = \mathbf{0} \in \mathbb{R}^n, \quad K_\mathbf{0}(\mathbf{x}', \mathbf{x}'') = \mathbf{x}'^T \mathbf{x}'', \quad K_\mathbf{0}(\mathbf{x}, \mathbf{x}) = \mathbf{x}^T \mathbf{x} = \|\mathbf{x}\|^2 > 0 \text{ if } \mathbf{x} \neq \mathbf{0},$$
$$\rho^2(\mathbf{x}', \mathbf{x}'') = K_\mathbf{0}(\mathbf{x}', \mathbf{x}') + K_\mathbf{0}(\mathbf{x}'', \mathbf{x}'') - 2K_\mathbf{0}(\mathbf{x}', \mathbf{x}'') = (\mathbf{x}' - \mathbf{x}'')^T (\mathbf{x}' - \mathbf{x}'') \geq 0, \tag{27}$$
$$K_\mathbf{0}(\mathbf{x}', \mathbf{x}'') = \frac{1}{2} \left[\rho^2(\mathbf{x}', \mathbf{x}') + \rho^2(\mathbf{x}'', \mathbf{x}'') - \rho^2(\mathbf{x}', \mathbf{x}'') \right].$$

The equality (25) is biunique, and its additive inverse form defines the indefinite inner product as function of the squared distance:

$$K_\phi(\omega', \omega'') = \frac{1}{2} \left[\rho^2(\omega', \phi) + \rho^2(\omega'', \phi) - \rho^2(\omega', \omega'') \right] : \widetilde{\Omega} \times \widetilde{\Omega} \to \mathbb{R}. \tag{28}$$

However, as distinct from the always positive squared metric in a vector space (27), the squared distance in the pseudo-Euclidean linear space (25) and (26) may be negative for some pairs of points $\omega', \omega'' \in \widetilde{\Omega}$. The distances between such points and their norms can be imaginary numbers:

$$\rho^2(\omega', \omega'') < 0, \quad \rho(\omega', \omega'') = i|\rho^2(\omega', \omega'')|^{1/2},$$
$$\rho^2(\omega, \phi) < 0, \quad \rho(\omega, \phi) = \|\omega\| = i|\rho^2(\omega', \omega'')|^{1/2}. \tag{29}$$

Thus, the pseudo-Euclidean space we have defined is neither distance nor normed space.

But the squared distance remains always positive $\rho^2(\omega', \omega'') > 0$ within the bounds of the subset of original objects $\Omega \subset \widetilde{\Omega}$, which is a distance space by definition (8).

In the contemporary terminology, the similarity function (11) defined on the set of real-world objects of arbitrary kind is called indefinite kernel [11, 27]. It is said to reproduce the embedding pseudo-Euclidean linear space (Krein space). Table 1 summarizes the mathematical framework of linear embedding a distance space.

3.2   The Particular Case of a Proto-Euclidean Metric Space

Up to now we assumed that an arbitrary distance function $\rho(\omega', \omega'')$ is defined on the set of real-world objects Ω (8). Let us now assume the distance function (8) to be a metric on the set of real-world objects Ω (9). We shall use the notation $\mathbf{P}_n = \left[-\rho^2(\omega_i, \omega_j), \, i, j = 1, \ldots, n \right] (n \times n)$ for the matrix of the inverse squared distances of any finite collection of objects $\{\omega_1, \ldots, \omega_n\} \subset \Omega$. Recall that a matrix $\mathbf{P}_n$ is positive (non-negative) definite if the quadratic form induced by it remains always positive

Table 1. The continuum of (indefinite) kernels defined on the set of real-world objects by an arbitrary distance function

A set of real-world objects $\omega \in \Omega$

$\Downarrow$

Arbitrary distance $\rho(\omega', \omega''): \Omega \times \Omega \to \mathbb{R}^+$

$\Downarrow$

Choice of the center $\phi \in \Omega$

$\Downarrow$

Indefinite kernel $K_\phi(\omega', \omega'') = \dfrac{1}{2}\left[\rho^2(\omega', \phi) + \rho^2(\omega'', \phi) - \rho^2(\omega', \omega'')\right]$

$\Downarrow$

Embedding the set of real-world objects Ω into a linear space $\tilde{\Omega} \supset \Omega$ with the null element $\phi \in \Omega$, specific linear operations, and the indefinite inner product

$$K_\phi(\omega', \omega''): \tilde{\Omega} \times \tilde{\Omega} \to \mathbb{R}$$

$\Downarrow$

The continuum of kernels defined by the continuum of null elements $\tilde{\phi} \in \bar{\Omega}$

$$K_{\tilde{\phi}}(\omega', \omega'') = K_\phi(\omega', \omega'') - K_\phi(\omega', \tilde{\phi}) - K_\phi(\omega'', \tilde{\phi}) + K_\phi(\tilde{\phi}, \tilde{\phi})$$

$\Downarrow$

Any distance on Ω defines a continuum of linear spaces $\tilde{\Omega} \supseteq \Omega$ with different null elements $\tilde{\phi} \in \tilde{\Omega}$, different linear operations, different inner products $K_{\tilde{\phi}}(\omega', \omega'')$, but with the same squared distance $\rho^2(\omega', \omega'')$, which may be negative $\rho^2(\omega', \omega'') < 0$ for some hypothetical points if $\omega' \notin \Omega$, $\omega'' \notin \Omega$, or both $\omega', \omega'' \notin \Omega$, but remains always positive $\rho^2(\omega', \omega'') \geq 0$ if both $\omega', \omega'' \in \Omega$.

(non-negative): $\mathbf{a}^T \mathbf{P}_n \mathbf{a} \geq 0$ for all $\mathbf{a} \in \mathbb{R}^n$. But we shall confine ourselves to an essentially weaker requirement – the quadratic form must be positive only for vectors whose elements sum up to zero:

$$\mathbf{P}_n = \left[-\rho^2(\omega_j, \omega_l), \; j, l = 1, \ldots, n\right], \mathbf{a}^T \mathbf{P}_n \mathbf{a} \geq 0, \mathbf{1}^T \mathbf{a} = \sum_{j=1}^{N} a_j = 0, \quad (30)$$

for all the finite sets of objects $\{\omega_1, \ldots, \omega_n\} \subseteq \Omega$. Such matrices are said to be conditionally positive (non-negative) definite [23].

We shall call metric of such a kind proto-Euclidean metric on an arbitrary set of objects, as distinct from the standard notion of Euclidean metric, which is defined in a linear space by choice of inner product. However, this is a small generalization. We shall see in Sect. 4 (Theorem 2) that any proto-Euclidean metric allows for embedding the set of real-world objects Ω into a Euclidean linear space $\tilde{\Omega} \supset \Omega$ with the usual (positive definite) inner product $K_\phi(\omega', \omega'')$ possessing the property of non-negativity when the arguments coincide $K_\phi(\omega,\omega) \geq 0$ (24), as distinct from indefinite inner product in the case of a distance function, even if it is a metric of general kind.

The property of conditional positive definiteness (30) guaranties the positive values of all squared distances in the entire embedding space $\rho^2(\omega',\omega'') \geq 0$, just as in the vector space (27), in contrast to their possible negativeness in the case of distance or even metric of general kind (29).

4 Mathematical Justification of Embedding the Set of Real-World Objects into a Linear Space

This Sect. 4 may seem boring, because it contains only detailed mathematical justification of the statements already formulated in the previous Sect. 3. The reader may skip it and go to the following Sect. 5 without loss of understanding the general logic of how Braverman's ideas have affected the contemporary Machine Learning methodology.

4.1 The Embedding Linear Space

So well, we are considering the idea of embedding an arbitrary distance space of some real-world objects into a linear space (Sect. 2).

Let us assume for simplicity sake that the set of all objects Ω is finite $|\Omega| = M$, $\Omega = \{\omega_1,\ldots,\omega_M\}^2$, where the numbering is of no significance, and consider the square symmetric matrix of similarity values with respect to the assigned center $\mathbf{K}_\phi = \left[K_\phi(\omega_i,\omega_j),\ i,j = 1,\ldots,M\right]$. In what follows, this matrix will play only a conceptual role. The number M of its rows and column may be arbitrarily huge, but there will be no need to deal with the full set in any computations.

Since matrix $\mathbf{K}_\phi(M \times M)$ is symmetric, its eigenvalues are real valued, and the corresponding real eigenvectors are pairwise orthogonal. Without loss of generality, it is always possible to consider eigenvectors of unity norm. Of course, the eigenvalues and eigenvectors will depend on the choice of the center of the distance space $\phi \in \Omega$:

$$\xi_{\phi,j} \in \mathbb{R}, \mathbf{z}_{\phi,j} = \begin{pmatrix} z_{\phi,11} \\ \vdots \\ z_{\phi,1i} \\ \vdots \\ z_{\phi,1M} \end{pmatrix} \in \mathbb{R}^M, \mathbf{z}_{\phi,i}^T \mathbf{z}_{\phi,j} = \begin{cases} 1,\ i = j, \\ 0,\ i \neq j, \end{cases} j = 1,\ldots,M. \qquad (31)$$

---

[2] Otherwise, we would be forced to assume that the distance is a metric, and the respective metric space is separable, i.e., contains a countable everywhere dense subset. All the mathematical construction would become much more complicated without any gain in generalization of the resulting class of dependence models.

Matrix $\mathbf{K}_\phi$ can always be represented as sum of elementary matrices

$$\mathbf{K}_\phi = \sum_{i=1}^{M} \xi_{\phi,i} \mathbf{z}_{\phi,i} \mathbf{z}_{\phi,i}^T = \begin{pmatrix} \sum_{i=1}^{M} \xi_{\phi,i} z_{\phi,i1} z_{\phi,i1} & \cdots & \sum_{i=1}^{M} \xi_{\phi,i} z_{\phi,i1} z_{\phi,iM} \\ \vdots & \ddots & \vdots \\ \sum_{i=1}^{M} \xi_{\phi,i} z_{\phi,iM} z_{\phi,i1} & \cdots & \sum_{i=1}^{M} \xi_{\phi,i} z_{\phi,iM} z_{\phi,iM} \end{pmatrix}. \quad (32)$$

For an arbitrary distance function, matrix $\mathbf{K}_\phi$ may be not positive definite, so, it may have both positive and negative eigenvalues. Let us rank all the eigenvalues in descending order $\xi_{\phi,1} \geq 0, \ldots, \xi_{\phi,p_\phi} \geq 0, \xi_{\phi,p_\phi+1} < 0, \ldots, \xi_{\phi,M} < 0$, and consider them as nonnegative numbers using special notations:

$$\xi_{\phi,1} \geq 0, \ldots, \xi_{\phi,p_\phi} \geq 0, \bar{\xi}_{\phi,p_\phi+1} = -\xi_{\phi,p_\phi+1} > 0, \ldots, \bar{\xi}_{\phi,M} = -\xi_{\phi,M} > 0. \quad (33)$$

The pair of whole numbers $p_\phi + q_\phi = M$ is said the signature of matrix $\mathbf{K}_\phi$ [28], in our case, similarity matrix of the given distance function.

Theorem 2. The signature of matrix $\mathbf{K}_\phi$ does not depend on the choice of the center of the distance space $\phi \in \Omega$.

Proof. In Appendix 2.

Generally speaking, the eigenvalues themselves $\xi_{\phi,1}, \ldots, \xi_{\phi,M}$ depend on the choice of the center, but after placing them in descending order the numbers of positive p and negative q eigenvalues remain the same on the force of Theorem 2, therefore, the second lower index in notations of eigenvalues (33) may be omitted:

$$\xi_{\phi,1} \geq 0, \ldots, \xi_{\phi,p_\phi} \geq 0, \bar{\xi}_{\phi,p+1} = -\xi_{\phi,p+1} > 0, \ldots, \bar{\xi}_{\phi,M} = -\xi_{\phi,M} > 0. \quad (34)$$

This means that the signature $p + q = M$ is a characteristic of the given distance function $\rho(\omega', \omega'') : \Omega \times \Omega \to \mathbb{R}$ (8) but not of the choice of the center $\phi \in \Omega$.

In accepted notations equality (32) may be written as

$$\mathbf{K}_\phi = \sum_{i=1}^{p} \xi_{\phi,i} \mathbf{z}_{\phi,i} \mathbf{z}_{\phi,i}^T + \sum_{i=p+1}^{M} \xi_{\phi,i} \mathbf{z}_{\phi,i} \mathbf{z}_{\phi,i}^T =$$
$$\begin{pmatrix} \sum_{i=1}^{p} \xi_{\phi,i} z_{\phi,i1} z_{\phi,i1} & \cdots & \sum_{i=1}^{p} \xi_{\phi,i} z_{\phi,i1} z_{\phi,iM} \\ \vdots & \ddots & \vdots \\ \sum_{i=1}^{p} \xi_{\phi,i} z_{\phi,iM} z_{\phi,i1} & \cdots & \sum_{i=1}^{p} \xi_{\phi,i} z_{\phi,iM} z_{\phi,iM} \end{pmatrix} -$$
$$\begin{pmatrix} \sum_{i=p+1}^{M} \bar{\xi}_{\phi,i} z_{\phi,i1} z_{\phi,i1} & \cdots & \sum_{i=p+1}^{M} \bar{\xi}_{\phi,i} z_{\phi,i1} z_{\phi,iM} \\ \vdots & \ddots & \vdots \\ \sum_{i=p+1}^{M} \bar{\xi}_{\phi,i} z_{\phi,iM} z_{\phi,i1} & \cdots & \sum_{i=p+1}^{M} \bar{\xi}_{\phi,i} z_{\phi,iM} z_{\phi,iM} \end{pmatrix}.$$

Here all numbers $\xi_{\phi,i}$ and $\bar{\xi}_{\phi,i}$ are nonnegative and allow for extracting the roots. Using the notations

$$\mathbf{K}_\phi = \begin{pmatrix} \sum_{i=1}^{p} \underbrace{\left(\sqrt{\xi_{\phi,i}}z_{\phi,i1}\right)}_{x_{\phi,1i}}\underbrace{\left(\sqrt{\xi_{\phi,i}}z_{\phi,i1}\right)}_{x_{\phi,1i}} & \cdots & \sum_{i=1}^{p} \underbrace{\left(\sqrt{\xi_{\phi,i}}z_{\phi,i1}\right)}_{x_{\phi,1i}}\underbrace{\left(\sqrt{\xi_{\phi,i}}z_{\phi,iM}\right)}_{x_{\phi,Mi}} \\ \vdots & \ddots & \vdots \\ \sum_{i=1}^{p} \underbrace{\left(\sqrt{\xi_{\phi,i}}z_{\phi,iM}\right)}_{x_{\phi,Mi}}\underbrace{\left(\sqrt{\xi_{\phi,i}}z_{\phi,i1}\right)}_{x_{\phi,1i}} & \cdots & \sum_{i=1}^{p} \underbrace{\left(\sqrt{\xi_{\phi,i}}z_{\phi,iM}\right)}_{x_{\phi,Mi}}\underbrace{\left(\sqrt{\xi_{\phi,i}}\right)z_{\phi,iM}}_{x_{\phi,Mi}} \end{pmatrix} -$$

$$\begin{pmatrix} \sum_{i=p+1}^{M} \underbrace{\left(\sqrt{\bar{\xi}_{\phi,i}}z_{\phi,i1}\right)}_{x_{\phi,1i}}\underbrace{\left(\sqrt{\bar{\xi}_{\phi,i}}z_{\phi,i1}\right)}_{x_{\phi,1i}} & \cdots & \sum_{i=p+1}^{M} \underbrace{\left(\sqrt{\bar{\xi}_{\phi,i}}z_{\phi,i1}\right)}_{x_{\phi,1i}}\underbrace{\left(\sqrt{\bar{\xi}_{\phi,i}}z_{\phi,iM}\right)}_{x_{\phi,Mi}} \\ \vdots & \ddots & \vdots \\ \sum_{i=p+1}^{M} \underbrace{\left(\sqrt{\bar{\xi}_{\phi,i}}z_{\phi,iM}\right)}_{x_{\phi,Mi}}\underbrace{\left(\sqrt{\bar{\xi}_{\phi,i}}z_{\phi,i1}\right)}_{x_{\phi,1i}} & \cdots & \sum_{i=p+1}^{M} \underbrace{\left(\sqrt{\bar{\xi}_{\phi,i}}z_{\phi,iM}\right)}_{x_{\phi,Mi}}\underbrace{\left(\sqrt{\bar{\xi}_{\phi,i}}z_{\phi,iM}\right)}_{x_{\phi,Mi}} \end{pmatrix}, \tag{35}$$

we shall have

$$\mathbf{K}_\phi = \begin{pmatrix} \sum_{i=1}^{p} x_{\phi,1i}x_{\phi,1i} & \cdots & \sum_{i=1}^{p} x_{\phi,1i}x_{\phi,Mi} \\ \vdots & \ddots & \vdots \\ \sum_{i=1}^{p} x_{\phi,Mi}x_{\phi,1i} & \cdots & \sum_{i=1}^{p} x_{\phi,Mi}x_{\phi,Mi} \end{pmatrix}$$
$$- \begin{pmatrix} \sum_{i=p+1}^{M} x_{\phi,1i}x_{\phi,1i} & \cdots & \sum_{i=p+1}^{M} x_{\phi,1i}x_{\phi,Mi} \\ \vdots & \ddots & \vdots \\ \sum_{i=p+1}^{M} x_{\phi,Mi}x_{\phi,1i} & \cdots & \sum_{i=p+1}^{M} x_{\phi,Mi}x_{\phi,Mi} \end{pmatrix}.$$

If we form matrix

$$\mathbf{G}_\phi = \left(\mathbf{z}_{\phi,1}\cdots\mathbf{z}_{\phi,M}\right) = \begin{pmatrix} z_{\phi,1,1} & \cdots & z_{\phi,1,M} \\ \vdots & \ddots & \vdots \\ z_{\phi,M,1} & \cdots & z_{\phi,M,M} \end{pmatrix} (M \times M) \tag{36}$$

whose columns are eigenvectors of matrix $\mathbf{K}_\phi$, then, in accordance with notations (35), vectors

$$\mathbf{x}_{\phi,i} = \left(x_{\phi,1,i}, \ldots, x_{\phi,M,i}\right)^T$$
$$= \left(\sqrt{\xi_{\phi,i}}z_{\phi,i,1}, \ldots, \sqrt{\xi_{\phi,i}}z_{\phi,i,p}, \sqrt{\bar{\xi}_{\phi,i}}z_{\phi,i,p+1}, \ldots, \sqrt{\bar{\xi}_{\phi,i}}z_{\phi,i,M}\right)^T \in \mathbb{R}^M \tag{37}$$

will be its rows with coefficients $\sqrt{\xi_{\phi,i}}$ and $\sqrt{\bar{\xi}_{\phi,i}}$.

We have tied each object of an arbitrary finite distance space $\omega \in \Omega = \{\omega_1, \ldots, \omega_M\}$, in which the central element $\phi \in \Omega$ is chosen, to a real-valued vector $\mathbf{x}_{\phi,\omega_i} = \mathbf{x}_{\phi,i} \in \mathbb{R}^M$ defined by M eigenvectors (36) of the similarity matrix $\mathbf{K}_\phi$ of this finite set and M eigenvalues, among which p ones are nonnegative $\xi_{\phi,1} \geq 0, \ldots, \xi_{\phi,p} \geq 0$, and the remaining $M - p$ are negative $\xi_{\phi,p+1} < 0, \ldots, \xi_{\phi,M} < 0$ (31). Owing to the latter fact, it will be convenient to consider each vector (37) as consisting of two parts:

$$
\begin{cases}
\mathbf{x}_{\phi,i} = \begin{pmatrix} \mathbf{u}_{\phi,i} \\ \mathbf{v}_{\phi,i} \end{pmatrix} \in \mathbb{R}^M, \quad \mathbf{u}_{\phi,i} = \left(\sqrt{\xi_{\phi,i}} z_{\phi,i,1}, \ldots, \sqrt{\xi_{\phi,i}} z_{\phi,i,p} \right), \quad \mathbf{v}_{\phi,i} = \left(\sqrt{\xi_{\phi,i}} z_{\phi,i,p+1}, \ldots, \sqrt{\xi_{\phi,i}} z_{\phi,i,M} \right). \\
\text{It is clear that the central element } \phi \in \Omega \text{ is tied to zero vector } \mathbf{x}_\phi = \mathbf{0} \in \mathbb{R}^M.
\end{cases}
$$

$$(38)$$

As to the continuum of other points of the M-dimensional linear space $\mathbf{x} \in \mathbb{R}^M$, there exist no real-world objects to which these points are tied. It appears natural to imagine the set of missing hypothetical objects $\omega \in \widetilde{\Omega}$ that could be meant as corresponding to all the points of the representing linear space.

This set may be viewed as a linear space $\widetilde{\Omega} \supset \Omega$ with null point $\phi \in \Omega \subset \widetilde{\Omega}$, which is spanned over the initial finite distance space $\Omega = \{\omega_1, \ldots, \omega_M\}$ of real-world objects and is isomorphic to $\mathbb{R}^M$.

4.2 Indefinite Inner Product in the Embedding Linear Space

Let us define square matrix

$$
\mathbf{J}_p = \begin{pmatrix} \mathbf{I}_{p \times p} & \mathbf{0}_{p \times (M-p)} \\ \mathbf{0}_{(M-p) \times p} & -\mathbf{I}_{(M-p) \times (M-p)} \end{pmatrix} (M \times M), \tag{39}
$$

which will be said unity matrix of signature p. In these terms, the similarity matrix of elements of the distance space of real-world objects $\Omega = \{\omega_1, \ldots, \omega_M\}$ will have the form

$$
\mathbf{K}_\phi = \begin{pmatrix} \mathbf{x}_{\phi,1}^T \mathbf{J}_p \mathbf{x}_{\phi,1} & \cdots & \mathbf{x}_{\phi,1}^T \mathbf{J}_p \mathbf{x}_{\phi,M} \\ \vdots & \ddots & \vdots \\ \mathbf{x}_{\phi,M}^T \mathbf{J}_p \mathbf{x}_{\phi,1} & \cdots & \mathbf{x}_{\phi,M}^T \mathbf{J}_p \mathbf{x}_{\phi,M} \end{pmatrix} =
$$

$$
\begin{pmatrix} \mathbf{u}_{\phi,1}^T \mathbf{u}_{\phi,1} & \cdots & \mathbf{u}_{\phi,1}^T \mathbf{u}_{\phi,M} \\ \vdots & \ddots & \vdots \\ \mathbf{u}_{\phi,M}^T \mathbf{u}_{\phi,1} & \cdots & \mathbf{u}_{\phi,M}^T \mathbf{u}_{\phi,M} \end{pmatrix} - \begin{pmatrix} \mathbf{v}_{\phi,1}^T \mathbf{v}_{\phi,1} & \cdots & \mathbf{v}_{\phi,1}^T \mathbf{v}_{\phi,M} \\ \vdots & \ddots & \vdots \\ \mathbf{v}_{\phi,M}^T \mathbf{v}_{\phi,1} & \cdots & \mathbf{v}_{\phi,M}^T \mathbf{v}_{\phi,M} \end{pmatrix}, \quad \mathbf{x}_{\phi,i} = \begin{pmatrix} \mathbf{u}_{\phi,i} \\ \mathbf{v}_{\phi,i} \end{pmatrix} \in \mathbb{R}^M,
$$

$$(40)$$

where vectors $\mathbf{u}_{\phi,i} \in \mathbb{R}^p$ and $\mathbf{v}_{\phi,i} \in \mathbb{R}^{M-p}$ are parts of vectors $\mathbf{x}_{\phi,i}$.

Let's notice, that the similarity function $K(\mathbf{x}', \mathbf{x}'') : \mathbb{R}^M \times \mathbb{R}^M \to \mathbb{R}$ can be naturally continued onto the entire embedding space with respect to previously defined notation (39):

$$K_0(\mathbf{x}', \mathbf{x}'') = (\mathbf{x}' - \underbrace{\mathbf{x}_\phi}_{=0})^T \mathbf{J}_p(\mathbf{x}'' - \underbrace{\mathbf{x}_\phi}_{=0}) = \mathbf{x}'^T \mathbf{J}_p \mathbf{x}'' = \mathbf{u}'^T \mathbf{u}'' - \mathbf{v}'^T \mathbf{v}'',$$

$$K_0(\mathbf{x}, \mathbf{x}) = \mathbf{x}^T \mathbf{J}_p \mathbf{x} = (\mathbf{u}^T \mathbf{v}^T) \begin{pmatrix} \mathbf{I}_{p \times p} & \mathbf{0}_{p \times (M-p)} \\ \mathbf{0}_{(M-p) \times p} & -\mathbf{I}_{(M-p) \times (M-p)} \end{pmatrix} \begin{pmatrix} \mathbf{u} \\ \mathbf{v} \end{pmatrix} = \mathbf{u}^T \mathbf{u} - \mathbf{v}^T \mathbf{v}.$$

$$(41)$$

The lower index $\mathbf{0} \in \mathbb{R}^M$ is here a trivial reminder that the linear operations are understood with respect to the null point of $\mathbb{R}^M$, into which the central object $\phi \in \Omega$ is mapped (38).

Thus, we have spanned an M-dimensional linear space $\mathbb{R}^M$ over the finite distance space of real-world objects $\Omega = \{\omega_1, \ldots, \omega_M\}$ with the given distance function $\rho(\omega', \omega'') : \Omega \times \Omega \to \mathbb{R}$ (8). Further, we have defined a two-argument real-valued function (41) in $\mathbb{R}^M$ (41). Let us consider its properties:

(1) $K_0(\mathbf{x}', \mathbf{x}'') = K_0(\mathbf{x}'', \mathbf{x}') -$ symmetry; $\qquad\qquad\qquad\qquad$ (42)

(2) $K_0(c'\mathbf{x}' + c''\mathbf{x}'', \ \mathbf{x}''') = c'K_0(\mathbf{x}', \mathbf{x}''') + c''K_0(\mathbf{x}'', \mathbf{x}''') -$ bilinearity. $\quad$ (43)

These two properties coincide with the properties of inner product in a linear space, but the third extremely important property is absent – there is no non-negativity when both arguments are equal to each other.

$\qquad\qquad$ (3') $K_0(\mathbf{x}, \mathbf{x}) \geq 0$ as in the usual vector space (27). $\qquad\qquad\qquad$ (44)

Instead, we have:

$\qquad\qquad$ (3) $K_0(\mathbf{x}, \mathbf{x}) \geq 0$ if $\mathbf{u}^T \mathbf{u} \geq \mathbf{v}^T \mathbf{v}$, but $K_0(\mathbf{x}, \mathbf{x}) < 0$ if $\mathbf{u}^T \mathbf{u} < \mathbf{v}^T \mathbf{v}$. $\qquad$ (45)

The two-argument function $K(\mathbf{x}', \mathbf{x}'') : \mathbb{R}^M \times \mathbb{R}^M \to \mathbb{R}$ having properties (42)–(45) is referred to as indefinite inner product of signature $p + q = M$, and the linear space with indefinite inner product is called pseudo-Euclidean linear space [8, 9, 11, 25, 26].

In particular, in the case of signature $p = M$, $q = 0$, we have the usual Euclidean linear space with usual inner product $K(\mathbf{x}, \mathbf{x}) \geq 0$.

As a consequence, the two-argument function

$$r^2(\mathbf{x}', \mathbf{x}'') = (\mathbf{x}' - \mathbf{x}'')^T \mathbf{J}_p(\mathbf{x}' - \mathbf{x}'') = (\mathbf{u}' - \mathbf{u}'')^T(\mathbf{u}' - \mathbf{u}'') - (\mathbf{v}' - \mathbf{v}'')^T(\mathbf{v}' - \mathbf{v}'')$$

$$(46)$$

defined in the pseudo-Euclidean space should be referred to as "squared distance", which may be positive $r^2(\mathbf{x}', \mathbf{x}'') > 0$, zero $r^2(\mathbf{x}', \mathbf{x}'') = 0$ or negative $r^2(\mathbf{x}', \mathbf{x}'') < 0$.

$$r^2(\mathbf{x}',\mathbf{x}'') \begin{cases} > 0 \text{ if } (\mathbf{u}'-\mathbf{u}'')^T(\mathbf{u}'-\mathbf{u}'') > (\mathbf{v}'-\mathbf{v}'')^T(\mathbf{v}'-\mathbf{v}''), \\ = 0 \text{ if } (\mathbf{u}'-\mathbf{u}'')^T(\mathbf{u}'-\mathbf{u}'') = (\mathbf{v}'-\mathbf{v}'')^T(\mathbf{v}'-\mathbf{v}''), \\ < 0 \text{ if } (\mathbf{u}'-\mathbf{u}'')^T(\mathbf{u}'-\mathbf{u}'') < (\mathbf{v}'-\mathbf{v}'')^T(\mathbf{v}'-\mathbf{v}''). \end{cases} \quad (47)$$

The distance in the initial sense

$$r(\mathbf{x}',\mathbf{x}'') = \sqrt{(\mathbf{u}'-\mathbf{u}'')^T(\mathbf{u}'-\mathbf{u}'') - (\mathbf{v}'-\mathbf{v}'')^T(\mathbf{v}'-\mathbf{v}'')} \quad (48)$$

is defined only for the pairs of vectors, for which the radicand is non-negative, otherwise the distance is an imaginary number.

Thus, the pseudo-Euclidean space spanned over a distance space $\{\omega_1,\ldots,\omega_M;\phi\}$ is, generally speaking, no distance space, because the distance $r(\mathbf{x}',\mathbf{x}'')$ is defined not for all vector pairs $(\mathbf{x}',\mathbf{x}'') \in \mathbb{R}^M \times \mathbb{R}^M$, whereas its squared value $r^2(\mathbf{x}',\mathbf{x}'')$ (47) is defined for all pairs but may be negative for some of them.

But any choice of the center $\phi \in \Omega = \{\omega_1,\ldots,\omega_M\}$ in the initial finite set of real-world objects defines a mapping of this set into its finite image of the entire continuum pseudo-Euclidean embedding space. The next Section will show that the squared distances between images of real-world objects coincide with squared distanced between original objects and, so, are always positive.

4.3   Isodistant Image of a Distance Space in Its Pseudo-Euclidean Embedding Space

Eigenvalues and eigenvectors of the symmetric similarity matrix $\mathbf{K}_\phi$ (37) and (40) jointly determine coordinates of the finite number of points

$$\mathbb{R}^M_{\Omega,\phi} = \{\mathbf{x}_{\phi,\omega_1},\ldots,\mathbf{x}_{\phi,\omega_M}\} \subset \mathbb{R}^M \quad (49)$$

that are images of real-world objects $\omega \in \Omega \Rightarrow \mathbf{x}_{\phi,\omega} \in \mathbb{R}^M$. We shall refer to this set as the image of the distance space in the pseudo-Euclidean linear space. Of course, the image of the distance space $\Omega \xrightarrow{\phi \in \Omega} \mathbb{R}^M_{\Omega,\phi}$ essentially depends on the choice of the center.

As we stated above (47), the distance is defined in the pseudo-Euclidean space $\mathbb{R}^M$ only for vector pairs having non-negative squared distance $r^2(\mathbf{x}',\mathbf{x}'') \geq 0$. But it is just this property that possess all the vectors in the image of the distance space in the pseudo-Euclidean linear space (49). Indeed,

$$\begin{aligned} r^2(\mathbf{x}_{\phi,\omega_k},\mathbf{x}_{\phi,\omega_l}) &= (\mathbf{x}_{\phi,\omega_k} - \mathbf{x}_{\phi,\omega_l})^T \mathbf{J}_p(\mathbf{x}_{\phi,\omega_k} - \mathbf{x}_{\phi,\omega_l}) \\ &= \mathbf{x}_{\phi,\omega_k}^T \mathbf{J}_p \mathbf{x}_{\phi,\omega_k} + \mathbf{x}_{\phi,\omega_l}^T \mathbf{J}_p \mathbf{x}_{\phi,\omega_l} - 2\mathbf{x}_{\phi,\omega_k}^T \mathbf{J}_p \mathbf{x}_{\phi,\omega_l}, \end{aligned}$$

whence in accordance with (40) and (15) we have

$$\begin{aligned} r^2(\mathbf{x}_{\phi,\omega_k},\mathbf{x}_{\phi,\omega_l}) &= K_\phi(\omega_{\phi,k},\omega_{\phi,k}) + K_\phi(\omega_{\phi,l},\omega_{\phi,l}) - 2K_\phi(\omega_{\phi,k},\omega_{\phi,l}) \\ &= \rho^2(\omega_k,\omega_l) \geq 0, \omega_k,\omega_l \in \Omega. \end{aligned}$$

It is to be emphasized that, despite the pseudo-Euclidean linear space $\mathbb{R}^M$ spanned over the distance space $(\Omega = \{\omega_1, \ldots, \omega_M\}, \rho(\omega', \omega''))$ does not depend on the center $\phi \in \Omega$, the embedding (49) essentially depends on it, but this embedding remains isodistant for any choice of the center.

4.4   An Important Particular Case: Embedding of a Proto-Euclidean Metric Space into the Euclidean Linear Space

Up to now we assumed that an arbitrary distance function $\rho(\omega', \omega'')$ is defined on the set of real-world objects Ω (8). Let us now assume the distance function to obey, first, the triangle inequality (9), i.e. to be a metric, and, additionally, the requirement of conditional non-negative definiteness $\mathbf{c}^T \mathbf{P}_N \mathbf{c} \geq 0$, $\mathbf{1}^T \mathbf{c} = \sum_{j=1}^{N} c_j = 0$ (30) of matrices $\mathbf{P}_N = \left[-\rho^2(\omega_j, \omega_l), j, l = 1, \ldots, N \right]$ for all the finite sets of objects $\{\omega_1, \ldots, \omega_N\} \subseteq \Omega = \{\omega_1, \ldots, \omega_M\}$, $N \leq M$. In Sect. 3.2 we called metrics of such a kind proto-Euclidean.

Theorem 3. In the case of proto-Euclidean metric (30), matrix $\mathbf{K}_\phi = \left[K_\phi(\omega_i, \omega_j), i, j = 1, \ldots, M \right]$ of pair-wise similarities in a finite metric space $K_\phi(\omega_i, \omega_j) = (1/2)[\rho^2(\omega_i, \phi) + \rho^2(\omega_j, \phi) - \rho^2(\omega_i, \omega_j)]$ (11) is non-negative definite for any $\phi \in \Omega$.

Proof. In Appendix 3.

On the force of this Theorem, all the eigenvalues of matrix $\mathbf{K}_\phi$ are non-negative $\xi_{\phi,1} \geq 0, \ldots, \xi_{\phi,M} \geq 0$, and signature of this matrix is $(p = M, q = 0)$ (35). Thus, in accordance with (35)–(40), matrix $\mathbf{K}_\phi$ can be represented as that of usual inner products

$$
\mathbf{K}_\phi =
\begin{pmatrix}
\mathbf{x}_{\phi,1}^T \mathbf{x}_{\phi,1} & \cdots & \mathbf{x}_{\phi,1}^T \mathbf{x}_{\phi,M} \\
\vdots & \ddots & \vdots \\
\mathbf{x}_{\phi,M}^T \mathbf{x}_{\phi,1} & \cdots & \mathbf{x}_{\phi,M}^T \mathbf{x}_{\phi,M}
\end{pmatrix}
$$

of vectors $\mathbf{x}_{\phi,i} \in \mathbb{R}^M$ formed by rows of matrix (36) multiplied by coefficients $\sqrt{\xi_{\phi,i}}$.

Just as in the general case (37), objects of an arbitrary proto-Euclidean metric space $\Omega = \{\omega_1, \ldots, \omega_M\}$ with assigned center $\phi \in \Omega$ can be mapped into a Euclidean linear space as real-valued vectors $\mathbf{x}_{\phi,1} = \mathbf{x}_{\phi,\omega_1} \in \mathbb{R}^M, \ldots, \mathbf{x}_{\phi,M} = \mathbf{x}_{\phi,\omega_M} \in \mathbb{R}^M$, of a huge (if not infinite) dimension M. The distinction is that the metric space is embedded now into a Euclidean linear space with a usual inner product that possesses all three properties (42), (43) and (44). Respectively, the squared metric (particular kind of distance) is always non-negative $r^2(\mathbf{x}', \mathbf{x}'') \geq 0$ (47), and the metric itself (48) is defined for each pair of points $\mathbf{x}', \mathbf{x}'' \in \mathbb{R}^M$.

5 The Generalized Distance Based Linear Approach to Dependence Estimation

5.1 The Generalized Linear Dependence Model

Our adoption of the term "Generalized Liner Model (GLM)", introduced by John Nelder and his colleagues as applied to mathematical statistics [18, 19], is meant here to call attention to the fact that the essence of this utterly pertinent notion had been, actually, anticipated by Braverman's idea of the rectifying linear space. Since a set of objects of any kind Ω is embedded in a linear space $\widetilde{\Omega} \supset \Omega$ with null point $\phi \in \widetilde{\Omega}$ and inner product $K_\phi(\omega', \omega'') : \widetilde{\Omega} \times \widetilde{\Omega} \to \mathbb{R}$, maybe, an indefinite one, it is enough to consider, at least theoretically, the parametric family of real-valued generalized linear features of each object $\omega \in \Omega$

$$z(\omega|\phi, \vartheta, b) = K_\phi(\omega, \vartheta) + b : \Omega \xrightarrow{(\phi, \vartheta, b)} \mathbb{R}. \tag{50}$$

Here the null point and direction vector in the embedding space $\vartheta \in \widetilde{\Omega}$ along with the real bias $b \in \mathbb{R}$ are free parameters to be adjusted. But it should be marked that the null point $\phi \in \widetilde{\Omega}$ can be free chosen in accordance with the rule of its translation (14).

As we see, the generalized linear feature of an object depends not only on the direction vector $\vartheta \in \widetilde{\Omega}$ but also on the null point $\phi \in \widetilde{\Omega}$. However, we shall see in Sect. 5.2 that the generalized linear feature (50) depends only on the difference $\vartheta - \phi$ in terms of linear operations defined in $\widetilde{\Omega}$, and it is enough to consider only this difference that should be named direction vector. In Sect. 5.2, we shall get rid of the dependence on the null point by a natural parametric representation of the difference $\vartheta - \phi \in \widetilde{\Omega}$ in terms of a basic set of real-world objects, to be strict, their images in the embedding space $\widetilde{\Omega}$.

As to the relation of the linear feature $z(\omega|\phi, \vartheta, b)$ to the object's goal characteristic $y \in \mathbb{Y}$, it will be established, just as in John Nelder's GLM, by a link function

$$q(y, z) : \mathbb{Y} \times \mathbb{R} \to \mathbb{R}^+, \tag{51}$$

which is to be chosen by the observer (Sect. 5.1.2).

Actually, the generalized linear feature (50) along with the link function $q(y, z)$ jointly convey the idea of the generalized linear model of a dependence. However, the point ϑ, which is the main parameter of a hypothetical linear space, cannot be represented in a computer, because the extremely high-dimensional linear space $\widetilde{\Omega}$ is an abstract notion that exists only in the imagination of the observer.

5.1.1 The First Heuristic of the Observer – Distance Function

The generalized linear feature (50) is completely defined by the indefinite inner product $K_\phi(\omega', \omega'') : \widetilde{\Omega} \times \widetilde{\Omega} \to \mathbb{R}$ in the pseudo-Euclidean linear space $\widetilde{\Omega}$ in which the initial distance space of objects is embedded $\widetilde{\Omega} \supset \Omega$ (Table 1). As it is stated in Sect. 3 and mathematically proved in Sect. 4, the inner product coincides with the similarity function (11) (synonyms – potential function, indefinite kernel) within the initial set of really existing objects

$$K_\phi(\omega', \omega'') = \frac{1}{2}\left[\rho^2(\omega', \phi) + \rho^2(\omega'', \phi) - \rho^2(\omega', \omega'')\right], \omega', \omega'' \in \Omega, \phi \in \Omega, \quad (52)$$

where all the squared distances are positive $\rho^2(\omega', \omega'') \geq 0$ including $\rho^2(\omega, \phi) \geq 0$. As to other hypothetical (imagined by the observer) points $\omega \in \widetilde{\Omega}\backslash\Omega$, the function $K_\phi(\omega', \omega'')$ is naturally extended on them in accordance with the reasons outlined in Sect. 4. However, the squared $\rho^2(\omega', \omega'')$ and $\rho^2(\omega, \phi)$ may occur to be negative.

Thus, for all the pairs of points in the embedding space $\widetilde{\Omega}$, the inner product $K_\phi(\omega', \omega'') : \widetilde{\Omega} \times \widetilde{\Omega} \to \mathbb{R}$, maybe, an indefinite one, is completely defined by the distance function (8) chosen by the observer:

$$\rho(\omega', \omega'') : \Omega \times \Omega \to \mathbb{R}^+.$$

We call this choice his *first heuristic* of the observer.

The observer is meant to have made this choice in the hope that it meets Braverman's *compactness hypothesis* (Sect. 1.2). All the forthcoming properties of the dependence model the observer craves to build will crucially and irreversibly depend on this choice.

5.1.2   The Second Heuristic of the Observer – Link Function

In our terms, John Neldor's Generalized Linear Model (GLM) [18, 19] generalizes linear regression by allowing the linear model $z(\omega|\vartheta, b) \in \mathbb{R}$ (50) to be related to the response variable of any kind $y \in \mathbb{Y}$, in our case, the hidden characteristic of the respective real-world object $\omega \in \Omega$.

The link function $q(y, z) \geq 0$ (51) serves as a means of formalizing this relationship. This is a kind of loss function that is meant to express the observer's suggestion on how the Nature would penalize the estimate of unknown y for an object $\omega \in \Omega$ represented by its generalized linear feature $z(\omega|\phi, \vartheta, b)$. When the null point in the hypothetical embedding linear space $\phi \in \widetilde{\Omega}$, direction point $\vartheta \in \widetilde{\Omega}$, and numerical bias $b \in \mathbb{R}$ are chosen in some way, the value $q(y, z(\omega|\phi, \vartheta, b))$ is an appropriate measure of the individual risk of error when making decision on the goal characteristic of a single object. The decision rule that minimizes this risk would have the form

$$\hat{y}(\omega|\phi, \vartheta, b) = \underset{y \in \mathbb{Y}}{\arg\min}\, q(y, z(\omega|\phi, \vartheta, b)) : \quad \Omega \xrightarrow{(\phi, \vartheta, b)} \mathbb{Y}. \quad (53)$$

We consider the choice of the link function $q(y, z) \geq 0$ (51) as the *second heuristic* of the observer. It is clear that this choice is as crucial for the properties of the desired dependence model as the choice of distance function (Sect. 5.1.1).

5.1.3   Empirical Risk Minimization

Once both heuristics are accepted, the null point $\phi \in \widetilde{\Omega}$ (not obligatory $\phi \in \Omega \subset \widetilde{\Omega}$), direction point $\vartheta \in \widetilde{\Omega}$ and bias $b \in \mathbb{R}$ are the only parameters that define the decision rule applicable to any object $\omega \in \Omega$. Generally speaking, these two heuristics are sufficient for formulating the criterion of adjusting the model parameters to the given training set $\{(\omega_j, y_j), j = 1, \ldots, N\}$. The requirement of finding a compromise between individual risks at different training-set objects $q\big(y_j, z(\omega_j|\phi, \vartheta, b)\big)$ leads to an analog of the well-known principle of empirical risk minimization [29, 30]:

$$\frac{1}{N}\sum_{j=1}^{N} q\big(y_j, z(\omega_j|\phi, \vartheta, b)\big) \rightarrow \min(\phi, \vartheta \in \widetilde{\Omega},\ b \in \mathbb{R}), z(\omega_j|\phi, \vartheta, b) = K_\phi(\omega_j, \vartheta) + b.$$

$$(54)$$

However, since the dimension of the embedding linear space $\widetilde{\Omega}$ is assumed to be extremely large, the too high capacity of the class of decision rules (too large freedom of varying the abstract pair $\phi, \vartheta \in \widetilde{\Omega}$) will result, as a rule, in the danger of overfitting (overtraining) [30]. As a result, the hypothetical "average" risk of the learned model $(\widehat{\phi}, \widehat{\vartheta}, \widehat{b})$ over the "entire" set of real world objects $\bar{q}\big(y(\omega), z(\omega|\widehat{\phi}, \widehat{\vartheta}, \widehat{b})\big)$ may be much larger than the empirical risk on the training set $(1/N)\sum_{j=1}^{N} q\big(y_j, z(\omega_j|\widehat{\phi}, \widehat{\vartheta}, \widehat{b})\big)$ (54). This is the effect well known in Machine Learning as low generalization performance.

Of course, this is only a rough explanation of the notion of generalization performance. The strict theory based on probabilistic assumptions on the set of real-world objects [3, 4] is beyond the topic of this paper.

5.1.4   The Third Heuristic of the Observer – Regularization Function

The commonly adopted way of enhancing the generalization performance is adding a penalty on model complexity to the empirical risk (54). This penalty usually has the form of a function of the main model parameter, in our case of the generalized linear model this is the null point and direction point in the embedding linear space $\phi, \vartheta \in \widetilde{\Omega}$. The function to be minimized $V(\phi, \vartheta) \rightarrow \min(\phi, \vartheta \in \widetilde{\Omega})$ is meant to express the observer's intuitive a priori knowledge on which models are less likely than the others. As a rule, it is enough to penalize the direction vector, i.e., the difference $\vartheta - \phi$:

$$V(\vartheta - \phi) \rightarrow \min(\phi, \vartheta \in \widetilde{\Omega}). \tag{55}$$

We shall call the choice of regularization function the ***third heuristic*** of the observer. This choice is not less important for the success of learning than the first and second heuristics (Sects. 5.1.1 and 5.1.2).

Some typical kinds of regularization functions will be briefly considered below in Sect. 5.2.3 in term of a parametric representation of the direction vector $\vartheta - \phi$.

5.1.5   The Generalized Training Criterion: The Principle of Regularized Empirical Risk Minimization

Once the observer has done all three heuristic choices (Sects. 5.1.1, 5.1.2, and 5.1.4), the generalized training principle is defined as demand to find a compromise between minimization of both empirical risk (54) and regularization function (55). It is commonly adopted to replace this two-criteria optimization problem by fusing the mutually contradicting requirements into a unified weighted criterion:

$$
V(\vartheta - \phi) + C \frac{1}{N} \sum_{j=1}^{N} q\left(y_j, z(\omega_j | \phi, \vartheta, b)\right) \to \min(\phi, \vartheta \in \widetilde{\Omega},\ b \in \mathbb{R}),
$$

$$
z(\omega_j | \phi, \vartheta, b) = K_\phi(\omega_j, \vartheta) + b.
$$

(56)

This is just the generalized version of the commonly adopted criterion of *regularized empirical risk minimization* [29, 30]. Since the generalized linear feature $z(\omega | \phi, \vartheta, b) = K_\phi(\omega, \vartheta) + b$ (50) is linear function of $\phi, \vartheta \in \widetilde{\Omega}$ on the force of bilinearity of the (indefinite) inner product (21), the training criterion (56) will be convex in $\widetilde{\Omega} \times \mathbb{R}$, if each of link function $q(y, z)$ and regularization function $V(\phi, \vartheta)$ is chosen as convex relative to, respectively, $z \in \mathbb{R}$ and $\vartheta - \phi \in \widetilde{\Omega}$.

In accordance with (50) and (53), the minimum point of the training criterion $(\widehat{\phi}, \widehat{\vartheta}, \widehat{b})$ yields the decision rule, which is applicable to any object independently of whether it is represented in the training set or not:

$$
\hat{y}(\omega | \widehat{\phi}, \widehat{\vartheta}, \widehat{b}) = \underset{y \in \mathbb{Y}}{\arg\min}\, q\left[y,\ \left(K_{\widehat{\phi}}(\omega, \widehat{\vartheta}) + \widehat{b}\right)\right] :\ \Omega \to \mathbb{Y}.
$$

(57)

The trade-off coefficient $C > 0$ in the generalized training criterion (56) is an inevitable hyperparameter, whose increasing from 0 to ∞ provides smooth switch from the extremely introvertive to the completely extrovertive strategy of the observer. Both of these extremes are awful. The coefficient C is to be adjusted in the course of learning by maximizing an appropriate index of generalization performance [3, 4, 23].

Thus, to completely determine the training problem (56), whose solution will result in the decision making rule (53) applicable to any real-world object, represented or not represented in the training set, it is enough that the observer makes three heuristic decisions. He has to specify, first, the distance function on the set of real-world objects $\rho(\omega', \omega'') : \Omega \times \Omega \to \mathbb{R}^+$ (8), second, the link function $q(y, z) : \mathbb{Y} \times \mathbb{R} \to \mathbb{R}^+$ (51), and, third, the regularization function $V(\vartheta - \phi) : \widetilde{\Omega} \times \widetilde{\Omega} \to \mathbb{R}^+$ (55).

Some ways of choosing the distance and the regularization functions will be considered below in Sect. 5.2 in terms of parametric representation of the generalized linear model. As to the link function, in the next Section we consider three simplest versions of it.

5.1.6   Some Link Functions

As stated above, the link function is meant to express the observer's view of how the Nature would penalize the suggestion $y \in \mathbb{Y}$ on the unknown hidden characteristic of an object whose generalized linear feature had taken the value $z \in \mathbb{R}$ (50). We restrict ourselves here to regression estimation and two-class pattern recognition (2).

Regression Estimation implies that the hidden goal characteristic of objects is a real number $y \in \mathbb{Y} = \mathbb{R}$. The generalized linear feature is also a real number $z \in \mathbb{R}$, so, it is enough to penalize the absolute value of the difference $z - y$. The simplest symmetric penalty is a quadratic function

$$q(y, z) = (z - y)^2, \tag{58}$$

as it is shown in Fig. 1(a).

In the case of *two-class pattern recognition*, it appears natural to treat positive values of the generalized feature $z > 0$ as evidence in favor of, say, class $y = +1$ and, vice versa, negative values $z < 0$ as voting for the opposite class $y = -1$. Then the penalty $q(y = +1, z)$ should be a decreasing function of $z \in \mathbb{R}$, whereas $q(y = -1, z)$ has to decrease along the z axis, as in Fig. 1(b):

$$\lim_{z \to -\infty} q(y = +1, z) = \infty, \quad \lim_{z \to \infty} q(y = +1, z) = 0,$$
$$\lim_{z \to -\infty} q(y = -1, z) = 0, \quad \lim_{z \to \infty} q(y = -1, z) = \infty. \tag{59}$$

Any symmetric pair of such functions, which always can be chosen as convex, as it is shown in Fig. 1,b, meets the idea of two-class pattern recognition.

In particular, if symmetric convex functions $q(y = +1, z)$ and $q(y = -1, z)$ are defined by the equality

$$q(y, z) = \ln[1 + \exp(-yz)], \tag{60}$$

we obtain the well known *Logistic Regression* method of dichotomous pattern recognition [31]. The plot of the respective link function shown in Fig. 1(c) is a more specific instantiation of the general requirement (56),

The greater the generalized feature $z(\omega_j | \phi, \vartheta, b)$ of a training-set object $\omega_j \in \Omega$ of class $y_j = +1$, the smaller the logistic-regression penalty (60), the weaker the influence of this object on the value of the training criterion (56), and, hence, on the resulting decision rule (53). The exactly inverse happens also with objects of the opposite class $y_j = +1$, but in both cases the penalty never becomes zero. Thus, all the objects of the training set affect the result of training.

The specificity of the convex link function shown in Fig. 1,d

$$q(y, z) = \max(0, \ 1 - yz) \tag{61}$$

is that it stops penalizing the generalized feature z, when its absolute value exceeds some threshold, say, $|z| \geq 1$, and the sign of the feature is "correct" – $z > 0$ if $y = +1$, and vice versa, $z < 0$ in the case $y = -1$. The link function of this kind results in the

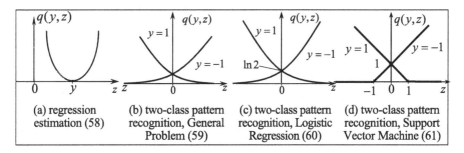

| (a) regression estimation (58) | (b) two-class pattern recognition, General Problem (59) | (c) two-class pattern recognition, Logistic Regression (60) | (d) two-class pattern recognition, Support Vector Machine (61) |

Fig. 1. Some link functions.

dissimilarity based version of the extremely popular Support Vector Machine (SVM) [3, 32].

This name refers to the fact that only the training-set objects for which the respective summands of the training criterion (56) differ from zero $q(y_j, z(\omega_j | \phi, \vartheta, b)) \geq 0$, i.e., $y_j z(\omega_j | \phi, \vartheta, b) \leq 1$, as it is seen from Fig. 1, d, "support" the finally inferred parameters $(\hat{\vartheta}, \hat{b})$ of the decision rule.

5.2 Parametric Representation of the Generalized Linear Model and Numerical Versions of the Training Criterion

5.2.1 A Finite Basis in the Hypothetical Linear Embedding Space

However, the most general training criterion (56) is not more than a universal conceptual principle of organizing the training process in terms of Braverman's compactness hypothesis coupled with his idea of a rectifying linear space, i.e., in terms of the distance-based generalized linear model of the dependence of any kind. It cannot be immediately realized as a computer-implementable algorithm, and the obstacle is the fact that at least one parameter $\vartheta \in \widetilde{\Omega}$ of the generalized linear feature $z(\omega | \phi, \vartheta, b) = K_\phi(\omega, \vartheta) + b$ (50) is element of the hypothetical linear space, which exists only in the imagination of the observer.

Before going to the idea of finite-dimensional parametric representation of the generalized linear feature, let the reader pay attention to the fact that the linear operations (17)-(19) and inner product $K_\phi(\omega, \vartheta)$ (20)-(24) in the embedding linear space will be different for different choices of the null point $\phi \in \widetilde{\Omega}$. A translation of the null point $\phi \to \phi'$ results in the change of the inner product $K_\phi(\omega, \vartheta) \to K_{\phi'}(\omega, \vartheta)$ (22). But the generalized linear feature of any object $z(\omega | \phi, \vartheta, b)$ depends only on the vector difference $\vartheta - \phi \in \widetilde{\Omega}$, and it is always possible to simultaneously change the direction vector and the null point, so that this difference remains the same $\vartheta' - \phi' = \vartheta - \phi$. Thus, in accordance with the rule of null point translation (22), the linear model $z(\omega | \phi, \vartheta, b) = K_\phi(\omega, \vartheta) + b$ (50) can be equivalently represented by a continuum of pairs $(\phi' \in \widetilde{\Omega}, \vartheta \in \widetilde{\Omega})$, where ϕ' can be different from the fixed center $\phi \in \Omega \subset \widetilde{\Omega}$:

$$K_{\phi'}(\omega, \vartheta) = K_\phi(\omega, \vartheta) - K_\phi(\omega, \phi') - K_\phi(\vartheta, \phi') + K_\phi(\phi', \phi'). \qquad (62)$$

Introduction of a finite basis of relatively small size in the embedding linear space $\widetilde{\Omega}$ may serve as a way to make the general training principle (56) computationally realizable. Only the really existing objects are observable elements of this space $\omega \in \Omega \subset \widetilde{\Omega}$, therefore, the basic set should be formed by some finite number of real-world objects $\{\omega_1, \ldots, \omega_n\} \subset \Omega$. In what follows, we restrict the search for the parameters $(\phi \in \widetilde{\Omega}, \vartheta \in \widetilde{\Omega})$ to finite affine combinations of the basic objects:

$$\phi = \sum_{i=1}^{n} c_{\phi,i}\omega_i, \ \ \vartheta = \sum_{i=1}^{n} c_{\vartheta,i}\omega_i \in \widetilde{\Omega}, \ \ \{\omega_1, \ldots, \omega_n\} \subset \Omega, \ \ c_{\phi,i}, c_{\vartheta,i} \in \mathbb{R}, \ \ \sum_{i=1}^{n} c_{\phi,i} = \sum_{i=1}^{n} c_{\vartheta,i} = 1.$$
$$(63)$$

5.2.2 The Parametric Family of Generalized Linear Features and Parametric Representation of the Regularized Empirical Risk

In accordance with (21) and (62), we shall have:

$$z(\omega|\phi', \vartheta, b) = K_\phi\left(\omega, \sum_{i=1}^{n} c_{\vartheta,i}\omega_i\right) - K_\phi\left(\omega, \sum_{i=1}^{n} c_{\phi,i}\omega_i\right) - K_\phi\left(\sum_{i=1}^{n} c_{\vartheta,i}\omega_i, \sum_{j=1}^{n} c_{\phi,j}\omega_j\right) +$$

$$K_\phi\left(\sum_{i=1}^{n} c_{\phi,i}\omega_i, \sum_{j=1}^{n} c_{\phi,j}\omega_j\right) + b = \sum_{i=1}^{n} K_\phi(\omega, \omega_i)(c_{\vartheta,i} - c_{\phi,i})$$

$$\underbrace{- \sum_{i=1}^{n}\left(\sum_{j=1}^{n} K_\phi(\omega_i, \omega_j)c_{\phi,j}\right)c_{\vartheta,i} + \sum_{i=1}^{n}\left(\sum_{j=1}^{n} K_\phi(\omega_i, \omega_j)c_{\phi,j}\right)c_{\phi,i} + b}.$$

const relative to the given entity ω, a new bias b

If we additionally denote

$$a_i = (c_{\vartheta,i} - c_{\phi,i}), \text{ so that } \sum_{i=1}^{n} a_i = 0 \text{ on the force of (63)}, \qquad (64)$$

we come to the parametric representation of the direction vector in accordance with (63)

$$\vartheta - \phi = \sum_{i=1}^{n} a_i\omega_i, \qquad (65)$$

as well as to the parametric family of generalized linear features:

$$z(\omega|a_1,\ldots,a_n,b) = \sum_{i=1}^{n} a_i K_\phi(\omega,\omega_i) + b, \sum_{i=1}^{n} a_i = 0. \tag{66}$$

This family should be named similarity-based (kernel-based) family because the similarity function (indefinite kernel) $K_\phi(\omega',\omega'') : \Omega \times \Omega \xrightarrow{\phi} \mathbb{R}$ (11) is assumed to be defined on the set of real-world objects. Despite the fact that similarity depends on the choice of the "central" object $\phi \in \Omega$, it is easy to see that generalized feature (66) does not depend on it.

Indeed, in accordance with (11)

$$z(\omega|a_1,\ldots,a_n,b) = \frac{1}{2}\sum_{i=1}^{n} a_i \left[\rho^2(\omega,\phi) + \rho^2(\omega_i,\phi) - \rho^2(\omega,\omega_i)\right] + b =$$

$$\frac{1}{2}\left[\underbrace{\left(\sum_{i=1}^{n} a_i\right)\rho^2(\omega,\phi)}_{=0} + \underbrace{\sum_{i=1}^{n} a_i\rho^2(\omega_i,\phi) - \sum_{i=1}^{n} a_i\rho^2(\omega,\omega_i)}_{const}\right] + b.$$

Coefficient $(1/2)$ is out of significance here, and $(1/2)const + b$ is an equivalent bias, for which we retain the same notation b. Thus, we come to the parametric family of generalized linear features as the sum equivalent to (66):

$$z(\omega|a_1,\ldots,a_n,b) = \sum_{i=1}^{n} \left(-\rho^2(\omega,\omega_i)\right)a_i + b, \sum_{i=1}^{n} a_i = 0. \tag{67}$$

This is the final form of the family of generalized linear features applicable to any real-world object $\omega \in \Omega$. It may seem that the square sign $\rho^2(\omega,\omega_i)$ is redundant here, because the squared distance remains to be a distance. But this sign will be significant in the particular case of a metric, especially, proto-Euclidean metric, which doesn't remain to be a metric when having been squared.

The minus sign is also out of significance, but we retain it as a reminder on its inheritance from the initial similarity (kernel) function (11).

Thus, the search of the direction vector $\vartheta - \phi \in \tilde{\Omega}$ in the hypothetical extremely high-dimensional space boils down to the search of a real valued vector $\mathbf{a} \in \mathbb{R}^n$ under the additional condition $\mathbf{1}^T\mathbf{a} = 0$, where $\mathbf{1} \in \mathbb{R}^n$ is all-ones vector (64).

Application of (65) and (67) to the training criterion of regularized empirical risk minimization (56) yields its parametric distance-based form:

$$\begin{cases} V(a_1,\ldots,a_n) + C\frac{1}{N}\sum_{j=1}^{N} q\left(y_j, z(\omega_j|a_1,\ldots,a_n,b)\right) \to \min(a_1,\ldots,a_n,b), \\ z(\omega|a_1,\ldots,a_n,b) = \sum_{i=1}^{n}\left(-\rho^2(\omega,\omega_i)\right)a_i + b, \quad \sum_{i=1}^{n} a_i = 0. \end{cases} \tag{68}$$

It is clear that this training criterion is convex when the link function $q(y, z)$ and regularization function $V(a_1, \ldots, a_n)$ are convex relative to, respectively, $z \in \mathbb{R}$ and $\mathbf{a} = (a_1, \ldots, a_n) \in \mathbb{R}^n$.

Just like the general case (56) and on the force of (67), the minimum point of the parametric training criterion $\widehat{\mathbf{a}} = (\hat{a}_1, \ldots, \hat{a}_n) \in \mathbb{R}^n$ defines the distance-based decision rule applicable to any object:

$$\hat{y}(\omega \,|\, \hat{a}_1, \ldots, \hat{a}_n) = \operatorname*{argmin}_{y \in \mathbb{Y}} q\left[y, \ \left(\sum_{i=1}^{n} \left(-\rho^2(\omega, \omega_i) \right) \hat{a}_i + \hat{b} \right) \right] : \quad \Omega \to \mathbb{Y}. \qquad (69)$$

As we see, the basic set of objects $\{\omega_1, \ldots, \omega_n\} \subset \Omega$ is integral part of the decision rule and must be kept in memory along with the resulting coefficients $(\hat{a}_1, \ldots, \hat{a}_n)$ and link function $q(y, z) : \mathbb{Y} \times \mathbb{R} \to \mathbb{R}^+$, because it implies comparison of the given object $\omega \in \Omega$ to all the basic objects. To overcome this extremely burdensome circumstance, it is required to choose the link function and (or) regularization function $V(a_1, \ldots, a_n) : \mathbb{R}^n \to \mathbb{R}^+$ in such a way, that the majority of coefficients $\hat{a}_i$ would turn into zeros or become very small in result of training. Some ways of providing this property of the training criterion will be considered below in Sect. 5.2.3.

5.2.3   Some Parametric Regularization Functions

Ridge and Bridge Regularization
The idea of regularization goes back to Tikhonov's mathematical methodology of solving incorrectly formulated problems [33–35]. It turned out to be exceptionally useful in mathematical statistics, primarily, for estimation of regression dependences from noisy data. The simplest kind of quadratic regularization is well known under the name of *ridge regularization* [36, 37].

As applied to the generalized linear approach to dependence estimation, the quadratic (ridge) regularization may be defined as function of some norm of the direction vector in the embedding linear space.

Since the squared norm of the direction vector may occur to be negative for an arbitrary distance (Sects. 3 and 4), its modulus should be used for measuring the length of the direction vector $\left| \|\vartheta - \phi\|^2 \right|^{1/2}$. In most cases, it is enough to consider the squared absolute norm of the direction vector:

$$V(\vartheta - \phi) = \left| \|\vartheta - \phi\|^2 \right| \to \min.$$

In terms of the pseudo-Euclidean embedding when the squared norm is always defined, the ridge regularization function should be defined as squared distance of the direction point to the zero point:

$$V(\vartheta - \phi) = \|\vartheta - \phi\|^2 = \rho^2(\vartheta, \phi) = K_\phi(\vartheta, \vartheta) \to \min. \qquad (70)$$

This, perhaps, most popular kind of regularization can be treated, at least, in two ways. The respective versions fundamentally differ from each other by their properties in the embedding linear space.

Absence of any a Priory Preferences on Direction Vector's Orientation. In parametric terms (65), the quadratic (ridge) regularization function is a quadratic form of the parameter vector in accordance with (26):

$$V'(\vartheta - \phi) = V'(a_1, \ldots, a_n) = \left\| \sum_{i=1}^{n} a_i \omega_i \right\|^2 = K_\phi \left(\sum_{i=1}^{n} a_i \omega_i, \sum_{l=1}^{n} a_l \omega_l \right) =$$

$$\sum_{i=1}^{n} \sum_{l=1}^{n} K_\phi(\omega_i, \omega_l) a_i a_l \rightarrow \min, \quad \sum_{i=1}^{n} a_i = 0.$$
(71)

The matrix $\left[K_\phi(\omega_i, \omega_j), \ i,j = 1, \ldots, n \right]$ may be not positive definite in a pseudo-Euclidean space, then function $V(\vartheta - \phi)$ will not be convex. Thus, this kind of regularization may be incorrect in the general case.

On the force of (28),

$$V'(a_1, \ldots, a_n) = \rho^2(\vartheta, \phi) = \sum_{i=1}^{n} \sum_{l=1}^{n} K_\phi(\omega_i, \omega_l) a_i a_l =$$

$$\frac{1}{2} \sum_{i=1}^{n} \sum_{l=1}^{n} [\rho^2(\omega_i, \phi) + \rho^2(\omega_l, \phi) - \rho^2(\omega_i, \omega_l)] a_i a_l =$$

$$\frac{1}{2} \underbrace{\left(\sum_{l=1}^{n} a_l \right)}_{=0} \sum_{i=1}^{n} \rho^2(\omega_i, \phi) a_i + \frac{1}{2} \underbrace{\left(\sum_{i=1}^{n} a_i \right)}_{=0} \sum_{l=1}^{n} \rho^2(\omega_l, \phi) a_l - \frac{1}{2} \sum_{i=1}^{n} \sum_{l=1}^{n} \rho^2(\omega_i, \omega_l) a_i a_l, \text{ and, finally,}$$

$$V'(a_1, \ldots, a_n) =$$

$$K_\phi \left(\sum_{i=1}^{n} a_i \omega_i, \sum_{i=1}^{n} a_i \omega_i \right) = \sum_{i=1}^{n} \sum_{l=1}^{n} K_\phi(\omega_i, \omega_l) a_i a_l = \frac{1}{2} \sum_{i=1}^{n} \sum_{l=1}^{n} \left(-\rho^2(\omega_i, \omega_l) \right) a_i a_l, \quad \sum_{i=1}^{n} a_i = 0.$$
(72)

Hence, it is necessary and sufficient for the convexity of this ridge regularization function for any basic set of objects $\{\omega_1, \ldots, \omega_n\} \subset \Omega$ that the matrix $\mathbf{P}_n = \left[-\rho^2(\omega_j, \omega_l), \ j,l = 1, \ldots, n \right]$ would be conditionally positive definite (30), i.e. distance $\rho(\omega', \omega'')$ would be proto-Euclidean metric, see above (30) in Sect. 3.2.

Thus, the simplest ridge regularization (71) and (72) is applicable only in the case of the Euclidean embedding space. All orientations of the direction vector $\vartheta - \phi$ in the embedding space are equally preferable, because only its length is penalized.

Pronounced preference in favor of direction vectors oriented along the major inertia axis of the basic assembly of objects in the embedding space [38]. Equality (71) defines the norm of the direction vector as quadratic form whose matrix depends on inner products between basic objects. However, such a quadratic form remains to be a norm with any positive definite matrix, which may be constant:

$$V''(a_1, \ldots, a_n) = \sum_{i=1}^{n} \sum_{l=1}^{n} \beta_{il} a_i a_l = \mathbf{a}^T \mathbf{B} \to \min. \tag{73}$$

In contrast to (72), this regularization function always remains to be convex, because $\mathbf{B}$ does not depend on the distance function.

In particular, in the case of unity matrix $\mathbf{B} = \mathbf{I}$, we obtain the simplest squared norm of the coefficient vector $\mathbf{a} = (a_1, \ldots, a_n) \in \mathbb{R}^n$ that parametrically represents the direction vector $\vartheta - \phi \in \widetilde{\Omega}$:

$$V''(a_1, \ldots, a_n) = \|\mathbf{a}\|^2 = \mathbf{a}^T \mathbf{a} = \sum_{i=1}^{n} a_i^2 \to \min, \sum_{i=1}^{n} a_i = 0, \tag{74}$$

which, in contrast to (72), remains convex for any distance function. It is just this regularization function that is usually considered in the literature as ridge regularization, primarily for regression estimation [36, 37].

A generalization of this idea, when the model coefficients are raised to an arbitrary power, is known under the name of bridge regularization [39, 40], Fig. 2(a):

$$V''(a_1, \ldots, a_n) = \sum_{i=1}^{n} |a_i|^p \to \min, p \geq 2. \tag{75}$$

In this work, we concentrate on the quadratic version (74). This regularization possesses a special property that may be very useful in practice of dependence estimation.

Let $\Theta = \left\{ \vartheta \in \widetilde{\Omega} : \rho^2(\vartheta, \phi) = const \right\} \subset \widetilde{\Omega}$ be the sphere of fixed radius around the null point in the embedding space. The value of the first regularization function (72) remains the same on this sphere $V'(\vartheta - \phi) = const$, $\vartheta \in \Theta$, and depends only on its radius.

At the same time, the sphere in the embedding space $\Theta \subset \widetilde{\Omega}$ will map as an ellipsoid into the finite-dimensional space of coefficient vectors $\mathbf{a} = (a_1, \ldots, a_n) \in \mathbb{R}^n$ (72):
$$\rho^2(\vartheta, \phi) = const \Rightarrow \mathbf{a}^T \mathbf{K}_\phi \mathbf{a} = const, \qquad \text{where}$$
$\mathbf{K}_\phi = [K_\phi(\omega_i, \omega_l), i, l = 1, \ldots, n] (n \times n)$.

It is still assumed here that the distance $\rho(\omega', \omega'')$ is Euclidean metric and, hence, matrix $\mathbf{K}_\phi$ is nonnegative definite.

Let us consider the second regularization function $V''(\vartheta - \phi) = \mathbf{a}^T \mathbf{a}$ (74), $\vartheta \in \Theta$, and find its conditional minimum point on this ellipsoid:

$$V''(\mathbf{a}) = \mathbf{a}^T \mathbf{a} \to \min, \mathbf{a}^T \mathbf{K}_\phi \mathbf{a} = const. \tag{76}$$

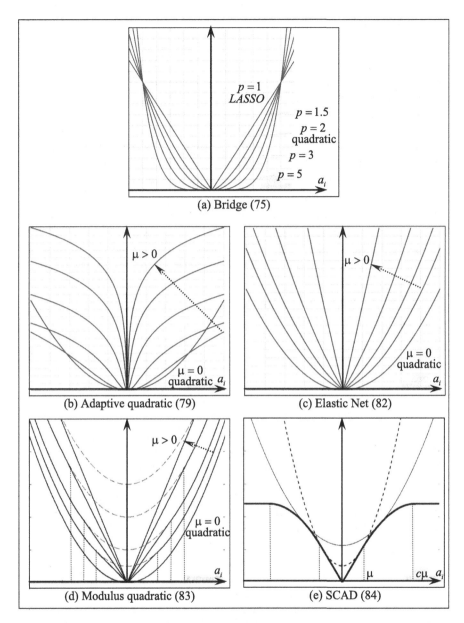

Fig. 2. Some regularization functions.

It is easy to see that, when minimizing this regularization function, we maximize the sum of inner products of basic objects with the normalized direction vector in the embedding Euclidean space.

$$\sum_{i=1}^{n} K_{\phi}(\omega_i, \vartheta(a)) \rightarrow \max. \tag{77}$$

This means that minimizing the second regularization function is equivalent to the a priori requirement to choose the direction vector close to the main inertia axis of the "cloud" of basic points in the Euclidean embedding space.

Indeed, it is easy to see that this condition is met only if $\mathbf{a} \in \mathbb{R}^n$ is the eigenvector of matrix $\mathbf{K}_{\phi}$ corresponding to its greatest eigenvalue $\mathbf{K}_{\phi}\mathbf{a} = \lambda\mathbf{a}$ (the main eigenvector). Further, the main eigenvector of matrix $\mathbf{K}_{\phi}$ meets the condition $\mathbf{a}^T\mathbf{K}_{\phi}\mathbf{K}_{\phi}\mathbf{a} \rightarrow \max$ under the equality constraint $\mathbf{a}^T\mathbf{K}_{\phi}\mathbf{a} = const$, which is equivalent to the condition $\mathbf{a}^T\mathbf{a} \rightarrow \min$ under constraint $\mathbf{a}^T\mathbf{K}_{\phi}\mathbf{a} = const$.

On the other hand, $\mathbf{a}^T\mathbf{K}_{\phi} = \mathbf{a}^T(\mathbf{x}_1 \cdots \mathbf{x}_n)$ where $\mathbf{x}_i \in \mathbb{R}^n$ are columns of the symmetric matrix $\mathbf{K}_{\phi}$, and $\mathbf{a}^T\mathbf{K}_{\phi} = (\mathbf{K}_{\phi}\mathbf{a})^T$, so that $\sum_{i=1}^{n} \mathbf{a}^T\mathbf{x}_i\mathbf{x}_i^T\mathbf{a} = \sum_{i=1}^{n}(\mathbf{x}_i^T\mathbf{a})^2$. In its turn, $\mathbf{x}_i^T\mathbf{a} = \sum_{l=1}^{n} a_l K_{\phi}(\omega_i, \omega_l) = K_{\phi}(\omega_i, \sum_{l=1}^{n} a_l\omega_l) = K_{\phi}(\omega_i, \vartheta(\mathbf{a}))$ and $\mathbf{a}^T\mathbf{K}_{\phi}\mathbf{K}_{\phi}\mathbf{a} = K_{\phi}(\omega_i, \vartheta(\mathbf{a}))$, whence it follows that (76) involves (77).

Adaptive Quadratic Regularization

A particular case of quadratic form (73) is a diagonal matrix with non-negative diagonal elements:

$$V(a_1,\ldots,a_n) = \sum_{i=1}^{n} V(a_i) \rightarrow \min, \ V(a_i) = \beta_i a_i^2, \ \beta_i > 0, \tag{78}$$

Let us assume, so far, that the weights β_i are assigned by the observer. The greater β_i, the stronger the criterion penalizes the estimated deflection $\hat{a}_i^2$ of the i-th component of the direction vector from zero. If $\beta_i \rightarrow \infty$ then $\hat{a}_i^2 \rightarrow 0$, and the respective basic object gets practically excluded from the participation in the decision rule (69).

Actually, the observer can control the size of active basic set by excluding some basic objects and retaining only those of them which he will hold as most informative. However, the choice of informative basic objects (secondary features) is just the main issue, and the a priori choice the weights $(\beta_1,\ldots,\beta_n)$ is extremely problematic.

On the force of this circumstance, the regularization function (78) hardly may have independent value. A way is needed to automatically rank the basic objects by their relevance for the sought-for dependence model.

Since the principle of machine learning is to let the algorithm learn by itself the best parameters from data, the generic idea is to consider the weights in the regularization function β_i as supplementary parameters to be adjusted in the course of training in addition to the main parameters a_i (68). However, the straightforwardly combined criterion

$$\sum_{i=1}^{n} \beta_i a_i^2 + C\frac{1}{N}\sum_{j=1}^{N} q\left[y_j, \ \left(\sum_{i=1}^{n}(-\rho^2(\omega,\omega_i))a_i + b\right)\right]$$
$$\rightarrow \min(a_1,\ldots,a_n,b,\beta_1,\ldots,\beta_n)$$

would always turn a_i into zeros to gain more freedom in minimizing the empirical risk.

We omit here the statistical interpretation of the selectivity methodology considered in [41, 42], but its optimization essence boils down to replacing the weighted quadratic regularization function (78) by a more sophisticated one:

$$V(a_i|\mu) =$$
$$\min_{\beta_i}\left[\beta_i\left(a_i^2 + \frac{1}{\mu}\right) - \left(1 + \frac{1}{\mu}\right)\ln\beta_i\right] - \min_{a_i}\min_{\beta_i}\left[\beta_i\left(a_i^2 + \frac{1}{\mu}\right) - \left(1 + \frac{1}{\mu}\right)\ln\beta_i\right],$$

$$\text{or, finally,} \quad V(a_i) = \frac{\mu+1}{\mu}\ln\left(\mu a_i^2 + 1\right).$$

$$(79)$$

Here $\mu > 0$ is selectivity parameter that controls the propensity of the optimization to suppress irrelevant basic objects. If $\mu \to 0$ and, so, selectivity is absent, it is easy to prove that

$$\lim_{\mu\to 0} V(a_i|\mu) = a_i^2,$$

and we have the simplest quadratic regularization (74). On the contrary, the growth of selectivity leads to growing sharpness of the regularization function in the vicinity of zero and increasingly penalizing even small positive values of a_i^2, as it is seen from Fig. 2 (b).

Training by adaptive criterion with regularization function (79)

$$\frac{\mu+1}{\mu}\sum_{i=1}^{n}\ln\left(\mu a_i^2 + 1\right) + C\frac{1}{N}\sum_{j=1}^{N}q\left[y_j, \left(\sum_{i=1}^{n}\left(-\rho^2(\omega_j, \omega_i)\right)a_i + b\right)\right] \rightarrow$$

$$(80)$$

$$\min(a_1, \ldots, a_n, b)$$

yields estimated coefficients at basic objects $\{\omega_1, \ldots, \omega_n\}$.

When the training is over, the resulting coefficients $(\hat{a}_1, \ldots, \hat{a}_n)$ define the found adaptive weights in the adapted quadratic regularization function $V(a_i) = \hat{\beta}_i a_i^2$ (78), as it is seen from the first summand in (79):

$$\hat{\beta}_i = \min_{\beta_i}\left[\beta_i\left(\hat{a}_i^2 + \frac{1}{\mu}\right) - \left(1 + \frac{1}{\mu}\right)\ln\beta_i\right] = \frac{1 + 1/\mu}{\hat{a}_i^2 + 1/\mu}. \quad (81)$$

The found weights may be sorted in ascending order $\hat{\beta}_1 < \hat{\beta}_2 < \ldots < \hat{\beta}_n$ what is equivalent to descending sorting of coefficients at basic objects $\hat{a}_1 > \hat{a}_2 > \ldots > \hat{a}_n$. It is up to the observer to decide how to use this succession. In particular, if selectivity is set at a moderate level $\mu = 1 \div 10$, he can select most relevant basic objects (secondary features) by a threshold adjusted, say, using cross-validation, and drop the others as redundant ones. What is important, this method allows to not partition subsets of secondary features having approximately equal relevance. If selectivity is high, the adaptive training suppresses almost all secondary features, retaining one or two of them.

Singular Regularization

Adaptive quadratic regularization (Sect. 5.2.3.2) emphasizes some basic objects (secondary features) and relatively suppresses the others, but does not completely remove the redundant ones. The cause of this soft behavior is smoothness of the regularization penalty in the vicinity of the zero point, which acts as almost absence of penalty on small non-zero values of coefficients a_i.

However, it may appear to be desirable to endow the regularization function with pronounced ability to completely exclude the redundant basic objects. The four kinds of regularization functions we mention here are based on the same generic idea of a singularity at zero produced by the modulus function $\mathrm{mod}(a) = |a|$ employed by each of them in combination with the quadratic function $\mathrm{quad}(a) = a^2$.

Elastic Net [43, 44], Fig. 2(c). The idea is to combine the modulus and quadratic function as a linear combination:

$$V(a_i|\mu) = a_i^2 + \mu|a_i|. \tag{82}$$

We shall call the coefficient $\mu \geq 0$ the selectivity parameter, like in (79). If $\mu = 0$, this is the initial quadratic regularization, which displays no ability to select features. On the contrary, if $\mu \to \infty$, the combined regularization function turns into the pure LASSO penalty [45, 46] providing the maximal selectivity. Both modulus and quadratic modes are active at each value of argument a_i.

Modulus Quadratic regularization [21, 47], Fig. 2(d). In this version, two kinds of regularization are smoothly clipped. When $|a_1| \leq \mu$, only the modulus function is active, whereas the values $|a_1| > \mu$ are purely quadratically penalized:

$$V(a_i|\mu) = \begin{cases} 2\mu|a_i|, & \text{if } |a_i| \leq \mu \text{ (LASSO)}, \\ \mu^2 + a_i^2, & \text{if } |a_i| > \mu \text{ (quadratic)}. \end{cases} \tag{83}$$

By its behavior, this regularization function is analogous to Elastic Net (82). Its importance is that an additional sophistication of this kind of penalty results in the following well known regularization.

Smoothly Clipped Absolute Deviation (SCAD) [48], Fig. 2(e):

$$V(a_i|\mu) = \begin{cases} \mu|a_i|, & |a_i| \leq \mu, \\ -(a_i^2 - 2c\mu|a_i| + \mu^2)/(2(c-1)), & \mu < |a_i| \leq c\mu, \quad c > 1. \\ (c+1)\mu^2/2, & |a_i| > c\mu, \end{cases} \tag{84}$$

If $|a_i| \leq \mu$, this penalty coincides with Modulus Quadratic and LASSO – the penalty is linearly growing. But further increase of $|a_i|$ in the interval $\mu < |a_i| \leq c\mu$ results in inverse quadratic growth with linearly diminishing rate. Finally, the penalty remains constant at the level $(c+1)\mu^2/2$ when $|a_i| > c\mu$.

We don't consider here advantages and drawbacks of these different kinds of regularization. But one important circumstance is worth being mentioned. As we see from Fig. 2, adaptive quadratic regularization function (b) and SCAD (e) are non-convex, whereas the others (a), (c) and (d) are convex. In both of these two cases, for any fixed value of selectivity $\mu = const$, the growth rate of the penalty in (79) and (84) tends to zero when $|a_i|$ is increasing, moreover, in SCAD (84) it becomes strict zero for sufficiently large values of $|a_i|$:

$$
\begin{cases}
\lim\limits_{|a_i| \to \infty} \frac{\partial}{\partial (a_i^2)} V''(a_i|\mu) = \lim\limits_{|a_i| \to \infty} \frac{\partial}{\partial (a_i^2)} \frac{\mu+1}{\mu} ln\left(\mu a_i^2 + 1\right) = \frac{\mu+1}{\mu} \lim\limits_{|a_i| \to \infty} \frac{1}{\mu a_i^2 + 1} = 0 \\
\hspace{8cm} \text{(adaptive quadratic),} \\
\frac{\partial}{\partial (a_i^2)} V''(a_i|\mu) = 0 \quad \text{when} \quad |a_i| > c\mu \quad \text{(SCAD).}
\end{cases}
$$

$$(85)$$

Despite the fact that non-convexity will inevitably complicate the optimization procedure required for numerical implementation of the training criterion (68), the latter specificity may occur to be of much greater importance. At least in the case of regression estimation, the equalities (85) guarantee the *"Oracle Property"* of the training criterion [48].

We took the latter term in quotation marks because it is applicable, strictly speaking, only under respective probabilistic assumptions on the data source and data model. We don't consider here the statistical interpretation of the dependence estimation problem, but the essence of the oracle property can be spelled out clearly in simple words.

If there would be an "oracle" who knew the "actual" model parameters and, so, the "actual" set of active features, he could see that the training criterion estimates them correctly, when the size of a "sufficiently good" training set grows unboundedly and the data model is "regular". In terms of the traditional mathematical statistics this means that the estimates of the model parameters are asymptotically consistent.

The statistical oracle properties of the regression parameter estimators using SCAD regularization (84) are outlined in detail in [48]. These properties of parameter estimators for some other kinds of dependences with Adaptive Quadratic regularization (79) are studied in [49, 50].

5.3   Distance-Based Dependence Models of Growing Complexity

5.3.1   Distance Transformation via Exponential Potential Function

It may happen that the observer, being not sure which exactly distance function $\rho(\omega', \omega'')$ (8) will be consistent with the Nature's design, accepts a parametric family:

$$\rho(\omega', \omega''|\alpha) : \Omega \times \Omega \xrightarrow{\alpha} \mathbb{R}^+, \ \alpha \in A. \tag{86}$$

In this case, the respective parametric family of similarity functions should be considered $K_\phi(\omega', \omega''|\alpha)$ instead of a fixed similarity $K_\phi(\omega', \omega'')$ (52).

Let us consider an example of such a parametric family based on Braverman's exponential potential function (7)

$$K_\phi(\omega', \omega''|\alpha) = \frac{1+\alpha}{\alpha} + \frac{1+\alpha}{2\alpha} \exp\left(-\alpha \rho^2(\omega', \omega'')\right) \tag{87}$$

where $\rho(\omega', \omega'')$ is assumed here to be distance function of general kind (8) instead of Euclidean metric in a finite-dimensional linear space as Braverman originally assumed (7). The two-argument function on the set of real-world objects

$$\rho(\omega', \omega''|\alpha) = \left\{ \frac{1+\alpha}{\alpha} \left[1 - \exp\left(-\alpha\rho^2(\omega', \omega'')\right)\right] \right\}^{1/2} \tag{88}$$

remains to be a distance function, which is different from the initial distance function $\rho(\omega', \omega'')$.

Theorem 4. Transformation (88) is identical if $\alpha \to 0$

$$\lim_{\alpha \to 0} \rho(\omega', \omega''|\alpha) \equiv \rho(\omega', \omega''), \tag{89}$$

and it results in the degenerate distance if $\alpha \to \infty$

$$\lim_{\alpha \to \infty} \rho(\omega', \omega''|\alpha) \equiv \begin{cases} 0, & \omega' = \omega'', \\ const = 1, & \omega' \neq \omega''. \end{cases} \tag{90}$$

Proof. In Appendix 4.

The following two theorems state that in the particular case when $\rho(\omega', \omega'')$ is metric (9), the new function $\rho(\omega', \omega'')|\alpha)$ inherits its properties.

Theorem 5. If $\rho(\omega', \omega'')$ is metric, the function $\rho(\omega', \omega''|\alpha)$ is metric too. If $\rho(\omega', \omega'')$ is proto-Euclidean metric (Sect. 3.2), the metric $\rho(\omega', \omega''|\alpha)$ remains proto-Euclidean.

Proof. In Appendix 5.

Theorem 6. If $\rho(\omega', \omega'')$ is proto-Euclidean metric, the metric $\rho(\omega', \omega''|\alpha)$ remains to be proto-Euclidean.

Proof. In Appendix 6.

More over, the transformation with increasing parameter α turns any distance into a proto-Euclidean metric.

Theorem 7. For any distance $\rho(\omega', \omega'')$, the transformed metric $\rho(\omega', \omega''|\alpha)$ is proto-Euclidean with sufficiently large value of parameter $\alpha > 0$.

Proof. In Appendix 7.

Since $\rho(\omega', \omega''|\alpha)$ (88) is a distance function, everything said in previous sections remains valid. Immediate substitution of (88) in the parametric family of generalized linear features (67) gives

$$z(\omega|a_1, \ldots, a_n, b) = \sum_{i=1}^{n} \left(-\rho^2(\omega, \omega_i|\alpha)\right)a_i + b, \quad \sum_{i=1}^{n} a_i = 0. \tag{91}$$

Further substitution of this equality into (68) results in the training criterion of regularized empirical risk minimization based on Braverman's potential function in the set of real-world objects:

$$
\begin{cases}
V(a_1,\ldots,a_n) + C\frac{1}{N}\sum_{j=1}^{N} q\big(y_j, z(\omega_j|a_1,\ldots,a_n,b)\big) \rightarrow \min(a_1,\ldots,a_n,b), \\
z(\omega_j|a_1,\ldots,a_n,b) = \sum_{i=1}^{n}\big(-\rho^2(\omega,\omega_i|\alpha)\big)a_i + b, \quad \sum_{i=1}^{n} a_i = 0.
\end{cases}
\tag{92}
$$

When this optimization problem is solved, the resulting decision rule applicable to any object (69) will have the form

$$
\hat{y}(\omega|\hat{a}_1,\ldots,\hat{a}_n) = \operatorname*{argmin}_{y \in \mathbb{Y}} q\left[y, \ \left(\sum_{i=1}^{n}\big(-\rho^2(\omega,\omega_i|\alpha)\big)\hat{a}_i + \hat{b}\right)\right] : \ \Omega \rightarrow \mathbb{Y}.
\tag{93}
$$

In particular, in the case of two-class pattern recognition $y \in \{-1, 1\}$, we have $q(y,z) = ln[1 + \exp(-yz)]$ (60) or $q(y,z) = \max(0, \ 1 - yz)$ (61), what results in the decision rule

$$
\begin{aligned}
\hat{y}(\omega|\hat{a}_1,\ldots,\hat{a}_n,\hat{b}) &= \operatorname*{argmin}_{y \in \{-1, 1\}} \max(0, \ 1 - yz) \\
&= \begin{cases}
-1, \ \sum_{i=1}^{n}\big(-\rho^2(\omega,\omega_i|\alpha)\big)a_i + b < 0, \\
1, \ \sum_{i=1}^{n}\big(-\rho^2(\omega,\omega_i|\alpha)\big)a_i + b > 0.
\end{cases}
\end{aligned}
\tag{94}
$$

It is appropriate to consider α as one more hyperparameter of the data model in addition to the trade-off coefficient C that balances the empirical risk and regularization in (68).

5.3.2   Fusion of Several Distance Functions

It is natural to expect that different experts skilled in specific knowledge areas will propose to the observer different distance functions $\{\rho_k(\omega', \omega''), \ k = 1,\ldots,m\}$ (8), which may put him in hesitation between them. On the force of reasoning outlined in Sect. 3 and, in more details, in Sect. 4, each of the alternative distance functions independently embeds the initial set of objects Ω into a specific, generally speaking, indefinite Euclidean linear space $\widetilde{\Omega}_k$ with a specific null point $\phi_k \in \widetilde{\Omega}_k$ and specific inner product $K_{k,\phi_k}(\omega', \omega'')$.

Respectively, each embedding will produce its own family of generalized linear features (50) $z_k(\omega|\phi_k, \vartheta_k, b_k) = K_{k,\phi_k}(\omega, \vartheta_k) + b_k : \Omega \xrightarrow{k} \mathbb{R}$, or, in parametric terms (67),

$$
z_k(\omega|a_{k1},\ldots,a_{kn},b_k) = \sum_{i=1}^{n}\big(-\rho_k^2(\omega,\omega_i|\alpha_k)\big)a_{ki} + b_k, \quad \sum_{i=1}^{n} a_{ki} = 0,
\tag{95}
$$

where all the features jointly form a vector.

$$\mathbf{z}(\omega|\phi_k, \vartheta_k, b_k; \ k = 1, \ldots, m) : \Omega \ \to \ \mathbb{R}^m. \tag{96}$$

As previously, it is up to the observer to accept his second heuristic – to choose the link function (Sect. 5.1.2). But now this choice will be much harder, because the link function has become a function of several variables $q(y, z_1, \ldots, z_m) = q(y, \mathbf{z})$. It is always possible to represent the link function of such a kind as a usual one-feature function $q(y(\omega), z(\omega)) : \mathbb{Y} \times \mathbb{R} \to \mathbb{R}^+$ (51), in which the "global" linear feature $z(\omega) \in \mathbb{R}$ is a function of "preliminary" features $z_k(\omega) \in \mathbb{R}$ (97):

$$\begin{aligned} q(y(\omega), z(\omega)) &= q(y(\omega), f(\mathbf{z}(\omega))), \\ z(\omega) &= f(\mathbf{z}(\omega)) = f(z_k(\omega|a_{k1}, \ldots, a_{kn}, b_k), \ k = 1, \ldots, m) = \\ & \qquad f\left(\sum_{i=1}^{n} \left(-\rho_k^2(\omega, \omega_i|\alpha_k) \right) a_{ki} + b_k, \ k = 1, \ldots, m \right). \end{aligned} \tag{97}$$

Thus, the observer has to make an additional heuristic decision – he must choose the structure of this function $f(\bullet) : \mathbb{R}^m \to \mathbb{R}$, more likely, a parametric family of such functions. The training aimed at minimization of the regularized empirical risk (92) becomes much more complicated in the general case, because adjusting function $f(\bullet)$, most likely, in an extremely multidimensional space, is now integral part of training:

$$\left\{ \begin{aligned} & V(a_{ki}, \ i = 1, \ldots, n, \ k = 1, \ldots, m, \ f(\bullet)) + \\ & \qquad C \frac{1}{N} \sum_{j=1}^{N} q\left[y_j, f\left(z_k(\omega_j|a_{ki}, i = 1, \ldots, n), \ k = 1, \ldots, m \right) \right] \to \\ & \qquad\qquad\qquad\qquad\qquad \min(a_{ki}, \ i = 1, \ldots, n, \ k = 1, \ldots, m, \ f(\bullet)), \\ & z_k(\omega_j|a_1, \ldots, a_n) = \sum_{i=1}^{n} \left(-\rho_k^2(\omega, \omega_i|\alpha_k) \right) a_{ki} + b, \quad \sum_{i=1}^{n} a_{ki} = 0, \ k = 1, \ldots, m. \end{aligned} \right. \tag{98}$$

In the next Section, we consider some particular simple link and regularization functions, which have shown themselves to be quite sufficient for distance and kernel fusion in a series of dependence estimation problems [21, 47, 51–53].

5.3.3 Some Link and Regularization Functions for Distance Fusion

In this paper, we restrict our consideration to only one simplest particular version of function $f(z_k, \ k = 1, \ldots, m) : \mathbb{R}^m \to \mathbb{R}$ (97) – we assume it to be linear. Any scalar linear function of such a kind is linear combination $z = \sum_{k=1}^{m} c_k z_k$. Since the coefficients a_{ki} at basic objects in each particular feature in the vector $\mathbf{z} = (z_1, \ldots, z_m)$ (95) may be freely scaled by an arbitrary common coefficient, it is enough to consider the fixed vector of coefficients that consists of all units $\mathbf{c} = (1, \ldots, 1)$, so that $z = \sum_{k=1}^{m} z_k$. Thus, for this choice, the function $z(\omega) = f(\mathbf{z}(\omega))$ (97) is simply the sum that is determined by mn coefficients $(a_{ki}, \ k = 1, \ldots, m, \ i = 1, \ldots, n)$ $z(\omega|a_{ki}, \ k = 1, \ldots, m,$

$$i = 1, \ldots, n) = \sum_{k=1}^{m} \sum_{i=1}^{n} \left(-\rho_k^2(\omega, \omega_i|\alpha_k) \right) a_{ki} + b, \sum_{i=1}^{n} a_{ki} = 0, \ k = 1, \ldots, m.$$

Respectively, the regularization function $V(a_{ki}, i = 1, \ldots, n, k = 1, \ldots, m, f(\bullet))$ in (98) is also completely defined by this set of coefficients $V(a_{ki}, i = 1, \ldots, n, k = 1, \ldots, m)$, and training boils down to finding their values:

$$V(a_{ki}, i = 1, \ldots, n, k = 1, \ldots, m) + C\frac{1}{N}\sum_{j=1}^{N} q\left[y_j, \sum_{k=1}^{m}\sum_{i=1}^{n}\left(-\rho_k^2(\omega_j, \omega_i|\alpha_k)\right)a_{ki} + b\right] \rightarrow$$
$$\min(a_{ki}, i = 1, \ldots, n, k = 1, \ldots, m),$$

$$(99)$$

The trade-off coefficient $C > 0$ and coefficients $\alpha_k > 0$ in transformed distances (88) are hyperparameters to be adjusted by an appropriate validation method.

Do not forget that the size n of the basic set of objects is not less than the number of training objects N. Besides, there are m competing distances (modalities of object representation). Thus, even if the hyperparameters are fixed, the number nm of variables in the objective function (99) exceeds the size N of the training set not less than m times. In this situation, the experimenter inevitably encounters the problem of eliminating redundant modalities and basic objects.

Let $\mathbb{I} = \{1, \ldots, n\}$ denote the set of indices at basic objects, and symbol $\mathbb{K} = \{1, \ldots, m\}$ stand for the set of distance function indices. Then the Cartesian product $\mathbb{I} \times \mathbb{K}$ will be the set of all different modalities representing real-world objects $\omega \in \Omega$. In other words, $\mathbb{F} = \mathbb{I} \times \mathbb{K}$ is the set of all pairs $<i, k>$, i.e., $<$(basic enttity), (distance function)$>$ by which a new object $\omega \in \Omega$ will be compared with the basic set $\{\omega_1, \ldots, \omega_n\}$. Actually, both adaptive quadratic and singular (Sect. 5.2.3) regularization functions are aimed at finding the subset of most relevant distances $\hat{\mathbb{F}} \subset \mathbb{F}$. The final decision rule will refer only to this subset instead of (69):

$$\hat{y}\left(\omega|\hat{a}_{ki}, (k, i) \in \hat{\mathbb{F}}, \hat{b}\right) = \underset{y \in \mathbb{Y}}{\arg\min} \, q\left[y, \left(\sum_{(k,i)\in\hat{\mathbb{F}}} \left[\exp\left(-\alpha\rho^2(\omega, \omega_i)\right)\right]\hat{a}_{ki} + \hat{b}\right)\right] : \Omega \rightarrow \mathbb{Y}$$

$$(100)$$

However, the mechanisms of evaluating the relevance of competitive distances are different for these two methods of regularization.

5.3.4   Decision Rules of Growing Complexity

We have considered two ways of generalizing the generic distance-based criterion of training by the principle of regularized empirical risk minimization (68) and (69).

(1) *Distance transformation* (Sect. 5.3.1). Instead of one training criterion (68) and one parametric family of decision rules (69), we have obtained a parametric family of training criteria $\alpha \in \mathbb{R}^+$ (92) and, respectively, a parametric class of decision rules (93).

This generalization is not superfluous. On the force of Theorem 4, the initial training criterion and decision rule are a particular case in the parametric family of them with $\alpha \to 0$. Conversely, it is stated in the same Theorem that the distance function gets degenerate if $\alpha \to \infty$. Let's study how the degeneracy of the distance will affect the result of training (93).

The two-class pattern recognition problem $y \in \mathbb{Y} = \{-1, 1\}$ would be the simplest example to clarify this question.

Theorem 8. For an arbitrary initial nondegenerate distance (90) $\rho(\omega', \omega'') = \lim_{\alpha \to 0} \rho(\omega', \omega'' | \alpha)$, there exists the training assembly $\{(\omega_j, y_j), j = 1, \ldots, N\}$, which cannot be correctly classified by the decision rule (93) with any parameters $(\hat{a}_1, \ldots, \hat{a}_n, \hat{b})$.

Proof. In Appendix 8.

Theorem 9. Any training assembly $\{(\omega_j, y_j = \pm 1), j = 1, \ldots, N\}$ with an arbitrary initial distance function $\rho(\omega_j, \omega_l)$ can be completely separated by a decision rule

$$d(\omega | a_1, \ldots, a_N, b, \alpha) = \sum_{i=1}^n \left(-\rho^2(\omega, \omega_i | \alpha)\right) a_i + b \gtrless 0, \quad \sum_{i=1}^n a_i = 0, \qquad (101)$$

with sufficiently large value of parameter $\alpha > 0$.

Proof. In Appendix 9.

Thus, the zero value of the distance transformation parameter $\rho(\omega', \omega'' | \alpha)$ (87) retains the initial distance function, and its sufficiently large value provides full separability of two classes of objects in any training set. Of course, the separability of classes at the stage of training does not guarantee small recognition error in the universe, therefore, the choice of such a hyperparameter of the training criterion must be underlined by an appropriate validation procedure.

A simulated example of how distance transformation via Braverman's exponential potential function affects two-class pattern recognition is shown in Fig. 3. Objects are represented here by their two-dimensional feature vectors $x \in \mathbb{R}^2$, so that $\rho^2(\omega', \omega'') = (x' - x'')^T (x' - x'')$. Pictures like these are usually demonstrated in text books on Machine Learning. The analogous effect will have place for regression estimation.

(2) *Selective combination of several distance functions* (Sects. 5.3.2 and 5.3.3). Just as it was stated for distance transformation, we have, again, a parametric family of training criteria (99) and decision rules (100) instead of one pair (68)-(69). The selectivity μ (102) is an additional hyperparameter.

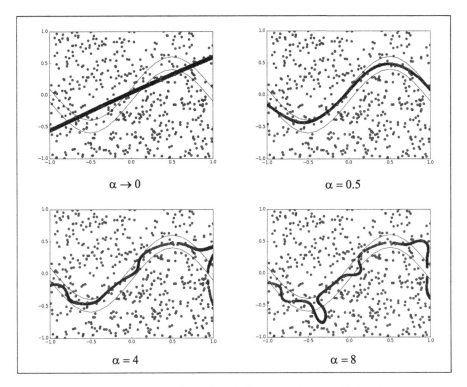

Fig. 3. Distance transformation via Braverman's potential function.

When training with the adaptive quadratic regularization (79) at a preset selectivity level $0 < \mu < \infty$

$$\frac{\mu+1}{\mu} \sum_{k=1}^{m} \sum_{i=1}^{n} \ln\left(\mu a_{ki}^2 + 1\right) + C \frac{1}{N} \sum_{j=1}^{N} q\left[y_j, \sum_{k=1}^{m} \sum_{i=1}^{n} \left(-\rho_k^2(\omega_j, \omega_i | \alpha_k)\right) a_{ki} + b\right] \rightarrow$$
$$\min(a_{ki}, \ i = 1, \ldots, n, \ k = 1, \ldots, m),$$
$$(102)$$

the result will be the values of all mn coefficients a_{ki} at the secondary features $\left(-\rho_k^2(\omega, \omega_i | \alpha_k)\right)$, none of which will turn into zero. But their squared values a_{ki}^2 (relevancies) will be as more different as greater is selectivity μ. The final selection may be done by ranking estimated coefficients in descending order $\hat{a}_{ki,1} > \hat{a}_{ki,2} > \ldots > \hat{a}_{ki,mn}$ (81) and thresholding $\hat{\mathbb{F}} \subset \mathbb{F}$, as it is shown in Fig. 4 [41, 42].

Singular regularization results in rigid selection of diminishing subsets of competitive distance functions $|\hat{\mathbb{F}}_1| > |\hat{\mathbb{F}}_2| > \ldots > |\hat{\mathbb{F}}_{\min}| = \emptyset$ as selectivity μ grows, however, strong nesting $\hat{\mathbb{F}}_1 \supset \hat{\mathbb{F}}_1 \supset \ldots \supset \hat{\mathbb{F}}_{\min} = \emptyset$ is not guaranteed. Our experience with Elastic Net (82) and Modulus Quadratic (83) regularization has shown their excellent performance in selecting pattern recognition modalities, in particular, distances [51–53]. The result is approximately the same as in Fig. 4 after thresholding.

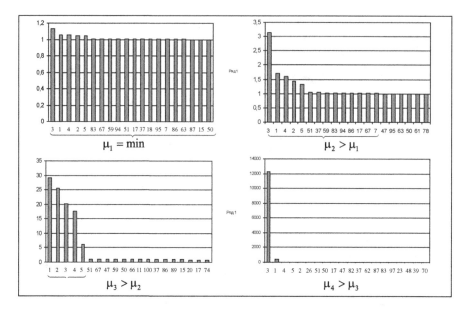

Fig. 4. Smoothly diminishing weights at object representation modalities.

Two axes of Complexity Growth. Thus, there are two hyperparameters that control the complexity of the class of dependence models. The growth of the distance transformation parameter $0 \rightarrow \alpha \rightarrow \infty$ in each of distance functions involves the increase in each of respective competing dependence models (Fig. 3) by increasing the "effective dimensionality" of each embedding linear space. Concurrently, the diminishing of the selectivity parameter $\infty \rightarrow \mu \rightarrow 0$ increases the number of elementary embedding linear spaces whose Cartesian product constitutes the combined dependence model space of proportionally growing dimensionality.

The values of these hyperparameters should be adapted in the learning phase in order to control the generalization performance. Methods of model verification are beyond the topics of this paper.

6 Conclusions

- In this paper we have outlined three generic ideas, which had been formulated by Emmanuel Braverman in the early sixties of the last century, and which constitute the core of the contemporary Machine Learning doctrine.
  - The first of them is the compactness hypothesis: The more alike the real-world entities are in some sense suggested by the Nature, the more their hidden labels are likely to coincide or be similar.

- The second idea, first suggested by Emmanuel Braverman and later essentially generalized by Vladimir Vapnik, is that it is up to the observer, armed with a computer, to choose a two-argument function on the set of "all" real-world entities in the hope that it would be a satisfactory evaluation of the Nature's understanding of similarity.
- The third generic idea is that of mentally embedding the set of real-world entities into an appropriate linear space, as a rule, of a huge or even infinite dimensionality. The purpose of the embedding is to harness the entire wealth of the linear mathematical methods for modeling and estimating dependences of any kind (rectifying linear space, in Braverman's original terminology).

- As a generalization of these three above-mentioned keystones, we have considered the entire linear framework of Machine Learning from the viewpoint of two mathematical ideas originally suggested by Lev Goldfarb, and later further refined by Robert Duin, Elzbieta Pękalska and Bernard Haasdonk.
  - A distance function on the set of real-world entities (not obligatory metric) is to be chosen by the observer as his/her guess on how to approximate the Nature's meaning of dissimilarity between them.
  - The resulting distance space allows for embedding it into a linear space with, generally speaking, indefinite inner product (pseudo-Euclidean linear space).
- Thus, it is enough to consider the embedding pseudo-Euclidean linear space as a mathematical model of the set of real-world entities along with a suggestion on their original dissimilarities. This will be Hilbert space with a usual inner product (positive definite kernel) and Euclidean metric, if the accepted distance on the initial set of real-world entities is metric of a special kind (proto-Euclidean metric). Such an embedding actually obliterates the difference between linear and non-linear decision rules, because a linear function in the embedding space produced by some distance function will be nonlinear in terms of another distance.
- To sketch out this generalized linear concept of Machine Learning, we have adopted the term "Generalized Liner Model", introduced by John Nelder and his colleagues as applied to mathematical statistics, however, we put aside its statistical origin. From this viewpoint, we have considered a wide class of dependence estimation problems including, at least, pattern recognition, numerical regression and ordinal regression.
- On the basis of Braverman's idea of potential function, we have considered a way of forming a parametric family of distances produced from the initially accepted distance function. This family embeds the distance space of entities into a succession of pseudo-Euclidean linear spaces and enables, therethrough, constructing dependence models of unrestrictedly growing complexity.
- It is natural to expect that different experts skilled in specific knowledge areas will propose to the observer different distance functions. Selective combining (fusion) of several competitive distances is an additional instrument of building more sophisticated dependence models. We have considered some ways of model complexity regularization to prevent the effect of overfitting the training set.

Acknowledgements. We would like to acknowledge support from grants of the Russian Foundation for Basic Research 14-07-00527, 16-57-52042, 17-07-00436, 17-07-00993, 18-07-01087, 18-07-00942, and from Tula State University within the framework of the scientific project № 2017-62PUBL.

Appendix. The Proofs of Theorems

Proof of Theorem 1.

Let $\omega', \omega'', \omega'''$ be three arbitrary objects. Due to (10), $h \geq \sup_{\tilde{\omega}'=\tilde{\omega}''=\tilde{\omega}''' \in \Omega} \{\rho(\tilde{\omega}', \tilde{\omega}''') - [\rho(\tilde{\omega}', \tilde{\omega}'') + \rho(\tilde{\omega}'', \tilde{\omega}''')]\} = 0$, thus, $r(\alpha, \beta)$ is nonnegative $r(\alpha, \beta) = \rho(\alpha, \beta) + h \geq 0$. Since $r(\omega', \omega'') = \rho(\omega', \omega'') + h$, we have $r(\omega', \omega'') + r(\omega', \omega''') - r(\omega', \omega''') = [\rho(\omega', \omega'') + h] + [\rho(\omega', \omega'') + h] - [\rho(\omega', \omega''') + h] = \rho(\omega', \omega'') + \rho(\omega', \omega'') - \rho(\omega', \omega''') + h$. Here, on the force of (10), $\rho(\omega', \omega''') + \rho(\omega', \omega''') - \rho(\omega', \omega''') \geq -h$, and $\rho(\omega', \omega'') + \rho(\omega'', \omega''') - \rho(\omega', \omega'') + h \geq -h + h = 0$, whence it follows that $r(\omega', \omega'') + r(\omega', \omega''') - r(\omega', \omega''') \geq 0$. ∎

Proof of Theorem 2.

Let $\{\omega_1, \ldots, \omega_M\} \subset \Omega$ be a finite collection of objects within a distance space, $\phi = \omega_k$ be chosen as the center, and $\mathbf{K}_\phi = \mathbf{K}_{\omega_k} = [K_{\omega_k}(\omega_i, \omega_j), i,j = 1, \ldots, M]$ be the matrix formed by similarity function (11). Let, further, another object be assigned as a new center $\varphi' = \omega_l$ and $\mathbf{K}_{\phi'} = \mathbf{K}_{\omega_l} = [K_{\omega_l}(\omega_i, \omega_j), i,j = 1, \ldots, M]$ be another similarity matrix. In accordance with (14),

$$K_{\phi'}(\omega', \omega'') = K_{\omega_l}(\omega_i, \omega_j) = K_{\omega_k}(\omega_i, \omega_j) - K_{\omega_k}(\omega_i, \omega_l) - K_{\omega_k}(\omega_j, \omega_l) + K_{\omega_k}(\omega_l, \omega_l).$$

Let notation $\mathbf{E}_{\omega_k \to \omega_l}(M \times M)$ stand for matrices in which all the elements are zeros except the l th row of units ($a_{li} = 1$, $i = 1, \ldots, M$) and one additional unit element $a_{kl} = 1$:

$$
\mathbf{E}_{\omega_k \to \omega_l} =
\begin{pmatrix}
0 & 0 & 0 & 0 & 0 & 0 & 0 \\
0 & 0 & 0 & 0 & 0 & 0 & 0 \\
0 & 0 & 0 & 0 & 1 & 0 & 0 \\
0 & 0 & 0 & 0 & 0 & 0 & 0 \\
1 & 1 & 1 & 1 & 1 & 1 & 1 \\
0 & 0 & 0 & 0 & 0 & 0 & 0 \\
0 & 0 & 0 & 0 & 0 & 0 & 0
\end{pmatrix}
\begin{matrix} 1 \\ \vdots \\ k \\ \vdots \\ l \\ \vdots \\ M \end{matrix}
$$

$$1 \quad \cdots \quad k \quad \cdots \quad l \quad \cdots \quad M$$

It is clear that the matrices $\mathbf{S}_{\omega_k \to \omega_l} = \mathbf{I} - \mathbf{E}_{\omega_k \to \omega_l}$ are nondegenerate.

Let us consider two quadratic forms $\mathbf{x}^T \mathbf{K}_{\omega_k} \mathbf{x}$ and $\mathbf{y}^T \mathbf{K}_{\omega_l} \mathbf{y}$, $\mathbf{x}, \mathbf{y} \in \mathbb{R}^M$. Here

$$\mathbf{y}^T \mathbf{K}_{\omega_l} \mathbf{y} = \mathbf{y}^T \mathbf{S}_{\omega_k \to \omega_l}^T \mathbf{K}_{\omega_k} \mathbf{S}_{\omega_k \to \omega_l} \mathbf{y} = (\mathbf{S}_{\omega_k \to \omega_l} \mathbf{y})^T \mathbf{K}_{\omega_k} (\mathbf{S}_{\omega_k \to \omega_l} \mathbf{y}).$$

As we see, the quadratic forms coincide after the one-to-one substitution $\mathbf{x} = \mathbf{S}_{\omega_k \to \omega_l} \mathbf{y}$. In accordance with Sylvester's law of inertia for quadratic forms, the numbers of positive, negative, and zero eigenvalues of matrices $\mathbf{K}_{\omega_k}$ and $\mathbf{K}_{\omega_l}$ coincide, so, their signatures coincide, too. ∎

Proof of Theorem 3.

Proof is based on the following lemma [54, p. 74].

Lemma 3.1. Function $K_\phi(\omega', \omega'') = \tilde{K}(\omega', \omega'') - \tilde{K}(\omega', \phi) - \tilde{K}(\omega'', \phi) + \tilde{K}(\phi, \phi)$, $\omega', \omega'', \phi \in \Omega$, is kernel if and only if $\tilde{K}(\vartheta', \vartheta'')$, $\vartheta', \vartheta'' \in \Omega$, is conditional kernel – matrix $[\tilde{K}(\vartheta_k, \vartheta_l), k, l = 1, \ldots, M]$ is conditionally positive definite (30) for any finite set $\{\vartheta_1, \ldots, \vartheta_M\}$.

Proof of the Theorem. In notations of this paper, the conditionally positive kernel is defined by the proto-Euclidean metric $\tilde{K}(\vartheta', \vartheta'') = -\rho^2(\vartheta', \vartheta'')$. The substitution of this equality in the assertion of Lemma 3.1 yields $K_\phi(\omega', \omega'') = -\rho^2(\omega', \omega'') + \rho^2(\omega', \phi) + \rho^2(\omega'', \phi) + \rho^2(\phi, \phi)$. Since the last summand is zero, we have $K_\phi(\omega_i, \omega_j) = (1/2)[\rho^2(\omega_i, \phi) + \rho^2(\omega_j, \phi) - \rho^2(\omega_i, \omega_j)]$. ∎

Proof of Theorem 4.

In accordance with (88), we have in the case (89)

$$\lim_{\alpha \to 0} \rho^2(\omega', \omega''|\alpha) = \lim_{\alpha \to 0} \frac{[1 - \exp(-\alpha\rho^2(\omega', \omega''))] - \alpha[1 - \exp(-\alpha\rho^2(\omega', \omega''))]}{\alpha} =$$

$$\lim_{\alpha \to 0} \left\{ \frac{1 - \exp(-\alpha\rho^2(\omega', \omega''))}{\alpha} - [1 - \exp(-\alpha\rho^2(\omega', \omega''))] \right\} = \lim_{\alpha \to 0} \frac{1 - \exp(-\alpha\rho^2(\omega', \omega''))}{\alpha} -$$

$$\lim_{\alpha \to 0} [1 - \exp(-\alpha\rho^2(\omega', \omega''))] = \lim_{\alpha \to 0} \frac{1 - \exp(-\alpha\rho^2(\omega', \omega''))}{\alpha} = \frac{0}{0},$$

$$\lim_{\alpha \to 0} \rho^2(\omega', \omega''|\alpha) = \lim_{\alpha \to 0} \frac{(\partial/\partial\alpha)[1 - \exp(-\alpha\rho^2(\omega', \omega''))]}{(\partial/\partial\alpha)\alpha} = \lim_{\alpha \to 0} \frac{\partial}{\partial\alpha}[1 - \exp(-\alpha\rho^2(\omega', \omega''))] =$$

$$-\lim_{\alpha \to 0} \frac{\partial}{\partial\alpha} \exp(-\alpha\rho^2(\omega', \omega'')) = -\lim_{\alpha \to 0} [(-\rho^2(\omega', \omega'')) \exp(-\alpha\rho^2(\omega', \omega''))] = \rho^2(\omega', \omega'').$$

Equality (90) immediately follows from (88). ∎

Proof of Theorem 5.

Let $\omega_1, \omega_2, \omega_3 \in \Omega$ be three objects, $\rho_{12} = \rho(\omega_1, \omega_2)$, $\rho_{23} = \rho(\omega_2, \omega_3)$, $\rho_{13} = \rho(\omega_1, \omega_3)$, and let $\rho_{12} \leq \rho_{23}$. Under notations (88) $\rho(\omega', \omega''|\alpha) = \rho_\alpha(\rho(\omega', \omega''))$ and $\rho(\omega', \omega'') = \rho$, the function $\rho_\alpha(\rho)$ is concave and increasing $(\partial/\partial\rho) \rho_\alpha(\rho) > 0$. Thus, the following inequalities hold:

$\rho_\alpha(\rho_{12} + \rho_{23}) \leq \rho_\alpha(\rho_{23}) + [(\partial/\partial\,\rho)\rho_\alpha(\rho)]_{\rho_{23}}\rho_{12}, \quad [(\partial/\partial\,\rho)\rho_\alpha(\rho)]_{\rho_{23}} \leq [(\partial/\partial\,\rho)\rho_\alpha(\rho)]_{\rho_{12}},$
$[(\partial/\partial\,\rho)\rho_\alpha(\rho)]_{\rho_{12}}\rho_{12} \leq \rho_\alpha(\rho_{12}).$

Here all the derivatives are positive, so, $\rho_\alpha(\rho_{12} + \rho_{23}) \leq \rho_\alpha(\rho_{12}) + \rho_\alpha(\rho_{23})$. Since the original metric satisfied the triangle inequality $\rho_{13} \leq \rho_{12} + \rho_{23}$, we have $\rho_\alpha(\rho_{13}) \leq \rho_\alpha(\rho_{12}) + \rho_\alpha(\rho_{23})$.   ∎

Proof of Theorem 6.
It is enough to prove that for any finite set of objects $\{\omega_1, \ldots, \omega_n\} \subset \Omega$ and any center $\phi \subset \Omega$ the matrix

$$\mathbf{K}_\phi(\alpha) = \frac{1}{2}\left[\rho^2(\omega_i, \phi|\alpha) + \rho^2(\omega_j, \phi|\alpha) - \rho^2(\omega_i, \omega_j|\alpha), \; i,j = 1,\ldots,n\right]$$

is positive definite if matrix

$$\mathbf{K}_\phi = \frac{1}{2}\left[\rho^2(\omega_i, \phi) + \rho^2(\omega_j, \phi) - \rho^2(\omega_i, \omega_j), \; i,j = 1,\ldots,n\right] \qquad (103)$$

is positive definite.
In accordance with Lemma 3.1 (proof of Theorem 3), it is necessary and sufficient for positive definiteness of matrix (103) that matrix

$$\mathbf{B}(\alpha) = \left[-\rho^2(\omega_i, \omega_j|\alpha), \; i,j = 1,\ldots,n\right]$$

would be conditionally positive definite.
At the same time, due to (85), $\rho^2(\omega_i, \omega_j|\alpha) \propto 1 - \exp\left(-\alpha\rho^2(\omega_i, \omega_j)\right)$, so,

$$\mathbf{B}(\alpha) \propto \mathbf{C}(\alpha) - \mathbf{D}, \mathbf{D} = \begin{pmatrix} 1 & 1 & \cdots & 1 \\ 1 & 1 & \cdots & 1 \\ \vdots & \vdots & \ddots & \vdots \\ 1 & 1 & \cdots & 1 \end{pmatrix} = \mathbf{1}\mathbf{1}^\mathrm{T}, \qquad (104)$$

$$\mathbf{C}(\alpha) = \begin{pmatrix} \exp(-\alpha\rho^2(\omega_1, \omega_1)) & \exp(-\alpha\rho^2(\omega_1, \omega_2)) & \cdots & \exp(-\alpha\rho^2(\omega_1, \omega_n)) \\ \exp(-\alpha\rho^2(\omega_2, \omega_1)) & \exp(-\alpha\rho^2(\omega_2, \omega_2)) & \cdots & \exp(-\alpha\rho^2(\omega_2, \omega_n)) \\ \vdots & & \ddots & \vdots \\ \exp(-\alpha\rho^2(\omega_n, \omega_1)) & \exp(-\alpha\rho^2(\omega_n, \omega_2)) & \cdots & \exp(-\alpha\rho^2(\omega_n, \omega_n)) \end{pmatrix}.$$

$$\qquad (105)$$

Here matrix $\mathbf{C}(\alpha)$ is always positive definite for any proto-Euclidean metric $\rho(\omega', \omega'')$ (103) on the force of Mercer's theorem (Sect. 1.4), i.e., $\boldsymbol{\beta}^\mathrm{T}\mathbf{C}(\alpha)\boldsymbol{\beta} > 0$ if $\boldsymbol{\beta} \neq \mathbf{0} \in \mathbb{R}^n$.

Let us consider the quadratic function $q(\boldsymbol{\beta}|\alpha) = \boldsymbol{\beta}^\mathrm{T}\mathbf{B}(\alpha)\boldsymbol{\beta} = \boldsymbol{\beta}^\mathrm{T}(\mathbf{C}(\alpha) - \mathbf{D})\boldsymbol{\beta} = \boldsymbol{\beta}^\mathrm{T}(\mathbf{C}(\alpha) - \mathbf{1}\mathbf{1}^\mathrm{T})\boldsymbol{\beta} = \boldsymbol{\beta}^\mathrm{T}\mathbf{C}(\alpha)\boldsymbol{\beta} - (\boldsymbol{\beta}^\mathrm{T}\mathbf{1})(\mathbf{1}^\mathrm{T}\boldsymbol{\beta})$ on the hyperplane $\mathbf{1}^\mathrm{T}\boldsymbol{\beta} = 0$. It is clear that $q(\boldsymbol{\beta}|\alpha)|_{\mathbf{1}^\mathrm{T}\boldsymbol{\beta}=0} = \boldsymbol{\beta}^\mathrm{T}\mathbf{C}(\alpha)\boldsymbol{\beta} > 0$. Thus, matrix $\mathbf{B}(\alpha)$ is conditionally positive definite.   ∎

Proof of Theorem 7.

Proof is facilitated by the following two Lemmas.

Lemma 7.1. Matrix $\mathbf{B}$ has the main eigenvalues $\lambda_1 = \ldots = \lambda_{n-1} = 1$ of multiplicity $n - 1$, the last one $\lambda_n = -(n - 1)$ of multiplicity 1, the main eigenvectors $\mathbf{z}_i, i = 1, \ldots, n - 1$ with zero sums of elements $\mathbf{1}^T \mathbf{z}_i = 0, i = 1, \ldots, n$, and the last eigenvector $\mathbf{z}_n = \mathbf{1} \in \mathbb{R}^n$.

$$\mathbf{B} = \begin{pmatrix} 0 & -1 & \cdots & -1 \\ -1 & 0 & \cdots & -1 \\ \cdots & \cdots & \ddots & \cdots \\ -1 & -1 & \cdots & 0 \end{pmatrix} \tag{106}$$

Indeed, the main eigenvalues and eigenvectors meet the equalities

$$\begin{pmatrix} 0 & -1 & \cdots & -1 \\ -1 & 0 & \cdots & -1 \\ \cdots & \cdots & \ddots & \cdots \\ -1 & -1 & \cdots & 0 \end{pmatrix} \begin{pmatrix} z_1 \\ z_2 \\ \vdots \\ z_n \end{pmatrix} = \begin{pmatrix} -\sum_{j=1, j\neq 1}^{n} z_j \\ -\sum_{j=1, j\neq 2}^{n} z_j \\ \vdots \\ -\sum_{j=1, j\neq n}^{n} z_j \end{pmatrix}$$

$$= \begin{pmatrix} z_1 \\ z_2 \\ \vdots \\ z_n \end{pmatrix}, \begin{pmatrix} z_1 = -\sum_{j=1, j\neq 1}^{n} z_j = -\sum_{j=1}^{n} z_j + z_1 \\ z_2 = -\sum_{j=1, j\neq 2}^{n} z_j = -\sum_{j=1}^{n} z_j + z_2 \\ \vdots \\ z_n = -\sum_{j=1, j\neq n}^{n} z_j = -\sum_{j=1}^{n} z_j + z_n \end{pmatrix},$$

$$\sum_{l=1}^{n} z_l = -n \sum_{j=1}^{n} z_j + \sum_{l=1}^{n} z_l, \quad n \sum_{l=1}^{n} z_l = \sum_{j=1}^{n} z_l - \sum_{l=1}^{n} z_l = 0, \text{ i.e., } \mathbf{1}^T \mathbf{z}_i = 0,$$

$$i = 1, \ldots, n.$$

The respective equalities for the last eigenvalue and eigenvector have the form

$$\begin{pmatrix} 0 & -1 & \cdots & -1 \\ -1 & 0 & \cdots & -1 \\ \cdots & \cdots & \ddots & \cdots \\ -1 & -1 & \cdots & 0 \end{pmatrix} \begin{pmatrix} z_1 \\ z_2 \\ \vdots \\ z_n \end{pmatrix} = -(n-1) \begin{pmatrix} z_1 \\ z_2 \\ \vdots \\ z_n \end{pmatrix}, \quad z_1 = z_2 = \ldots = z_n, \text{ i.e.}$$

$$\mathbf{z}_n = \mathbf{1} \in \mathbb{R}^n, \text{QED}.$$

Lemma 7.2. Quadratic form $q(\boldsymbol{\beta}) = \boldsymbol{\beta}^T \mathbf{B} \boldsymbol{\beta}$ is positive $q(\boldsymbol{\beta}) > 0$, when $\mathbf{1}^T \boldsymbol{\beta} = 0$ and $\boldsymbol{\beta} \neq \mathbf{0} \in \mathbb{R}^n$, hence, matrix $\mathbf{B}$ (106) is conditionally positive definite.

Indeed, due to Lemma 7.1, $\mathbf{B} = \sum_{i=1}^{n} \lambda_i \mathbf{z}_i \mathbf{z}_i^{\mathrm{T}} = \sum_{i=1}^{n-1} \mathbf{z}_i \mathbf{z}_i^{\mathrm{T}} - (n-1)\mathbf{z}_n \mathbf{z}_n^{\mathrm{T}} = \sum_{i=1}^{n-1} \mathbf{z}_i \mathbf{z}_i^{\mathrm{T}} - (n-1)\mathbf{1}\mathbf{1}^{\mathrm{T}}$. Let us consider the quadratic form $q(\boldsymbol{\beta}) = \boldsymbol{\beta}^{\mathrm{T}} \mathbf{B} \boldsymbol{\beta} = \sum_{i=1}^{n-1} \boldsymbol{\beta}^{\mathrm{T}} \mathbf{z}_i \mathbf{z}_i^{\mathrm{T}} \boldsymbol{\beta} - (n-1)\boldsymbol{\beta}^{\mathrm{T}} \mathbf{1}\mathbf{1}^{\mathrm{T}} \boldsymbol{\beta}$. On the hyperplane $\mathbf{1}^{\mathrm{T}} \boldsymbol{\beta} = 0$, it has the values $q(\boldsymbol{\beta})|_{\mathbf{1}^{\mathrm{T}}\boldsymbol{\beta}=0} = \boldsymbol{\beta}^{\mathrm{T}} \mathbf{B} \boldsymbol{\beta} = \boldsymbol{\beta}^{\mathrm{T}} \bar{\mathbf{B}} \boldsymbol{\beta}$. Here matrix $\bar{\mathbf{B}} = \sum_{i=1}^{n-1} \mathbf{z}_i \mathbf{z}_i^{\mathrm{T}}$ is positive definite, i.e., $q(\boldsymbol{\beta})|_{\mathbf{1}^{\mathrm{T}}\boldsymbol{\beta}=0} > 0$, $\boldsymbol{\beta} \neq \mathbf{0} \in \mathbb{R}^n$, QED.

Now we are ready **to prove Theorem 7.**

We have to prove that matrix $\mathbf{B}(\alpha)$ is conditionally positive definite if α is large enough. Let us consider the quadratic form

$$
\mathbf{B}(\alpha) = \begin{pmatrix}
0 & -\rho^2(\omega_1, \omega_2|\alpha) & \cdots & -\rho^2(\omega_1, \omega_n|\alpha) \\
-\rho^2(\omega_2, \omega_1|\alpha) & 0 & \cdots & -\rho^2(\omega_2, \omega_n|\alpha) \\
\cdots & & \ddots & \cdots \\
-\rho^2(\omega_n, \omega_1|\alpha) & -\rho^2(\omega_n, \omega_2|\alpha) & \cdots & 0
\end{pmatrix}
$$

$q(\boldsymbol{\beta}|\alpha) = \boldsymbol{\beta}^{\mathrm{T}} \mathbf{B}(\alpha) \boldsymbol{\beta}$ on the intersection of the hyperplane $\mathbf{1}^{\mathrm{T}} \boldsymbol{\beta} = 0$ and hypersphere $\boldsymbol{\beta}^{\mathrm{T}} \boldsymbol{\beta} = 1$. Since $q(\boldsymbol{\beta} \,|\, \alpha)$ is continuous function of α, $\lim_{\alpha \to \infty} \mathbf{B}(\alpha) = \mathbf{B}$ and, so, $\lim_{\alpha \to \infty} q(\boldsymbol{\beta} \,|\, \alpha) = \lim_{\alpha \to \infty} \boldsymbol{\beta}^{\mathrm{T}} \mathbf{B}(\alpha) \boldsymbol{\beta} = \boldsymbol{\beta}^{\mathrm{T}} \mathbf{B} \boldsymbol{\beta} = q(\boldsymbol{\beta})$. In accordance with Lemma 7.1, matrix $\mathbf{B}$ (106) is conditionally positive definite, hence, due to Lemma 7.2, there exists α_0, such that $q(\boldsymbol{\beta}|\alpha)|_{\mathbf{1}^{\mathrm{T}}\boldsymbol{\beta}=0} > 0$, $\boldsymbol{\beta} \neq \mathbf{0} \in \mathbb{R}^n$, i.e., $\mathbf{B}(\alpha)$ is conditionally positive definite if $\alpha > \alpha_0$. ∎

Proof of Theorem 8.

Let the training assembly consist of four objects $N = 4$:

$$
\rho(\omega_1, \omega_2) = \rho(\omega_3, \omega_4) = \sqrt{2}, \ \rho(\omega_1, \omega_3) = \rho(\omega_1, \omega_4) = \rho(\omega_2, \omega_3) = \rho(\omega_2, \omega_4) = 1,
$$
$$
y_1 = y_2 = 1, \ y_3 = y_4 = -1.
$$

(107)

Let us try to find to find a decision rule (68) that correctly classifies all the objects:

$$
\begin{cases}
\underbrace{(-\rho^2(\omega_1, \omega_1))}_{0} a_1 + \underbrace{(-\rho^2(\omega_2, \omega_1))}_{-2} a_2 + \underbrace{(-\rho^2(\omega_3, \omega_1))}_{-1} a_3 + \underbrace{(-\rho^2(\omega_4, \omega_1))}_{-1} a_4 + b > 0, \\
\underbrace{(-\rho^2(\omega_1, \omega_2))}_{-2} a_1 + \underbrace{(-\rho^2(\omega_2, \omega_2))}_{0} a_2 + \underbrace{(-\rho^2(\omega_3, \omega_2))}_{-1} a_3 + \underbrace{(-\rho^2(\omega_4, \omega_2))}_{-1} a_4 + b > 0, \\
\underbrace{(-\rho^2(\omega_1, \omega_3))}_{-1} a_1 + \underbrace{(-\rho^2(\omega_2, \omega_3))}_{-1} a_2 + \underbrace{(-\rho^2(\omega_3, \omega_3))}_{0} a_3 + \underbrace{(-\rho^2(\omega_4, \omega_3))}_{-2} a_4 + b < 0, \\
\underbrace{(-\rho^2(\omega_1, \omega_4))}_{-1} a_1 + \underbrace{(-\rho^2(\omega_2, \omega_4))}_{-1} a_2 + \underbrace{(-\rho^2(\omega_3, \omega_4))}_{-2} a_3 + \underbrace{(-\rho^2(\omega_4, \omega_4))}_{0} a_4 + b < 0.
\end{cases}
$$

(108)

Then, the numbers (a_1, a_2, a_3, a_4, b) have to meet the equalities

$$\begin{cases} -2a_2 - a_3 - a_4 + b > 0, \\ -2a_1 - a_3 - a_4 + b > 0, \\ -a_1 - a_2 - 2a_4 + b < 0, \\ -a_1 - a_2 - 2a_3 + b < 0, \end{cases} \text{ or, what is equivalent, } \begin{cases} -2a_2 - a_3 - a_4 + b > 0, \\ -2a_1 - a_3 - a_4 + b > 0, \\ a_1 + a_2 + 2a_4 - b > 0, \\ a_1 + a_2 + 2a_3 - b > 0. \end{cases}$$

$$(109)$$

Adding the left and right parts of the first two inequalities results in the inequality $-(a_1 + a_2) - a_3 - a_4 + b > 0$, i.e. $a_1 + a_2 + a_3 + a_4 - b < 0$, and the same operation applied to the second two inequalities gives $a_1 + a_2 + a_3 + a_4 - b > 0$. It is clear that these two pairs of inequalities are incompatible, hence, the inequalities (109) are incompatible, too. Thus, the decision rule of kind (69), which would correctly classify (108) the assembly (107), does not exist. ∎

Proof of Theorem 9.
Let us consider an arbitrary object of the training set $\omega_j \in \Omega$. In accordance with (69), the decision rule (101) can be represented as

$$d(\omega_j | a_1, \ldots, a_N, b, \alpha) = \frac{1+\alpha}{\alpha} \left[a_j + \sum_{i=1, i \neq j}^{n} \exp\left(-\alpha \rho^2(\omega_j, \omega_i)\right) a_i \right] + b.$$

Indeed,

$$\sum_{i=1}^{n} \left(-\rho^2(\omega_j, \omega_i | \alpha)\right) a_i + b = \frac{1+\alpha}{\alpha} \sum_{i=1}^{n} \left[\exp\left(-\alpha \rho^2(\omega_j, \omega_i)\right) - 1\right] a_i + b =$$

$$\frac{1+\alpha}{\alpha} \sum_{i=1}^{n} \exp\left(-\alpha \rho^2(\omega_j, \omega_i)\right) a_i - \frac{1+\alpha}{\alpha} \underbrace{\sum_{i=1}^{n} a_i}_{=0} + b =$$

$$\frac{1+\alpha}{\alpha} \underbrace{\exp\left(-\alpha \rho^2(\omega_j, \omega_j)\right)}_{=1} a_j + \frac{1+\alpha}{\alpha} \sum_{i=1, i \neq j}^{n} \exp\left(-\alpha \rho^2(\omega_j, \omega_i)\right) a_i + b.$$

Since $\lim_{\alpha \to \infty} [(1+\alpha)/\alpha] = 1$, we have

$$\lim_{\alpha \to \infty} d(\omega_j | a_1, \ldots, a_N, b, \alpha) = a_j + \sum_{i=1, i \neq j}^{n} \underbrace{\lim_{\alpha \to \infty} \exp\left(-\alpha \rho^2(\omega_j, \omega_i)\right)}_{=0} a_i + b = a_j + b.$$

Let $(a_j, j = 1, \ldots, N, b)$ be the parameter vector such that $a_j > 0$ if $y_j = 1$, $a_j < 0$ if $y_j = -1$, and $b = 0$. Then

$$\lim_{\alpha \to \infty} d(\omega_j | a_1, \ldots, a_N, b, \alpha) = a_j \begin{cases} > 0, & y_j = 1, \\ < 0, & y_j = -1, \end{cases}$$

i.e., parameter vector $(a_j, j = 1, \ldots, N, b)$ correctly separates the entire training set. ∎

References

1. Braverman, E.M.: Experiments on machine learning to recognize visual patterns. Autom. Remote Control **23**, 315–327 (1962). Translated from Russian Autimat. i Telemekh. 23, 349–364 (1962)
2. Arkad'ev, A.G., Braverman, E.M.: Computers and Pattern Recognition. Thompson Book Company, Washington (1967). 115 p.
3. Vapnik, V.: Statistical Learning Theory. Wiley, New York (1998)
4. Vapnik, V.: Estimation of Dependences Based on Empirical Data. Springer, New York (1982). https://doi.org/10.1007/0-387-34239-7
5. Duin, R.P.W.: Compactness and complexity of pattern recognition problems. In: Proceedings of International Symposium on Pattern Recognition "In Memoriam Pierre Devijver", Brussels, B, 12 February, Royal Military Academy, pp. 124–128 (1999)
6. Aizerman, M., Braverman, E., Rozonoer, L.: Theoretical foundations of the potential function method in pattern recognition learning. Autom. Remote Control **25**, 917–936 (1964)
7. Mercer, J.: Functions of positive and negative type and their connection with the theory of integral equations. Philos. Trans. Roy. Soc. A **209**, 415–446 (1909)
8. Goldfarb, L.: A unified approach to pattern recognition. Pattern Recogn. **17**, 575–582 (1984)
9. Goldfarb, L.: A New Approach to Pattern Recognition. Progress in Pattern Recognition, Elsevier Science Publishers BV **2**, 241–402 (1985)
10. Pȩkalska, E., Duin, R.P.W.: Dissimilarity representations allow for building good classifiers. Pattern Recogn. Lett. **23**(8), 943–956 (2002)
11. Pekalska, E., Duin, R.P.W.: The Dissimilarity Representation for Pattern Recognition: Foundations and Applications. World Scientific Publishing Co. Inc., River Edge (2005)
12. Haasdonk, B., Pekalska, E.: Indefinite kernel Fisher discriminant. In: Proceedings of the 19th International Conference on Pattern Recognition, Tampa, USA, 8–11 December 2008
13. Duin, R.P.W., Pȩkalska, E.: Non-Euclidean dissimilarities: causes and informativeness. In: Hancock, E.R., Wilson, R.C., Windeatt, T., Ulusoy, I., Escolano, F. (eds.) SSPR/SPR 2010. LNCS, vol. 6218, pp. 871–880. Springer, Heidelberg (2010). https://doi.org/10.1007/978-3-642-14980-1_31
14. Haasdonk, B.: Feature space interpretation of SVMs with indefinite kernels. TPAMI **25**, 482–492 (2005)
15. Pȩkalska, E., Harol, A., Duin, R.P.W., Spillmann, B., Bunke, H.: Non-Euclidean or non-metric measures can be informative. In: Yeung, D.-Y., Kwok, J.T., Fred, A., Roli, F., de Ridder, D. (eds.) SSPR/SPR 2006. LNCS, vol. 4109, pp. 871–880. Springer, Heidelberg (2006). https://doi.org/10.1007/11815921_96
16. Duin, R., Pekalska, E., De Ridder, D.: Relational discriminant analysis. Pattern Recogn. Lett. **20**, 1175–1181 (1999)

17. Maria-Florina Balcan, M.-F., Blum, A., Srebro, N.: A theory of learning with similarity functions. Mach. Learn. **72**, 89–112 (2008)
18. Nelder, J., Wedderburn, R.: Generalized linear models. J. Roy. Stat. Soc. Ser. A (Gen.) **135** (3), 370–384 (1972)
19. McCullagh, P., Nelder, J.: Generalized Linear Models, 511 p., 2nd edn. Chapman and Hall, London (1989)
20. Mottl, V., Krasotkina, O., Seredin, O., Muchnik, I.: Principles of multi-kernel data mining. In: Perner, P., Imiya, A. (eds.) MLDM 2005. LNCS (LNAI), vol. 3587, pp. 52–61. Springer, Heidelberg (2005). https://doi.org/10.1007/11510888_6
21. Tatarchuk, A., Urlov, E., Mottl, V., Windridge, D.: A support kernel machine for supervised selective combining of diverse pattern-recognition modalities. In: El Gayar, N., Kittler, J., Roli, F. (eds.) MCS 2010. LNCS, vol. 5997, pp. 165–174. Springer, Heidelberg (2010). https://doi.org/10.1007/978-3-642-12127-2_17
22. Gonen, M., Alpaydın, E.: Multiple kernel learning algorithms. J. Mach. Learn. Res. **12**, 2211–2268 (2011)
23. Schölkopf, B., Smola, A.: Learning with Kernels. MIT Press, Cambridge (2001)
24. Deza, M., Deza, E.: Encyclopedia of Distances. Springer, Heidelberg (2006). https://doi.org/10.1007/978-3-642-00234-2
25. Azizov, T.Y., Iokhvidov, I.S.: Linear Operators in Spaces with an Indefinite Metric. Wiley, Chichester (1989)
26. Langer, H.: Krein space. In: Hazewinkel, M. (ed.) Encyclopaedia of Mathematics (set). Springer, Netherlands (1994)
27. Ong, C.S., Mary, X., Canu, S., Smola, A.: Learning with non-positive kernels. In: Proceedings of the Twenty-First International Conference on Machine learning, ICML 2004, Banff, Alberta, Canada, 04–08 July 2004
28. Bugrov, S., Nikolsky, S.M.: Fundamentals of Linear Algebra and Analytical Geometry. Mir, Moscow (1982)
29. Vapnik, V.: The Nature of Statistical Learning Theory. Information Science and Statistics. Springer, New York (2000). https://doi.org/10.1007/978-1-4757-3264-1
30. Guyon, I., Vapnik, V.N., Boser, B.E., Bottou, L., Solla, S.A.: Structural risk minimization for character recognition. In: Advances in Neural Information Processing Systems, vol. 4. Morgan Kaufman, Denver (1992)
31. Wilson, J.R., Lorenz, K.A.: Short history of the logistic regression model. Modeling Binary Correlated Responses using SAS, SPSS and R. IBSS, vol. 9, pp. 17–23. Springer, Cham (2015). https://doi.org/10.1007/978-3-319-23805-0_2
32. Cortes, C., Vapnik, V.: Support-vector networks. Mach. Learning **20**, 273–297 (1995)
33. Tikhonov, A.N.: On the stability of inverse problems. Dokl. Akad. Nauk SSSR **39**(5), 195–198 (1943)
34. Tikhonov, A.N.: Solution of incorrectly formulated problems and the regularization method. Sov. Math. **4**, 1035–1038 (1963)
35. Tikhonov, A.N., Arsenin, V.Y.: Solution of Ill-Posed Problems. Winston & Sons, Washington (1977)
36. Hoerl, A.E., Kennard, D.J.: Application of ridge analysis to regression problems. Chem. Eng. Prog. **58**, 54–59 (1962)
37. Vinod, H.D., Ullah, A.: Recent advances in regression methods, vol. 41. In: Statistics: Textbooks and Monographs. Marcel Dekker Inc., New York (1981)
38. Mottl, V., Dvoenko, S., Seredin, O., Kulikowski, C., Muchnik, I.: Featureless pattern recognition in an imaginary Hilbert space and its application to protein fold classification. In: Perner, P. (ed.) MLDM 2001. LNCS (LNAI), vol. 2123, pp. 322–336. Springer, Heidelberg (2001). https://doi.org/10.1007/3-540-44596-X_26

39. Frank, I.E., Friedman, J.H.: A statistical view of some chemometrics regression tools. Technometrics **35**, 109–148 (1993)
40. Fu, W.J.: Penalized regression: the bridge versus the LASSO. J. Comput. Graph. Stat. **7**, 397–416 (1998)
41. Mottl, V., Seredin, O., Krasotkina, O., Muchnik, I.: Fusion of Euclidean metrics in featureless data analysis: an equivalent of the classical problem of feature selection. Pattern Recogn. Image Anal. **15**(1), 83–86 (2005)
42. Mottl, V., Seredin, O., Krasotkina, O., Mochnik, I.: Kernel fusion and feature selection in machine learning. In: Proceedings of the 8th IASTED International Conference on Intelligent Systems and Control, Cambridge, USA, 31 October–2 November, 2005, pp. 477–482
43. Zou, H., Hastie, T.: Regularization and variable selection via the elastic net. J. Roy. Stat. Soc. **67**, 301–320 (2005)
44. Wang, L., Zhu, J., Zou, H.: The doubly regularized support vector machine. Statistica Sinica **16**, 589–615 (2006)
45. Tibshirani, R.J.: Regression shrinkage and selection via the LASSO. J. Roy. Stat. Soc. Ser. B **58**, 267–288 (1996)
46. Tibshirani, R.J.: The LASSO method for variable selection in the Cox model. Stat. Med. **16**, 385–395 (1997)
47. Tatarchuk, A., Mottl, V., Eliseyev, A., Windridge, D.: Selectivity supervision in combining pattern-recognition modalities by feature- and kernel-selective support vector machines. In: Proceedings of the 19th International Conference on Pattern Recognition, ICPR-2008, vol. 1–6, pp. 2336–2339 (2008)
48. Fan, J., Li, R.: Variable selection via nonconcave penalized likelihood and its oracle properties. J. Am. Stat. Assoc. Theor. Methods **96**(456), 1348–1360 (2001)
49. Krasotkina, O., Mottl, V.A.: Bayesian approach to sparse Cox regression in high-dimensional survival analysis. In: Proceedings of the 11th International Conference on Machine Learning and Data Mining (MLDM 2015), Hamburg, Germany, 20–23 July 2015, pp. 425–437
50. Krasotkina, O., Mottl, V.A.: Bayesian approach to sparse learning-to-rank for search engine optimization. In: Proceedings of the 11th International Conference on Machine Learning and Data Mining (MLDM 2015), Hamburg, Germany, 20–23 July 2015, pp. 382–394
51. Tatarchuk, A., Sulimova, V., Windridge, D., Mottl, V., Lange, M.: Supervised selective combining pattern recognition modalities and its application to signature verification by fusing on-line and off-line kernels. In: Benediktsson, J.A., Kittler, J., Roli, F. (eds.) MCS 2009. LNCS, vol. 5519, pp. 324–334. Springer, Heidelberg (2009). https://doi.org/10.1007/978-3-642-02326-2_33
52. Razin, N., et al.: Application of the multi-modal relevance vector machine to the problem of protein secondary structure prediction. In: Shibuya, T., Kashima, H., Sese, J., Ahmad, S. (eds.) PRIB 2012. LNCS, vol. 7632, pp. 153–165. Springer, Heidelberg (2012). https://doi.org/10.1007/978-3-642-34123-6_14
53. Tatarchuk, A., Sulimova, V., Torshin, I., Mottl, V., Windridge, D.: Supervised selective kernel fusion for membrane protein prediction. In: Comin, M., Käll, L., Marchiori, E., Ngom, A., Rajapakse, J. (eds.) PRIB 2014. LNCS, vol. 8626, pp. 98–109. Springer, Cham (2014). https://doi.org/10.1007/978-3-319-09192-1_9
54. Berg, C., Christensen, J.P.R., Ressel, P.: Harmonic Analysis on Semigroups: Theory of Positive Definite and Related Functions. Springer, New York (1984). https://doi.org/10.1007/978-1-4612-1128-0

Conformal Predictive Distributions
with Kernels

Vladimir Vovk[(✉)], Ilia Nouretdinov, Valery Manokhin, and Alex Gammerman

Royal Holloway, University of London, Egham, Surrey, UK
{v.vovk,a.gammerman}@rhul.ac.uk, I.R.Nouretdinov@cs.rhul.ac.uk,
Valery.Manokhin.2015@live.rhul.ac.uk

Abstract. This paper reviews the checkered history of predictive distributions in statistics and discusses two developments, one from recent literature and the other new. The first development is bringing predictive distributions into machine learning, whose early development was so deeply influenced by two remarkable groups at the Institute of Automation and Remote Control. As result, they become more robust and their validity ceases to depend on Bayesian or narrow parametric assumptions. The second development is combining predictive distributions with kernel methods, which were originated by one of those groups, including Emmanuel Braverman. As result, they become more flexible and, therefore, their predictive efficiency improves significantly for realistic nonlinear data sets.

Keywords: Conformal prediction · Fiducial inference
Predictive distributions

1 Introduction

Prediction is a fundamental and difficult scientific problem. We limit the scope of our discussion by imposing, from the outset, two restrictions: we only want to predict one real number $y \in \mathbb{R}$, and we want our prediction to satisfy a reasonable property of validity (under a natural assumption). It can be argued that the fullest prediction for y is a probability measure on $\mathbb{R}$, which can be represented by its distribution function: see, e.g., [5,6,8]. We will refer to it as the predictive distribution. A standard property of validity for predictive distributions is being well-calibrated. Calibration can be defined as the "statistical compatibility between the probabilistic forecasts and the realizations" [8, Sect. 1.2], and its rough interpretation is that predictive distributions should tell the truth. Of course, truth can be uninteresting and non-informative, and there is a further requirement of efficiency, which is often referred to as sharpness [8, Sect. 2.3]. Our goal is to optimize the efficiency subject to validity [8, Sect. 1.2].

This paper is a very selective review of predictive distributions with validity guarantees. After introducing our notation and setting the prediction problem in Sect. 2, we start, in Sect. 3, from the oldest approach to predictive distributions,

© Springer Nature Switzerland AG 2018
L. Rozonoer et al. (Eds.): Braverman Readings in Machine Learning, LNAI 11100, pp. 103–121, 2018.
https://doi.org/10.1007/978-3-319-99492-5_4

Bayesian. This approach gives a perfect solution but under a very restrictive assumption: we need a full knowledge of the stochastic mechanism generating the data. In Sect. 4 we move to Fisher's fiducial predictive distributions.

The first recent development (in [27], as described in Sect. 5 of this paper) was to carry over predictive distributions to the framework of statistical machine learning as developed by two groups at the Institute of Automation and Remote Control (Aizerman's laboratory including Braverman and Rozonoer and Lerner's laboratory including Vapnik and Chervonenkis; for a brief history of the Institute and research on statistical learning there, including the role of Emmanuel Markovich Braverman, see [25], especially Chap. 5). That development consisted in adapting predictive distributions to the IID model, discussed in detail in the next section. The simplest linear case was considered in [27], with groundwork laid in [1].

The second development, which is this paper's contribution, is combination with kernel methods, developed by the members of Aizerman's laboratory, first of all Braverman and Rozonoer [25, p. 48]; namely, in Sect. 6 we derive the kernelized versions of the main algorithms of [27]. In the experimental section (Sect. 8), we demonstrate an important advantage of kernelized versions. The computational efficiency of our methods is studied theoretically in Sect. 6, where we show that pre-processing a training sequence of length n takes, asymptotically, the same time as inverting an $n \times n$ matrix (at most n^3) and, after that, processing a test object takes time $O(n^2)$. Their predictive efficiency is studied in Sect. 8 experimentally using an artificial data set, where we show that a universal (Laplacian) kernel works remarkably well.

The standard methods of probabilistic prediction that have been used so far in machine learning, such as those proposed by Platt [15] and Zadrozny and Elkan [29], are outside the scope of this paper for two reasons: first, they have no validity guarantees whatsoever, and second, they are applicable to classification problems, whereas in this paper we are interested in regression. A sister method to conformal prediction, Venn prediction, does have validity guarantees akin to those in conformal prediction (see, e.g., [26, Theorems 1 and 2]), but it is also applicable only to classification problems. Conformalized kernel ridge regression, albeit in the form of prediction intervals rather than predictive distributions, has been studied by Burnaev and Nazarov [2].

2   The Problem

In this section we will introduce our basic prediction problem. The training sequence consists of n observations $z_i = (x_i, y_i) \in \mathbf{X} \times \mathbf{Y} = \mathbf{X} \times \mathbb{R}$, $i = 1, \ldots, n$; given a test object x_{n+1} we are asked to predict its label y_{n+1}. Each observation $z_i = (x_i, y_i)$, $i = 1, \ldots, n+1$, consists of two components, the object x_i assumed to belong to a measurable space $\mathbf{X}$ that we call the *object space* and the label y_i that belongs to a measurable space $\mathbf{Y}$ that we call the *label space*. In this paper we are interested in the case of regression, where the object space is the real line, $\mathbf{Y} = \mathbb{R}$.

In the problem of probability forecasting our prediction takes the form of a probability measure on the label space $\mathbf{Y}$; since $\mathbf{Y} = \mathbb{R}$, this measure can be represented by its distribution function. This paper is be devoted to this problem and its modifications.

Our prediction problem can be tackled under different assumptions. In the chronological order, the standard assumptions are Bayesian (discussed in Sect. 3 below), statistical parametric (discussed in Sect. 4), and nonparametric, especially the IID model, standard in machine learning (and discussed in detail in the rest of this section and further sections). When using the method of conformal prediction, it becomes convenient to differentiate between two kinds of assumptions, hard and soft (to use the terminology of [24]). Our hard assumption is the IID model: the observations are generated independently from the same probability distribution. The validity of our probabilistic forecasts will depend only on the hard model. In designing prediction algorithms, we may also use, formally or informally, another model in hope that it will be not too far from being correct and under which we optimize efficiency. Whereas the hard model is a standard statistical model (the IID model in this paper), the soft model is not always even formalized; a typical soft model (avoided in this paper) is the assumption that the label y of an object x depends on x in an approximately linear fashion.

In the rest of this paper we will use a fixed parameter $a > 0$, determining the amount of regularization that we wish to apply to our solution to the problem of prediction. Regularization becomes indispensable when kernel methods are used.

3 Bayesian Solution

A very satisfactory solution to our prediction problem (and plethora of other problems of prediction and inference) is given by the theory that dominated statistical inference for more than 150 years, from the work of Thomas Bayes and Pierre-Simon Laplace to that of Karl Pearson, roughly from 1770 to 1930. This theory, however, requires rather strong assumptions.

Let us assume that our statistical model is linear in a feature space (спрямляемое пространство, in the terminology of Braverman and his colleagues) and the noise is Gaussian. Namely, we assume that $x_1, \ldots, x_{n+1}$ is a deterministic sequence of objects and that the labels are generated as

$$y_i = w \cdot F(x_i) + \xi_i, \quad i = 1, \ldots, n+1, \tag{1}$$

where $F : \mathbf{X} \to H$ is a mapping from the object space to a Hilbert space H, "$\cdot$" is the dot product in H, w is a random vector distributed as $N(0, (\sigma^2/a)I)$ (I being the identity operator on H), and ξ_i are random variables distributed as $N(0, \sigma^2)$ and independent of w and between themselves. Here a is the regularization constant introduced at the end of Sect. 2, and $\sigma > 0$ is another parameter, the standard deviation of the noise variables ξ_i.

It is easy to check that

$$\mathbb{E}\, y_i = 0, \qquad\qquad\qquad i = 1, \ldots, n,$$

$$\mathrm{cov}(y_i, y_j) = \frac{\sigma^2}{a}\mathcal{K}(x_i, x_j) + \sigma^2 1_{\{i=j\}}, \quad i, j = 1, \ldots, n, \tag{2}$$

where $\mathcal{K}(x, x') := F(x) \cdot F(x')$. By the theorem on normal correlation (see, e.g., [18, Theorem II.13.2]), the Bayesian predictive distribution for y_{n+1} given x_{n+1} and the training sequence is

$$N\left(k'(K + aI)^{-1}Y, \frac{\sigma^2}{a}\kappa + \sigma^2 - \frac{\sigma^2}{a}k'(K + aI)^{-1}k\right), \tag{3}$$

where k is the n-vector $k_i := \mathcal{K}(x_i, x_{n+1})$, $i = 1, \ldots, n$, K is the kernel matrix for the first n observations (the training observations only), $K_{i,j} := \mathcal{K}(x_i, x_j)$, $i, j = 1, \ldots, n$, $I = I_n$ is the $n \times n$ unit matrix, $Y := (y_1, \ldots, y_n)'$ is the vector of the n training labels, and $\kappa := \mathcal{K}(x_{n+1}, x_{n+1})$.

The weakness of the model (1) (used, e.g., in [23, Sect. 10.3]) is that the Gaussian measure $N(0, (\sigma^2/a)I)$ exists only when H is finite-dimensional, but we can circumvent this difficulty by using (2) directly as our Bayesian model, for a given symmetric positive semidefinite $\mathcal{K}$. The mapping F in not part of the picture any longer. This is the standard approach in Gaussian process regression in machine learning.

In the Bayesian solution, there is no difference between the hard and soft model; in particular, (2) is required for the validity of the predictive distribution (3).

4   Fiducial Predictive Distributions

After its sesquicentennial rule, Bayesian statistics was challenged by Fisher and Neyman, who had little sympathy with each other's views apart from their common disdain for Bayesian methods. Fisher's approach was more ambitious, and his goal was to compute a full probability distribution for a future value (test label in our context) or for the value of a parameter. Neyman and his followers were content with computing intervals for future values (prediction intervals) and values of a parameter (confidence intervals).

Fisher and Neyman relaxed the assumptions of Bayesian statistics by allowing uncertainty, in Knight's [11] terminology. In Bayesian statistics we have an overall probability measure, i.e., we are in a situation of risk without any uncertainty. Fisher and Neyman worked in the framework of parametric statistics, in which we do not have any stochastic model for the value of the parameter (a number or an element of a Euclidean space). In the next section we will discuss the next step, in which the amount of uncertainty (where we lack a stochastic model) is even greater: our statistical model will be the nonparametric IID model (standard in machine learning).

The available properties of validity naturally become weaker as we weaken our assumptions. For predicting future values, conformal prediction (to be discussed in the next section) ensures calibration in probability, in the terminology of [8, Definition 1]. It can be shown that Bayesian prediction satisfies a stronger conditional version of this property: Bayesian predictive distributions are calibrated in probability conditionally on the training sequence and test object (more generally, on the past). The property of being calibrated in probability for conformal prediction is, on the other hand, unconditional; or, in other words, it is conditional on the trivial σ-algebra. Fisher's fiducial predictive distributions satisfy an intermediate property of validity: they are calibrated in probability conditionally on what was called the σ-algebra of invariant events in [13], which is greater than the trivial σ-algebra but smaller than the σ-algebra representing the full knowledge of the past. Our plan is to give precise statements with proofs in future work.

Fisher did not formalize his fiducial inference, and it has often been regarded as erroneous (his "biggest blunder" [7]). Neyman's simplification, replacing probability distributions by intervals, allowed him to state suitable notions of validity more easily, and his approach to statistics became mainstream until the Bayesian approach started to reassert itself towards the end of the 20th century. However, there has been a recent revival of interest in fiducial inference: cf. the BFF (Bayesian, frequentist, and fiducial) series of workshops, with the fourth one held on 1–3 May 2017 in Cambridge, MA, right after the Braverman Readings in Boston. Fiducial inference is a key topic of the series, both in the form of confidence distributions (the term introduced by David Cox [4] in 1958 for distributions for parameters) and predictive distributions (which by definition [17, Definition 1] must be calibrated in probability).

Since fiducial inference was developed in the context of parametric statistics, it has two versions, one targeting computing confidence distributions and the other predictive distributions. Under nonparametric assumptions, such as our IID model, we are not interested in confidence distributions (the parameter space, the set of all probability measures on the observation space $\mathbf{X} \times \mathbb{R}$, is just too big), and concentrate on predictive distributions. The standard notion of validity for predictive distributions, introduced independently by Schweder and Hjort [16, Chap. 12] and Shen, Liu, and Xie [17], is calibration in probability, going back to Philip Dawid's work (see, e.g., [5, Sect. 5.3] and [6]).

5   Conformal Predictive Distributions

In order to obtain valid predictive distributions under the IID model, we will need to relax slightly the notion of a predictive distribution as given in [17]. In our definition we will follow [22,27]; see those papers for further intuition and motivation.

Let $U = U[0,1]$ be the uniform probability distribution on the interval $[0,1]$. We fix the length n of the training sequence. Set $\mathbf{Z} := \mathbf{X} \times \mathbb{R}$; this is our *observation space*.

A function $Q : \mathbf{Z}^{n+1} \times [0,1] \rightarrow [0,1]$ is a *randomized predictive system* (RPS) if:

R1a For each training sequence $(z_1, \ldots, z_n) \in \mathbf{Z}^n$ and each test object $x_{n+1} \in \mathbf{X}$, the function $Q(z_1, \ldots, z_n, (x_{n+1}, y), \tau)$ is monotonically increasing in both y and τ.

R1b For each training sequence $(z_1, \ldots, z_n) \in \mathbf{Z}^n$ and each test object $x_{n+1} \in \mathbf{X}$,

$$\lim_{y \to -\infty} Q(z_1, \ldots, z_n, (x_{n+1}, y), 0) = 0,$$

$$\lim_{y \to \infty} Q(z_1, \ldots, z_n, (x_{n+1}, y), 1) = 1.$$

R2 For any probability measure P on $\mathbf{Z}$, $Q(z_1, \ldots, z_n, z_{n+1}, \tau) \sim U$ when $(z_1, \ldots, z_{n+1}, \tau) \sim P^{n+1} \times U$.

The function

$$Q_n : (y, \tau) \in \mathbb{R} \times [0,1] \mapsto Q(z_1, \ldots, z_n, (x_{n+1}, y), \tau) \tag{4}$$

is the *randomized predictive distribution (function)* (RPD) output by the randomized predictive system Q on a training sequence $z_1, \ldots, z_n$ and a test object x_{n+1}.

A *conformity measure* is a measurable function $A : \mathbf{Z}^{n+1} \rightarrow \mathbb{R}$ that is invariant with respect to permutations of the first n observations. A simple example, used in this paper, is

$$A(z_1, \ldots, z_{n+1}) := y_{n+1} - \hat{y}_{n+1}, \tag{5}$$

$\hat{y}_{n+1}$ being the prediction for y_{n+1} computed from x_{n+1} and $z_1, \ldots, z_{n+1}$ as training sequence. The *conformal transducer* determined by a conformity measure A is defined as

$$Q(z_1, \ldots, z_n, (x_{n+1}, y), \tau) := \frac{1}{n+1} \Big(|\{i = 1, \ldots, n+1 \mid \alpha_i^y < \alpha_{n+1}^y\}|$$
$$+ \tau \, |\{i = 1, \ldots, n+1 \mid \alpha_i^y = \alpha_{n+1}^y\}| \Big), \tag{6}$$

where $(z_1, \ldots, z_n) \in \mathbf{Z}^n$ is a training sequence, $x_{n+1} \in \mathbf{X}$ is a test object, and for each $y \in \mathbb{R}$ the corresponding *conformity scores* α_i^y are defined by

$$\alpha_i^y := A(z_1, \ldots, z_{i-1}, z_{i+1}, \ldots, z_n, (x_{n+1}, y), z_i), \qquad i = 1, \ldots, n,$$
$$\alpha_{n+1}^y := A(z_1, \ldots, z_n, (x_{n+1}, y)). \tag{7}$$

A function is a *conformal transducer* if it is the conformal transducer determined by some conformity measure. A *conformal predictive system* (CPS) is a function which is both a conformal transducer and a randomized predictive system. A *conformal predictive distribution* (CPD) is a function Q_n defined by (4) for a conformal predictive system Q.

The following lemma, stated in [27], gives simple conditions for a conformal transducer to be an RPS; it uses the notation of (7).

Lemma 1. *The conformal transducer determined by a conformity measure A is an RPS if, for each training sequence $(z_1, \ldots, z_n) \in \mathbf{Z}^n$, each test object $x_{n+1} \in \mathbf{X}$, and each $i \in \{1, \ldots, n\}$:*

- $\alpha_{n+1}^y - \alpha_i^y$ *is a monotonically increasing function of $y \in \mathbb{R}$;*
- $\lim_{y \to \pm\infty} \left(\alpha_{n+1}^y - \alpha_i^y \right) = \pm\infty$.

6   Kernel Ridge Regression Prediction Machine

In this section we introduce the Kernel Ridge Regression Prediction Machine (KRRPM); it will be the conformal transducer determined by a conformity measure of the form (5), where $\hat{y}_{n+1}$ is computed using kernel ridge regression, to be defined momentarily. There are three natural versions of the definition, and we start from reviewing them. All three versions are based on (1) as soft model (with the IID model being the hard model).

Given a training sequence $(z_1, \ldots, z_n) \in \mathbf{Z}^n$ and a test object $x_{n+1} \in \mathbf{X}$, the *kernel ridge regression* predicts

$$\hat{y}_{n+1} := k'(K + aI)^{-1}Y$$

for the label y_{n+1} of x_{n+1}. This is just the mean in (3), and the variance is ignored. Plugging this definition into (5), we obtain the *deleted KRRPM*. Alternatively, we can replace the conformity measure (5) by

$$A(z_1, \ldots, z_{n+1}) := y_{n+1} - \widehat{\bar{y}}_{n+1}, \tag{8}$$

where

$$\widehat{\bar{y}}_{n+1} := \bar{k}'(\bar{K} + aI)^{-1}\bar{Y} \tag{9}$$

is the prediction for the label y_{n+1} of x_{n+1} computed using $z_1, \ldots, z_{n+1}$ as the training sequence. The notation used in (9) is: $\bar{k}$ is the $(n+1)$-vector $k_i := \mathcal{K}(x_i, x_{n+1})$, $i = 1, \ldots, n+1$, $\bar{K}$ is the kernel matrix for all $n+1$ observations, $\bar{K}_{i,j} := \mathcal{K}(x_i, x_j)$, $i, j = 1, \ldots, n+1$, $I = I_{n+1}$ is the $(n+1) \times (n+1)$ unit matrix, and $\bar{Y} := (y_1, \ldots, y_{n+1})'$ is the vector of all $n+1$ labels. In this context, $\mathcal{K}$ is any given *kernel*, i.e., symmetric positive semidefinite function $\mathcal{K} : \mathbf{X}^2 \to \mathbb{R}$. The corresponding conformal transducer is the *ordinary KRRPM*. The disadvantage of the deleted and ordinary KRRPM is that they are not RPSs (they can fail to produce a function increasing in y in the presence of extremely high-leverage objects).

Set

$$\bar{H} := (\bar{K} + aI)^{-1}\bar{K} = \bar{K}(\bar{K} + aI)^{-1}. \tag{10}$$

This *hat matrix* "puts hats on the ys": according to (9), $\bar{H}\bar{Y}$ is the vector $(\widehat{\bar{y}}_1, \ldots, \widehat{\bar{y}}_{n+1})'$, where $\widehat{\bar{y}}_i$, $i = 1, \ldots, n+1$, is the prediction for the label y_i of x_i computed using $z_1, \ldots, z_{n+1}$ as the training sequence. We will refer to the entries of the matrix $\bar{H}$ as $\bar{h}_{i,j}$ (where i is the row and j is the column of the entry), abbreviating $\bar{h}_{i,i}$ to $\bar{h}_i$. The usual relation between the residuals in (5) and (8) is

$$y_{n+1} - \hat{y}_{n+1} = \frac{y_{n+1} - \widehat{\hat{y}}_{n+1}}{1 - \bar{h}_{n+1}}. \tag{11}$$

This equality makes sense since the diagonal elements $\bar{h}_i$ of the hat matrix are always in the semi-open interval $[0, 1)$ (and so the numerator is non-zero); for details, see Appendix A. Equation (11) motivates using the *studentized residuals* $(y_{n+1} - \widehat{\hat{y}}_{n+1})(1 - \bar{h}_{n+1})^{-1/2}$, which are half-way between the deleted residuals in (5) and the ordinary residuals in (8). (We ignore a factor in the usual definition of studentized residuals, as in [14, (4.8)], that does not affect the value (6) of the conformal transducer.) The conformal transducer determined by the corresponding conformity measure

$$A(z_1, \ldots, z_{n+1}) := \frac{y_{n+1} - \widehat{\hat{y}}_{n+1}}{\sqrt{1 - \bar{h}_{n+1}}} \tag{12}$$

is the (studentized) *KRRPM*. Later in this section we will see that the KRRPM is an RPS. This is the main reason why this is the main version considered in this paper, with "studentized" usually omitted.

An Explicit Form of the KRRPM

According to (6), to compute the predictive distributions produced by the KRRPM (in its studentized version), we need to solve the equation $\alpha_i^y = \alpha_{n+1}^y$ (and the corresponding inequality $\alpha_i^y < \alpha_{n+1}^y$) for $i = 1, \ldots, n + 1$. Combining the Definition (7) of the conformity scores α_i^y, the Definition (12) of the conformity measure, and the fact that the predictions $\hat{y}_i$ can be obtained from $\bar{Y}$ by applying the hat matrix $\bar{H}$ (cf. (10)), we can rewrite $\alpha_i^y = \alpha_{n+1}^y$ as

$$\frac{y_i - \sum_{j=1}^n \bar{h}_{ij} y_j - \bar{h}_{i,n+1} y}{\sqrt{1 - \bar{h}_i}} = \frac{y - \sum_{j=1}^n \bar{h}_{n+1,j} y_j - \bar{h}_{n+1} y}{\sqrt{1 - \bar{h}_{n+1}}}.$$

This is a linear equation, $A_i = B_i y$, and solving it we obtain $y = C_i := A_i/B_i$, where

$$A_i := \frac{\sum_{j=1}^n \bar{h}_{n+1,j} y_j}{\sqrt{1 - \bar{h}_{n+1}}} + \frac{y_i - \sum_{j=1}^n \bar{h}_{ij} y_j}{\sqrt{1 - \bar{h}_i}}, \tag{13}$$

$$B_i := \sqrt{1 - \bar{h}_{n+1}} + \frac{\bar{h}_{i,n+1}}{\sqrt{1 - \bar{h}_i}}. \tag{14}$$

The following lemma, to be proved in Appendix A, allows us to compute (6) easily.

Lemma 2. *It is always true that $B_i > 0$.*

The lemma gives Algorithm 1 for computing the conformal predictive distribution (4). The notation i' and i'' used in line 6 is defined as $i' := \min\{j \mid C_{(j)} = C_{(i)}\}$ and $i'' := \max\{j \mid C_{(j)} = C_{(i)}\}$, to ensure that $Q_n(y, 0) = Q_n(y-, 0)$ and

Algorithm 1. Kernel Ridge Regression Prediction Machine

Require: A training sequence $(x_i, y_i) \in \mathbf{X} \times \mathbb{R}$, $i = 1, \ldots, n$.
Require: A test object $x_{n+1} \in \mathbf{X}$.
1: Define the hat matrix $\bar{H}$ by (10), $\bar{K}$ being the $(n+1) \times (n+1)$ kernel matrix.
2: **for** $i \in \{1, 2, \ldots, n\}$ **do**
3:    Define A_i and B_i by (13) and (14), respectively.
4:    Set $C_i := A_i / B_i$.
5: Sort $C_1, \ldots, C_n$ in the increasing order obtaining $C_{(1)} \leq \cdots \leq C_{(n)}$.
6: Return the following predictive distribution for y_{n+1}:

$$
Q_n(y, \tau) := \begin{cases} \frac{i+\tau}{n+1} & \text{if } y \in (C_{(i)}, C_{(i+1)}) \text{ for } i \in \{0, 1, \ldots, n\} \\ \frac{i'-1+\tau(i''-i'+2)}{n+1} & \text{if } y = C_{(i)} \text{ for } i \in \{1, \ldots, n\}. \end{cases} \tag{15}
$$

$Q_n(y, 1) = Q_n(y+, 1)$ at $y = C_{(i)}$; $C_{(0)}$ and $C_{(n+1)}$ are understood to be $-\infty$ and ∞, respectively. Notice that there is no need to apply Lemma 1 formally; Lemma 2 makes it obvious that the KRRPM is a CPS.

Algorithm 1 is not computationally efficient for a large test set, since the hat matrix $\bar{H}$ (cf. (10)) needs to be computed from scratch for each test object. To obtain a more efficient version, we use a standard formula for inverting partitioned matrices (see, e.g., [10, (8)] or [23, (2.44)]) to obtain

$$
\bar{H} = (\bar{K} + aI)^{-1}\bar{K} = \begin{pmatrix} K + aI & k \\ k' & \kappa + a \end{pmatrix}^{-1} \begin{pmatrix} K & k \\ k' & \kappa \end{pmatrix}
$$

$$
= \begin{pmatrix} (K+aI)^{-1} + d(K+aI)^{-1}kk'(K+aI)^{-1} & -d(K+aI)^{-1}k \\ -dk'(K+aI)^{-1} & d \end{pmatrix} \begin{pmatrix} K & k \\ k' & \kappa \end{pmatrix}
$$

$$
= \begin{pmatrix} H + d(K+aI)^{-1}kk'H - d(K+aI)^{-1}kk' \\ -dk'H + dk' \end{pmatrix} \tag{16}
$$

$$
\begin{pmatrix} (K+aI)^{-1}k + d(K+aI)^{-1}kk'(K+aI)^{-1}k - d\kappa(K+aI)^{-1}k \\ -dk'(K+aI)^{-1}k + d\kappa \end{pmatrix} \tag{17}
$$

$$
= \begin{pmatrix} H + d(K+aI)^{-1}kk'(H-I) & d(I-H)k \\ dk'(I-H) & -dk'(K+aI)^{-1}k + d\kappa \end{pmatrix} \tag{18}
$$

$$
= \begin{pmatrix} H - ad(K+aI)^{-1}kk'(K+aI)^{-1} & ad(K+aI)^{-1}k \\ adk'(K+aI)^{-1} & d\kappa - dk'(K+aI)^{-1}k \end{pmatrix}, \tag{19}
$$

where

$$
d := \frac{1}{\kappa + a - k'(K+aI)^{-1}k} \tag{20}
$$

(the denominator is positive by the theorem on normal correlation, already used in Sect. 3), the equality in line (18) follows from $\bar{H}$ being symmetric (which allows us to ignore the upper right block of the matrix (16)–(17)), and the equality in line (19) follows from

$$
I - H = (K+aI)^{-1}(K+aI) - (K+aI)^{-1}K = a(K+aI)^{-1}.
$$

We have been using the notation H for the training hat matrix

$$H = (K + aI)^{-1}K = K(K + aI)^{-1}. \tag{21}$$

Notice that the constant ad occurring in several places in (19) is between 0 and 1:

$$ad = \frac{a}{a + \kappa - k'(K + aI)^{-1}k} \in (0, 1] \tag{22}$$

(the fact that $\kappa - k'(K + aI)^{-1}k$ is nonnegative follows from the lower right entry $\bar{h}_{n+1}$ of the hat matrix (19) being nonnegative; the nonnegativity of the diagonal entries of hat matrices is discussed in Appendix A).

The important components in the expressions for A_i and B_i (cf. (13) and (14)) are, according to (19),

$$1 - \bar{h}_{n+1} = 1 + dk'(K + aI)^{-1}k - d\kappa = 1 + \frac{k'(K + aI)^{-1}k - \kappa}{\kappa + a - k'(K + aI)^{-1}k}$$

$$= \frac{a}{\kappa + a - k'(K + aI)^{-1}k} = ad, \tag{23}$$

$$1 - \bar{h}_i = 1 - h_i + ade_i'(K + aI)^{-1}kk'(K + aI)e_i$$

$$= 1 - h_i + ad(e_i'(K + aI)^{-1}k)^2, \tag{24}$$

where $h_i = h_{i,i}$ is the ith diagonal entry of the hat matrix (21) for the n training objects and e_i is the ith vector in the standard basis of $\mathbb{R}^n$ (so that the jth component of e_i is $1_{\{i=j\}}$ for $j = 1, \ldots, n$). Let $\hat{y}_i := e_i'HY$ be the prediction for y_i computed from the training sequence $z_1, \ldots, z_n$ and the test object x_i. Using (23) (but not using (24) for now), we can transform (13) and (14) as

$$A_i := \frac{\sum_{j=1}^n \bar{h}_{n+1,j}y_j}{\sqrt{1 - \bar{h}_{n+1}}} + \frac{y_i - \sum_{j=1}^n \bar{h}_{ij}y_j}{\sqrt{1 - \bar{h}_i}}$$

$$= (ad)^{-1/2} \sum_{j=1}^n ady_j k'(K + aI)^{-1}e_j$$

$$+ \frac{y_i - \sum_{j=1}^n h_{ij}y_j + \sum_{j=1}^n ady_j e_i'(K + aI)^{-1}kk'(K + aI)^{-1}e_j}{\sqrt{1 - \bar{h}_i}}$$

$$= (ad)^{1/2}k'(K + aI)^{-1}Y + \frac{y_i - \hat{y}_i + ade_i'(K + aI)^{-1}kk'(K + aI)^{-1}Y}{\sqrt{1 - \bar{h}_i}},$$

$$= \sqrt{ad}\hat{y}_{n+1} + \frac{y_i - \hat{y}_i + ad\hat{y}_{n+1}e_i'(K + aI)^{-1}k}{\sqrt{1 - \bar{h}_i}}, \tag{25}$$

where $\hat{y}_{n+1}$ is the Bayesian prediction for y_{n+1} (cf. the expected value in (3)), and

$$B_i := \sqrt{1 - \bar{h}_{n+1}} + \frac{\bar{h}_{i,n+1}}{\sqrt{1 - \bar{h}_i}} = \sqrt{ad} + \frac{adk'(K + aI)^{-1}e_i}{\sqrt{1 - \bar{h}_i}}. \tag{26}$$

Therefore, we can implement Algorithm 1 as follows. Preprocessing the training sequence takes time $O(n^3)$ (or faster if using, say, the Coppersmith–Winograd algorithm and its versions; we assume that the kernel $\mathcal{K}$ can be computed in time $O(1)$):

1. The $n \times n$ kernel matrix K can be computed in time $O(n^2)$.
2. The matrix $(K + aI)^{-1}$ can be computed in time $O(n^3)$.
3. The diagonal of the training hat matrix $H := (K + aI)^{-1}K$ can be computed in time $O(n^2)$.
4. All $\hat{y}_i$, $i = 1, \ldots, n$, can be computed by $\hat{y} := HY = (K + aI)^{-1}(KY)$ in time $O(n^2)$ (even without knowing H).

Processing each test object x_{n+1} takes time $O(n^2)$:

1. Vector k and number κ (as defined after (3)) can be computed in time $O(n)$ and $O(1)$, respectively.
2. Vector $(K + aI)^{-1}k$ can be computed in time $O(n^2)$.
3. Number $k'(K + aI)^{-1}k$ can now be computed in time $O(n)$.
4. Number d defined by (20) can be computed in time $O(1)$.
5. For all $i = 1, \ldots, n$, compute $1 - \bar{h}_i$ as (24), in time $O(n)$ overall (given the vector computed in 2).
6. Compute the number $\hat{y}_{n+1} := k'(K + aI)^{-1}Y$ in time $O(n)$ (given the vector computed in 2).
7. Finally, compute A_i and B_i for all $i = 1, \ldots, n$ as per (25) and (26), set $C_i := A_i/B_i$, and output the predictive distribution (15). This takes time $O(n)$ except for sorting the C_i, which takes time $O(n \log n)$.

7   Limitation of the KRRPM

The KRRPM makes a significant step forward as compared to the LSPM of [27]: our soft model (1) is no longer linear in x_i. In fact, using a universal kernel (such as Laplacian in Sect. 8) allows the function $x \in \mathbf{X} \mapsto w \cdot F(x)$ to approximate any continuous function (arbitrarily well within any compact set in $\mathbf{X}$). However, since we are interested in predictive distributions rather than point predictions, using the soft model (1) still results in the KRRPM being restricted. In this section we discuss the nature of the restriction, using the ordinary KRRPM as a technical tool.

The Bayesian predictive distribution (3) is Gaussian and (as clear from (1) and from the bottom right entry of (19) being nonnegative) its variance is at least σ^2. We will see that the situation with the conformal distribution is not as bad, despite the remaining restriction. To understand the nature of the restriction it will be convenient to ignore the denominator in (12), i.e., to consider the ordinary KRRPM; the difference between the (studentized) KRRPM and ordinary KRRPM will be small in the absence of high-leverage objects (an example will be given in the next section). For the ordinary KRRPM we have, in place of (13) and (14),

$$A_i := \sum_{j=1}^{n} \bar{h}_{n+1,j} y_j + y_i - \sum_{j=1}^{n} \bar{h}_{i,j} y_j,$$
$$B_i := 1 - \bar{h}_{n+1} + \bar{h}_{i,n+1}.$$

Therefore, (25) and (26) become

$$A_i = ad\hat{y}_{n+1} + y_i - \hat{y}_i + ad\hat{y}_{n+1}e_i'(K + aI)^{-1}k$$

and

$$B_i = ad + ade_i'(K + aI)^{-1}k,$$

respectively. For $C_i := A_i/B_i$ we now obtain

$$C_i = \hat{y}_{n+1} + \frac{y_i - \hat{y}_i}{ad + ade_i'(K + aI)^{-1}k}$$

$$= \hat{y}_{n+1} + \frac{\sigma_{\text{Bayes}}^2/\sigma^2}{1 + e_i'(K + aI)^{-1}k}(y_i - \hat{y}_i), \quad (27)$$

where $\hat{y}_{n+1}$ is, as before, the Bayesian prediction for y_{n+1}, and σ_{Bayes}^2 is the variance of the Bayesian predictive distribution (3) (cf. (22)).

The second addend $e_i'(K + aI)^{-1}k$ in the denominator of (27) is the prediction for the label of the test object x_{n+1} in the situation where all training labels are 0 apart from the ith, which is 1. For a long training sequence we can expect it to be close to 0 (unless x_i or x_{n+1} are highly influential); therefore, we can expect the shape of the predictive distribution output by the ordinary KRRPM to be similar to the shape of the empirical distribution function of the residuals $y_i - \hat{y}_i$. In particular, this shape does not depend (or depends weakly) on the test object x_{n+1}. This lack of sensitivity of the predictive distribution to the test object prevents the conformal predictive distributions output by the KRRPM from being universally consistent in the sense of [22]. The shape of the predictive distribution can be arbitrary, not necessarily Gaussian (as in (3)), but it is fitted to all training residuals and not just the residuals for objects similar to the test object. One possible way to get universally consistent conformal predictive distributions would be to replace the right-hand side of (5) by $\hat{F}_{n+1}(y_{n+1})$, where $\hat{F}_{n+1}$ is the Bayesian predictive distribution for y_{n+1} computed from x_{n+1} and $z_1, \ldots, z_{n+1}$ as training sequence for a sufficiently flexible Bayesian model (in any case, more flexible than our homoscedastic model (1)). This idea was referred to as de-Bayesing in [23, Sect. 4.2] and frequentizing in [28, Sect. 3]. However, modelling input-dependent (heteroscedastic) noise efficiently is a well-known difficult problem in Bayesian regression, including Gaussian process regression (see, e.g., [9,12,19]).

8   Experimental Results

In the first part of this section we illustrate the main advantage of the KRRPM over the LSPM introduced in [27], its flexibility: for a suitable kernel, it gets the location of the predictive distribution right. In the second part, we illustrate the limitation discussed in the previous section: while the KRRPM adapts to the shape of the distribution of labels, the adaptation is not conditional on the test object. Both points will be demonstrated using artificial data sets.

In our first experiment we generate a training sequence of length 1000 from the model

$$y_i = w_1 \cos x_{i,1} + w_2 \cos x_{i,2} + w_3 \sin x_{i,1} + w_4 \sin x_{i,2} + \xi_i, \qquad (28)$$

where $(w_1, w_2, w_3, w_4) \sim N(0, I_4)$ (I_4 being the unit 4×4 matrix), $(x_{i,1}, x_{i,2}) \sim U[-1, 1]^2$ ($U[-1, 1]$ being the uniform probability distribution on $[-1, 1]$), and $\xi_i \sim N(0, 1)$, all independent. This corresponds to the Bayesian ridge regression model with $a = \sigma = 1$. The true kernel is

$$\begin{aligned} \mathcal{K}&((x_1, x_2), (x_1', x_2')) \\ &= (\cos x_1, \cos x_2, \sin x_1, \sin x_2) \cdot (\cos x_1', \cos x_2', \sin x_1', \sin x_2') \\ &= \cos(x_1 - x_1') + \cos(x_2 - x_2'). \end{aligned} \qquad (29)$$

Remember that a kernel is *universal* [20] if any continuous function can be uniformly approximated (over each compact set) by functions in the corresponding reproducing kernel Hilbert space. An example of a universal kernel is the *Laplacian kernel*

$$\mathcal{K}(x, x') := \exp\left(-\|x - x'\|\right).$$

Laplacian kernels were introduced and studied in [21]; the corresponding reproducing kernel Hilbert space has the Sobolev norm

$$\|u\|^2 = 2 \int_{-\infty}^{\infty} u(t)^2 \mathrm{d}t + 2 \int_{-\infty}^{\infty} u'(t)^2 \mathrm{d}t$$

(see [21, Corollary 1]). This expression shows that Laplacian kernels are indeed universal. On the other hand, the *linear kernel* $\mathcal{K}(x, x') := x \cdot x'$ is far from being universal; remember that the LSPM [27] corresponds to this kernel and $a = 0$.

Figure 1 shows that, on this data set, universal kernels lead to better results. The parameter a in Fig. 1 is the true one, $a = 1$. In the case of the Bayesian predictive distribution, the parameter $\sigma = 1$ is also the true one; remember that conformal predictive distributions do not require σ. The right-most panel shows that, when based on the linear kernel, the conformal predictive distribution can get the predictive distribution wrong. The other two panels show that the true kernel and, more importantly, the Laplacian kernel (chosen independently of the model (28)) are much more accurate. Figure 1 shows predictive distributions for a specific test object, $(1, 1)$, but this behaviour is typical. The effect of using a universal kernel becomes much less pronounced (or even disappears completely) for smaller lengths of the training sequence: see Fig. 2 using 100 training observations (whereas Fig. 1 uses 1000).

We now illustrate the limitation of the KRRPM that we discussed in the previous section. An artificial data set is generated as follows: $x_i \in [0, 1]$, $i = 1, \ldots, n$, are chosen independently from the uniform distribution U on $[0, 1]$, and $y_i \in [-x_i, x_i]$ are then chosen independently, again from the uniform distributions $U[-x_i, x_i]$ on their intervals. Figure 3 shows the prediction for $x_{n+1} = 0$ on the left and for $x_{n+1} = 1$ on the right for $n = 1000$; there is no visible

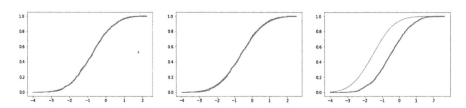

Fig. 1. The predictive distribution for the label of the test object $(1,1)$ based on a training sequence of length 1000 (all generated from the model (28)). The red line in each panel is the Bayesian predictive distribution based on the true kernel (29), and the blue line is the conformal predictive distribution based on: the true kernel (29) in the left-most panel; the Laplacian kernel in the middle panel; the linear kernel in the right-most panel.

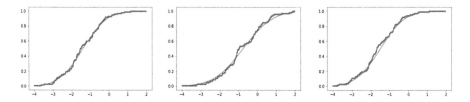

Fig. 2. The analogue of Fig. 1 for a training sequence of length 100.

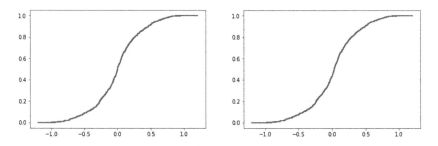

Fig. 3. Left panel: predictions of the KRRPM for a training sequence of length 1000 and $x_{1001} = 0$. Right panel: predictions for $x_{1001} = 1$. The data are described in the text.

difference between the studentized and ordinary versions of the KRRPM. The difference between the predictions for $x_{n+1} = 0$ and $x_{n+1} = 1$ is slight, whereas ideally we would like the former prediction to be concentrated at 0 whereas the latter should be close to the uniform distribution on $[-1, 1]$.

Fine details can be seen in Fig. 4, which is analogous to Fig. 3 but uses a training sequence of length $n = 10$. It shows the plots of the functions $Q_n(y, 0)$ and $Q_n(y, 1)$ of y, in the notation of (4). These functions carry all information about $Q_n(y, \tau)$ as function of y and τ since $Q_n(y, \tau)$ can be computed as the convex mixture $(1 - \tau)Q_n(y, 0) + \tau Q_n(y, 1)$ of $Q_n(y, 0)$ and $Q_n(y, 1)$.

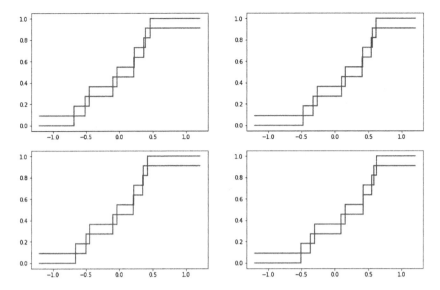

Fig. 4. Upper left panel: predictions of the (studentized) KRRPM for a training sequence of length 10 and $x_{11} = 0$. Upper right panel: analogous predictions for $x_{11} = 1$. Lower left panel: predictions of the ordinary KRRPM for a training sequence of length 10 and $x_{11} = 0$. Lower right panel: analogous predictions for $x_{11} = 1$.

In all experiments described in this section, the seed of the Python pseudo-random numbers generator was set to 0 for reproducibility.

9   Conclusion

In this section we list some directions of further research:

- An important problem in practice is choosing a suitable value of the parameter a; it deserves careful study in the context of conformal predictive distributions.
- It was shown in [27] that, under narrow parametric statistical models, the LSPM is almost as efficient as various oracles that are optimal (or almost optimal) under those models; it would be interesting to prove similar results in the context of this paper using (1) as the model and the Bayesian predictive distribution (3) as the oracle.
- On the other hand, it would be interesting to explore systematically cases where (1) is violated and this results in poor performance of the Bayesian predictive distributions (cf. [23, Sect. 10.3, experimental results]). One example of such a situation is described in Sect. 7: in the case of non-Gaussian homogeneous noise, the Bayesian predictive distribution (3) is still Gaussian, whereas the KRRPM adapts to the noise distribution.
- To cope with heterogeneous noise distribution (see Sect. 7), we need to develop conformal predictive systems that are more flexible than the KRRPM.

Acknowledgements. This work has been supported by the EU Horizon 2020 Research and Innovation programme (in the framework of the ExCAPE project under grant agreement 671555) and Astra Zeneca (in the framework of the project "Machine Learning for Chemical Synthesis").

A   Properties of the Hat Matrix

In the kernelized setting of this paper the hat matrix is defined as $H = (K + aI)^{-1}K$, where K is a symmetric positive semidefinite matrix whose size is denoted $n \times n$ in this appendix (cf. (10); in our current abstract setting we drop the bars over H and K and write n in place of $n + 1$). We will prove, or give references for, various properties of the hat matrix used in the main part of the paper.

Numerous useful properties of the hat matrix can be found in literature (see, e.g., [3]). However, the usual definition of the hat matrix is different from ours, since it is not kernelized; therefore, we start from reducing our kernelized definition to the standard one. Since K is symmetric positive semidefinite, it can be represented in the form $K = XX'$ for some matrix X, whose size will be denoted $n \times p$ (in fact, a matrix is symmetric positive semidefinite if and only if it can be represented as the Gram matrix of n vectors; this easily follows from the fact that a symmetric positive semidefinite K can be diagonalized: $K = Q'\Lambda Q$, where Q and Λ are $n \times n$ matrices, Λ is diagonal with nonnegative entries, and $Q'Q = I$). Now we can transform the hat matrix as

$$H = (K + aI)^{-1}K = (XX' + aI)^{-1}XX' = X(X'X + aI)^{-1}X'$$

(the last equality can be checked by multiplying both sides by $(XX' + aI)$ on the left). If we now extend X by adding $\sqrt{a}I_p$ on top of it (where $I_p = I$ is the $p \times p$ unit matrix),

$$\tilde{X} := \begin{pmatrix} \sqrt{a}I_p \\ X \end{pmatrix}, \tag{30}$$

and set

$$\tilde{H} := \tilde{X}(\tilde{X}'\tilde{X})^{-1}\tilde{X}' = \tilde{X}(X'X + aI)^{-1}\tilde{X}', \tag{31}$$

we will obtain a $(p + n) \times (p + n)$ matrix containing H in its lower right $n \times n$ corner. To find HY for a vector $Y \in \mathbb{R}^n$, we can extend Y to $\tilde{Y} \in \mathbb{R}^{p+n}$ by adding p zeros at the beginning of Y and then discard the first p elements in $\tilde{H}\tilde{Y}$. Notice that $\tilde{H}$ is the usual definition of the hat matrix associated with the data matrix $\tilde{X}$ (cf. [3, (1.4a)]).

When discussing (11), we used the fact that the diagonal elements of H are in $[0, 1)$. It is well-known that the diagonal elements of the usual hat matrix, such as $\tilde{H}$, are in $[0, 1]$ (see, e.g., [3, Property 2.5(a)]). Therefore, the diagonal elements of H are also in $[0, 1]$. Let us check that h_i are in fact in the semi-open interval $[0, 1)$ directly, without using the representation in terms of $\tilde{H}$.

Representing $K = Q'\Lambda Q$ as above, where Λ is diagonal with nonnegative entries and $Q'Q = I$, we have

$$H = (K + aI)^{-1}K = (Q'\Lambda Q + aI)^{-1}Q'\Lambda Q = (Q'(\Lambda + aI)Q)^{-1}Q'\Lambda Q$$
$$= Q^{-1}(\Lambda + aI)^{-1}(Q')^{-1}Q'\Lambda Q = Q'(\Lambda + aI)^{-1}\Lambda Q. \quad (32)$$

The matrix $(\Lambda + aI)^{-1}\Lambda$ is diagonal with the diagonal entries in the semi-open interval $[0, 1)$. Since $Q'Q = I$, the columns of Q are vectors of length 1. By (32), each diagonal element of H is then of the form $\sum_{i=1}^{n} \lambda_i q_i^2$, where all $\lambda_i \in [0, 1)$ and $\sum_{i=1}^{n} q_i^2 = 1$. We can see that each diagonal element of H is in $[0, 1)$.

The equality (11) itself was used only for motivation, so we do not prove it; for a proof in the non-kernelized case, see, e.g., [14, (4.11) and Appendix C.7].

Proof of Lemma 2

In our proof of $B_i > 0$ we will assume $a > 0$, as usual. We will apply the results discussed so far in this appendix to the matrix $\bar{H}$ in place of H and to $n + 1$ in place of n.

Our goal is to check the strict inequality

$$\sqrt{1 - \bar{h}_{n+1}} + \frac{\bar{h}_{i,n+1}}{\sqrt{1 - \bar{h}_i}} > 0; \quad (33)$$

remember that both $\bar{h}_{n+1}$ and $\bar{h}_i$ are numbers in the semi-open interval $[0, 1)$. The inequality (33) can be rewritten as

$$\bar{h}_{i,n+1} > -\sqrt{(1 - \bar{h}_{n+1})(1 - \bar{h}_i)} \quad (34)$$

and in the weakened form

$$\bar{h}_{i,n+1} \geq -\sqrt{(1 - \bar{h}_{n+1})(1 - \bar{h}_i)} \quad (35)$$

follows from [3, Property 2.6(b)] (which can be applied to $\tilde{H}$).

Instead of the original hat matrix $\bar{H}$ we will consider the extended matrix (31), where $\tilde{X}$ is defined by (30) with $\bar{X}$ in place of X. The elements of $\tilde{H}$ will be denoted $\tilde{h}$ with suitable indices, which will run from $-p + 1$ to $n + 1$, in order to have the familiar indices for the submatrix $\bar{H}$. We will assume that we have an equality in (34) and arrive at a contradiction. There will still be an equality in (34) if we replace $\bar{h}$ by $\tilde{h}$, since $\tilde{H}$ contains $\bar{H}$. Consider auxiliary "random residuals" $E := (I - \tilde{H})\epsilon$, where ϵ is a standard Gaussian random vector in $\mathbb{R}^{p+n+1}$; there are $p + n + 1$ random residuals $E_{-p+1}, \ldots, E_{n+1}$. Since the correlation between the random residuals E_i and E_{n+1} is

$$\mathrm{corr}(E_i, E_{n+1}) = \frac{-\tilde{h}_{i,n+1}}{\sqrt{(1 - \tilde{h}_{n+1})(1 - \tilde{h}_i)}}$$

(this easily follows from $I - \tilde{H}$ being a projection matrix and is given in, e.g., [3, p. 11]), (35) is indeed true. Since we have an equality in (34) (with $\tilde{h}$ in place of $\bar{h}$), E_i and E_{n+1} are perfectly correlated. Remember that neither row number i nor row number $n+1$ of the matrix $I - \bar{H}$ are zero (since the diagonal elements of $\bar{H}$ are in the semi-open interval $[0, 1)$), and so neither E_i nor E_{n+1} are zero vectors. Since E_i and E_{n+1} are perfectly correlated, the row number i of the matrix $I - \tilde{H}$ is equal to a positive scalar c times its row number $n + 1$. The projection matrix $I - \tilde{H}$ then projects $\mathbb{R}^{p+n+1}$ onto a subspace of the hyperplane in $\mathbb{R}^{p+n+1}$ consisting of the points with coordinate number i being c times the coordinate number $n + 1$. The orthogonal complement of this subspace, i.e., the range of $\tilde{H}$, will contain the vector $(0, \ldots, 0, -1, 0, \ldots, 0, c)$ (-1 being its coordinate number i). Therefore, this vector will be in the range of $\tilde{X}$ (cf. (31)). Therefore, this vector will be a linear combination of the columns of the extended matrix (30) (with $\bar{X}$ in place of X), which is impossible because of the first p rows $\sqrt{a} I_p$ of the extended matrix.

References

1. Burnaev, E., Vovk, V.: Efficiency of conformalized ridge regression. In: JMLR: Workshop and Conference Proceedings, COLT 2014, vol. 35, pp. 605–622 (2014)
2. Burnaev, E.V., Nazarov, I.N.: Conformalized Kernel Ridge Regression. Technical report arXiv:1609.05959 [stat.ML], arXiv.org e-Print archive, September 2016. Conference version: Proceedings of the Fifteenth International Conference on Machine Learning and Applications (ICMLA 2016), pp. 45–52
3. Chatterjee, S., Hadi, A.S.: Sensitivity Analysis in Linear Regression. Wiley, New York (1988)
4. Cox, D.R.: Some problems connected with statistical inference. Ann. Math. Stat. **29**, 357–372 (1958)
5. Dawid, A.P.: Statistical theory: the prequential approach (with discussion). J. Royal Stat. Soc. A **147**, 278–292 (1984)
6. Dawid, A.P., Vovk, V.: Prequential probability: principles and properties. Bernoulli **5**, 125–162 (1999)
7. Efron, B.: R. A. Fisher in the 21st century. Stat. Sci. **13**, 95–122 (1998)
8. Gneiting, T., Katzfuss, M.: Probabilistic forecasting. Ann. Rev. Stat. Appl. **1**, 125–151 (2014)
9. Goldberg, P.W., Williams, C.K.I., Bishop, C.M.: Regression with input-dependent noise: a Gaussian process treatment. In: Jordan, M.I., Kearns, M.J., Solla, S.A. (eds.) Advances in Neural Information Processing Systems 10, pp. 493–499. MIT Press, Cambridge (1998)
10. Henderson, H.V., Searle, S.R.: On deriving the inverse of a sum of matrices. SIAM Rev. **23**, 53–60 (1981)
11. Knight, F.H.: Risk, Uncertainty, and Profit. Houghton Mifflin Company, Boston (1921)
12. Le, Q.V., Smola, A.J., Canu, S.: Heteroscedastic Gaussian process regression. In: Dechter, R., Richardson, T. (eds.) Proceedings of the Twenty Second International Conference on Machine Learning, pp. 461–468. ACM, New York (2005)

13. McCullagh, P., Vovk, V., Nouretdinov, I., Devetyarov, D., Gammerman, A.: Conditional prediction intervals for linear regression. In: Proceedings of the Eighth International Conference on Machine Learning and Applications (ICMLA 2009), pp. 131–138 (2009). http://www.stat.uchicago.edu/~pmcc/reports/predict.pdf
14. Montgomery, D.C., Peck, E.A., Vining, G.G.: Introduction to Linear Regression Analysis, 5th edn. Wiley, Hoboken (2012)
15. Platt, J.C.: Probabilities for SV machines. In: Smola, A.J., Bartlett, P.L., Schölkopf, B., Schuurmans, D. (eds.) Advances in Large Margin Classifiers, pp. 61–74. MIT Press (2000)
16. Schweder, T., Hjort, N.L.: Confidence, Likelihood, Probability: Statistical Inference with Confidence Distributions. Cambridge University Press, Cambridge (2016)
17. Shen, J., Liu, R., Xie, M.: Prediction with confidence–A general framework for predictive inference. J. Stat. Plann. Infer. **195**, 126–140 (2018)
18. Shiryaev, A.N.: Вероятность (Probability), 3rd edn. МЦНМО, Moscow (2004)
19. Snelson, E., Ghahramani, Z.: Variable noise and dimensionality reduction for sparse Gaussian processes. In: Dechter, R., Richardson, T. (eds.) Proceedings of the Twenty Second Conference on Uncertainty in Artifical Intelligence (UAI 2006), pp. 461–468. AUAI Press, Arlington (2006)
20. Steinwart, I.: On the influence of the kernel on the consistency of support vector machines. J. Mach. Learn. Res. **2**, 67–93 (2001)
21. Thomas-Agnan, C.: Computing a family of reproducing kernels for statistical applications. Numer. Algorithms **13**, 21–32 (1996)
22. Vovk, V.: Universally consistent predictive distributions. Technical report. arXiv:1708.01902 [cs.LG], arXiv.org e-Print archive, August 2017
23. Vovk, V., Gammerman, A., Shafer, G.: Algorithmic Learning in a Random World. Springer, New York (2005)
24. Vovk, V., Nouretdinov, I., Gammerman, A.: On-line predictive linear regression. Ann. Stat. **37**, 1566–1590 (2009)
25. Vovk, V., Papadopoulos, H., Gammerman, A. (eds.): Measures of Complexity: Festschrift for Alexey Chervonenkis. Springer, Heidelberg (2015)
26. Vovk, V., Petej, I.: Venn-Abers predictors. In: Zhang, N.L., Tian, J. (eds.) Proceedings of the Thirtieth Conference on Uncertainty in Artificial Intelligence, pp. 829–838. AUAI Press, Corvallis (2014)
27. Vovk, V., Shen, J., Manokhin, V., Xie, M.: Nonparametric predictive distributions based on conformal prediction. In: Proceedings of Machine Learning Research, COPA 2017, vol. 60, pp. 82–102 (2017)
28. Wasserman, L.: Frasian inference. Stat. Sci. **26**, 322–325 (2011)
29. Zadrozny, B., Elkan, C.: Obtaining calibrated probability estimates from decision trees and naive Bayesian classifiers. In: Brodley, C.E., Danyluk, A.P. (eds.) Proceedings of the Eighteenth International Conference on Machine Learning (ICML 2001), pp. 609–616. Morgan Kaufmann, San Francisco (2001)

On the Concept of Compositional Complexity

Lev I. Rozonoer[(✉)]

Institute of Control Sciences RAS, Moscow, Russia
levroz@gmail.com

Abstract. This text is the main content of the presentation, prepared for the conference in honor of E.M. Braverman. In its results, the text does not exceed the 3rd chapter of the book of M.A. Aiserman, E.M. Braverman and L.I. Rozonoer, The Method of Potential Functions in Machine Learning Theory, Physical-Mathematical State Publishing, Moscow (1970) – a chapter dedicated to the choice of a potential function. I.B. Muchnik argued the need for the presentation of the results of this chapter from the long-published book by the fact that these results were never translated into English and remained not well known. Besides this, the accents in the exposition of this presentation have significantly shifted from the exposition given in the book – here only complexity is discussed, and concrete problems of machine learning theory are not touched upon at all. The concept of compositional complexity proves useful for the choice of a potential function, and this is essentially demonstrated in the 3rd chapter of the cited book. Benjamin Rozonoer translated, and Maxim Braverman edited the translation of this section of the book (c.f. list of main publications).

It is well known that the complexity theory is well developed in application to algorithmic and computational complexity, which discusses the complexity of computing functions. Problems of machine learning focus not on the computation of some defined function or the execution of some defined algorithm, but rather on the construction, the reproduction of some object – an image or function – by some means, i.e. the construction of an object from some elements. In particular, such elements can be systems of orthonormal functions (in particular, harmonic), from which with the help of series, reproducible functions are constructed. From the point of view of how well or poorly some function is reproduced by such series, we can discuss the compositional complexity of this function. The characteristic of complexity, suggested by Vapnik and Chervonenkis, the VC-dimension, though it is directly related to machine learning theory, characterizes the complexity of the learning algorithm used – its ability to break into groups the points that appear in space – and not the complexity of the function being restored in the process of learning. In this text the theory of compositional complexity is constructed for measurable functions, defined in metric spaces with measure. Specially distinguished is the class of the so-called symmetric metric spaces (examples of which are the many vertices of an

© Springer Nature Switzerland AG 2018
L. Rozonoer et al. (Eds.): Braverman Readings in Machine Learning, LNAI 11100, pp. 122–127, 2018.
https://doi.org/10.1007/978-3-319-99492-5_5

n-dimensional unit cube with Hammings metric, an n-dimensional sphere, whose metric is defined by the lengths of the arcs of large circles), with which machine learning theory often deals.

1 A Metric Space with a Measure; the Linear Space of L^2 Functions, Defined in the Metric Space, and the Theory of Representation of the Group of Isometric Transformations

We denote the distance between two points x and y of a metric space X by $\rho(x, y)$.

A transformation $A : x \mapsto Ax$ of a metrical space X into itself is called *isometric*, if $\rho(Ax, Ay) = \rho(x, y)$. The set of all isometric transformations clearly forms a group, in which the product of two transformations is their successive application, the unit element is the identity transformation, the inverse element is the inverse transformation.

We consider a metric space X with a measure and denote by $L^2(X)$ the linear space of square-integrable functions on X. In the case when X is a discrete set of points, the integration in the definition of scalar product is replaced with summation with respect to all points of the space. In this case the dimension of the space $L^2(X)$ is clearly equals the number of points of X. In practical application we can restrict ourselves to consider a metric space, consisting of a finite, but arbitrarily large number of points. From here on we assume this case. Thus the space $L^2(X)$ is finite-dimensional.

For an isometric transformation $A : X \to X$ we consider the operator $T_A : L^2(X) \to L^2(X)$, transforming the function $f(x)$ into $g(x) = T_A f(x) := f(A^{-1}x)$. It is easy to see that the operator T_A is a linear unitary operator and that

$$T_{A_1 \circ A_2} = T_{A_1} \circ T_{A_2}, \quad \text{and} \quad T_{A^{-1}} = T_A^{-1}.$$

Therefore, the set of all operators T_A is also a group, and the mapping $A \mapsto T_A$ is a unitary representation of the group of isometric transformations.

Recall that the representation of a group is called *irreducible*, if there does not exist a nontrivial (i.e. distinct from the zero subspace and from the whole space) invariant subspace for the mapping of all elements of the group. The representation $A \to T_A$ is reducible, since the constant function represents a nontrivial one-dimensional subspace $\Lambda_0 \subset L^2(X)$.

In accordance with the well-known theorem of representation theory, the finite-dimensional linear space $L^2(X)$ decomposes into a direct sum of orthogonal invariant subspaces

$$L^2(X) = \Lambda_0 \oplus \Lambda_1 \oplus \cdots \oplus \Lambda_n.$$

Each Λ_s is a unitary representation of the group of isometric transformations.

2 Classes of the Compositional Complexity of Functions

We will introduce a *hierarchy of compositional complexity* for the functions in $L^2(X)$: for two functions $f(x)$ and $g(x)$ we introduce the order relations: $f \preceq g$. This notation is read as: *"function f is not worse than function g"*. The relation $\preceq$ is transitive.

We will say: *"a function f is equivalent to a function g"* and denote $f \sim g$, if $f \preceq g$ and $g \preceq f$. We now postulate the following properties of the introduced order relations:

$$f(Ax) \sim f(x), \quad \text{for any isometric transformation } A; \tag{1}$$
$$\text{if } f \preceq g, \text{ then} \quad \lambda f(x) + \mu g(x) \preceq g(x). \tag{2}$$

The connection of the introduced hierarchy of complexity with the theory of representations of a group of isometric transformations is determined by the following theorem:

Theorem 1. *If two nonzero functions $f(x)$ and $g(x)$ belong to the same subspace Λ_s, then $f(x) \sim g(x)$.*

3 Functionals of Complexity in Symmetric Spaces

A metric space is called *symmetric*, if for any two pairs of points (x^1, y^1) and (x^2, y^2), such that $\rho(x^1, y^1) = \rho(x^2, y^2)$, there exists an isometric transformation A, such that $x^2 = Ax^1$ and $y^2 = Ay^1$.

In a symmetric space X we can introduce a *"functional of complexity"* – the indicator of the compositional complexity of functions. Namely, we consider the functional on $L^2(X)$ defined by the formula

$$\mathscr{F}\{f\} = \frac{\langle f, \Lambda f \rangle}{\|f\|^2}, \tag{3}$$

where $\langle \cdot, \cdot \rangle$ denotes the scalar product, and Λ is a linear operator. Specifically, let

$$\Lambda : f \mapsto g = \Lambda f$$

be the integral operator

$$g(x) = \int_X \Lambda(x, y) f(y) \mu(dy) \tag{4}$$

with integral kernel $\Lambda(x, y)$.

If functional (3) is an indicator of the complexity of functions, then it must satisfy requirement (1). Hence

$$\mathscr{F}\{Af(x)\} = \mathscr{F}\{f(x)\}, \tag{5}$$

for any isometric transformation A. The fulfillment of this requirement is guaranteed by the following theorem:

Theorem 2. *If functional* (3) *is defined in a symmetric space, then this functional is an indicator of complexity if and only if the integral kernel $\Lambda(x, y)$ of the operator* (4) *is the function of distance $\rho(x, y)$:*

$$\Lambda(x, y) = L(\rho(x, y)),\qquad(6)$$

where $L(\rho)$ is some function of distance.

It turns out that condition (6) guaranties that the functional $\mathscr{F}$ is indeed and indicator of complexity. This follows from Theorem 1 and the following

Theorem 3. *If the condition* (6) *is satisfied than the functional* (3) *takes the same values $\mathscr{F}_s$ on all functions $f \in \Lambda_s$.*

It turns out, that the value $\mathscr{F}_s$ depends only on the chosen function $L(\rho)$. Namely, it is the eigenvalue of the integral operator (4), and the functions $f \in \Lambda_s$ are the corresponding eigenfunctions.

It is easy to show that functional (3) is uniquely linked with functional

$$\mathscr{F}^*\{f\} = \iint \frac{L(\rho(x, y))\,(f(x) - f(y))^2\,\mu(dx)\mu(dy)}{\|f\|^2}.\qquad(7)$$

Namely, $\mathscr{F}^* = 2(C - \mathscr{F})$, where C is a constant, depending only on $L(\rho)$. Functional (7) is interpreted easier than (3). Suppose that function $L(\rho)$ is chosen in such a way that its values are equal to zero everywhere, except a small vicinity of $\rho = 0$ and is equal to 1 for small values of ρ. Then the integrand expression in (7) is distinct from zero only for sufficiently close points x and y, so that the value of the integral in (7) is determined by the difference of the values of f only in close points. Since we intuitively presuppose that the more "unsmooth" and "unusual" a function is, the more complex it is, then for a more complex function the differences of values in nearby points are greater than for the more "simple" functions. Therefore, with such a choice of $L(\rho)$, the value of functional (7) is, indeed, directly proportional to the "complexity" of the function in the intuitive sense.

In the case when the space X consists of a finite number of points let us choose $L(\rho)$ in such a way, that

$$L(\rho) = 1, \quad \text{if} \quad \rho = \rho_{\min}, \quad \text{and} \quad L(\rho) = 0 \quad \text{if} \quad \rho \neq \rho_{\min}.$$

Here $\rho_{\min}$ is the distance between two closest points (in a symmetric space it is the same for any pair of closest points). Here the kernel of functional $L(\rho(x, y))$ equals 1 for adjacent points x, y and equals 0 in the opposite case. For this reason the value of functional $\mathscr{F}^*$ is proportional to the sum of the squares of the differences of the values of functions in all adjacent points. Therefore the greater

$\mathscr{F}_s^*\{f\}$, the more "complex" the function f is. The values of $\mathscr{F}_s^*\{f\}$ thereby order the classes of equivalence $\Lambda_0, \ldots, \Lambda_n$ in order of "increasing complexity". With the appropriate normalization of the integral kernel of the functional the following inequalities hold:

$$0 \le \mathscr{F}_s^*\{f\} \le 1, \quad \text{and} \quad 0 \le \mathscr{F}_s^*\{f\} \le \frac{1}{2} \quad \text{if} \quad f(x) \ge 0.$$

4 Compositional Complexity of Functions and Its Decomposition into a Series. Non-complicating Operators

Our intuitive basis for the understanding of compositional complexity is the concept that the complexity of a function in problems of machine learning is determined by its "unsmoothness", "unusualness". A functional with the kernel examined above reflects precisely this concept. Intuitively it is also clear that the complexity understood in this way is reflected in the rate of the convergence of the series representing the decomposition of a function by some orthonormal system of functions – the system of "harmonic" functions. Theorem 4 below establishes the link between the coefficients of such a series and the value of the functional of complexity, discussed above. Let function $f(x)$ be represented by the series

$$f = \sum c_s \phi^s, \tag{8}$$

where $\{\phi^s\}$ is an orthonormal system of functions, $\langle \phi^s, \phi^t \rangle = \delta_{s,t}$. Here $f^s := c_s \phi^s$ is the projection of f onto the space Λ_s.

Theorem 4. *The following formula is valid:*

$$\mathscr{F}^*\{f\} = \frac{\sum \mathscr{F}_s^* |c_s|^2}{\sum |c_s|^2}. \tag{9}$$

Since the values of $\mathscr{F}_s^*$ are arranged in ascending order it follows from formula (9) that a function is the more complex, the slower the coefficients c_s decrease.

Let us now move on to series representing the kernels of operators which define the functionals of complexity. We will select a complete orthonormal system of functions $\{\phi_k^s\}$ in the space Λ_s and examine the kernel:

$$\Lambda^s(x,y) = \sum_k \phi_k^s(x)\, \phi_k^s(y).$$

Theorem 5. *(1) The integral kernel $\Lambda_s(x,y)$ does not depend on the choice of the system of functions ϕ_k^s and there is a function of distance $L^s(\rho)$ such that*

$$\Lambda^s(x,y) = L^s\big(\rho(x,y)\big). \tag{10}$$

(2) The functions $L^s(\rho)$ comprise the complete system of functions over the set of distance functions, i.e. any distance function $L(\rho)$ is representable by a series:

$$L(\rho) = \sum_s \mu_s L^s \tag{11}$$

with some coefficients μ_s.

The operator with kernel $\Lambda^s(x, y)$ projects any function from L^2 onto subspace Λ^s.

We will call kernel $L(\rho)$ *non-complicating* or *complicating*, if the application of the operator with this kernel to any function from $L^2(X)$ does not complicate, or, respectively, complicates, the value of functional (7).

Theorem 6. *A kernel $L(\rho)$ is non-complicating if and only if in its representation by series (11) the coefficients μ_s do not increase in absolute value:* $|\mu_{s+1}| \leq |\mu_s|$.

In conclusion I will note that a natural generalization of the theory of compositional complexity would be such a theory of complexity as would also be applicable in the situation when "elements" would be not the members of the series which represents the dividing function in the feature space, but rather the fragments of the image itself.

On the Choice of a Kernel Function
in Symmetric Spaces

M. A. Aizerman, E. M. Braverman, and Lev I. Rozonoer$^{(\boxtimes)}$

Institute of Control Sciences RAS, 65 Profsoyuznaya Street,
117997 Moscow, Russian Federation
levroz@gmail.com

Abstract. The book *"Pattern recognition: the method of potential functions"* was published in Soviet Union more than 45 years ago. It was never translated to English and many ideas of the book remained unknown outside of Russia. We present here a sligtly abridged version of Chapter III.3 of the book. Technical details of some proofs are omitted and replaced by a short sketch of the main steps of the proof. An interested reader can either fill those details or consult the original Russian edition.

1 Symmetric Space

In this section we consider a metric space of special type, called *symmetric spaces*. An m-dimensional cube, which often appears in problems of machine learning, is an important particular instance of such spaces. For symmetric spaces it is possible to give more precise choice of a system of functions $\phi_i(x)$ and the kernel function $K(x, y)$, and also to justify the usefulness of the choice of a kernel function as a function of distance.

The definition of a symmetric space is based on the concept of a isometric transformation of a metric space into itself. A transformation of a metric space into itself is called *isometric* if it preserves the distance between any pair of points. If A is an isometric transformation, such that point x transforms to point Ax as a result of the transformation, then, in accordance with the definition, for any pair of points $x, y \in X$,

$$\rho(Ax, Ay) \;=\; \rho(x, y). \tag{1}$$

It is clear that the successive application of several isometric transformations is an isometric transformation and that the inverse transformation exists and is also isometric. thus the set of isometric transformations forms a group, where, as usual, the product of two elements (transformations) means successive application of these transformations; the identical transformation can be considered

---

The chapter was translated by Benjamin Rozonoer. The translation was edited by Maxim Braverman.

L. Rozonoer et al. (Eds.): Braverman Readings in Machine Learning, LNAI 11100, pp. 128–147, 2018.
https://doi.org/10.1007/978-3-319-99492-5_6

as the groups unit, and the inverse transformation plays the role of the inverse element.

A metric space X is called *symmetric* if, for any two pairs of points x', y' and x'', y'', located at the same distance $\rho(x', y') = \rho(x'', y'')$, there exists an isometric transformation A of the space X into itself, such that $x'' = Ax'$ and $y'' = Ay'$.

An example of a symmetric space is the above-mentioned set of vertices of an m-dimensional cube, containing $N = 2m$ points, if we use Hammings definition of distance. In this case the isometric transformations are rotations and reflections of the cube (c.f. article [5] below for more details).

Another example of a symmetric space is the set, consisting of N points, evenly distributed along a circle in such a way that the lengths of the shortest arcs between any pair of neighboring points is the same. In this space we set the distance between any two points to be the length of the shortest arc which connects them. Then the isometric transformations are rotations and reflections relative to the corresponding diameters.

From now on, when talking about symmetric space, we will assume that it contains only a finite number of points.

2   Quadratic Quality Functionals on Symmetric Spaces

Next we define functionals of type $\mathscr{F}\{f(x)\}$, which evaluates the "quality of the function $f(x)$, i.e. its "smoothness, simplicity etc. From the point of view of intuitive notions about the quality of functions, it is natural to demand for the functional $\mathscr{F}\{f(x)\}$ to possess the following properties:

$$\mathscr{F}\{\lambda f(x)\} = \mathscr{F}\{f(x)\}; \tag{2}$$
$$\mathscr{F}\{Af(x)\} = \mathscr{F}\{f(x)\}, \tag{3}$$

where λ is any non-zero constant, and A is any isometric transformation of the symmetric space X, on which f(x) is defined. Really, the multiplication of a function by a non-zero constant does not change its "spectrum composition", which is what defines the quality of the function. The second requirement is justified by the fact that the function $f(Ax)$ is simply "shift" of the function $f(x)$.

In this chapter we consider quality functionals of the form

$$\mathscr{F}\{f(x)\} := \frac{\sum_{x,y \in X} L(x, y)\, f(x)\, f(y)}{\|f\|^2}, \tag{4}$$

where

$$\|f\| := \sqrt{\sum_{x \in X} |f(x)|^2}.$$

Without loss of generality we can assume that the kernel $L(x, y)$ is symmetric

$$L(x, y) = L(y, x).$$

The specific form of the functional is determined by the choice of the kernel $L(x, y)$. The requirement (2) is satisfied automatically due to the appearance of the quantity $\|f\|^2$ in the denominator. The requirement (3) considerably limits the potential form of the kernel $L(x, y)$. Namely, the following theorem holds.

Theorem 1. *Let X be a symmetric space. Then, the functional (4) satisfies (3) for every function $f(x)$ and every isometric transformation A if and only if the kernel $L(x, y)$ is a function of distance between points* x *and* y*:*

$$L(x, y) = L\big(\rho(x, y)\big).$$

We precede the proof of Theorem 1 with the following

Lemma 1. *For a function $\psi(x, y)$ defined on a finite symmetric space X to be a function of distance $\rho(x, y)$, it is both necessary and sufficient that for any isometric transformation A*

$$\psi(Ax, Ay) = \psi(x, y).$$

Proof (Proof of Lemma 1). The necessity of the lemmas condition immediately follows from the definition an isometry since

$$\psi\big(\rho(Ax, Ay)\big) = \psi\big(\rho(x, y)\big).$$

Let us prove the sufficiency of the of the lemmas conditions. Let x, y and x', y' be two pares of points. We must prove, that if $\rho(x, y) = \rho(x', y')$, then it follows from the assumptions of the lemma that $\psi(x, y) = \psi(x', y')$. Since the space X is symmetric there exists an isometry A such that

$$x' = Ax, \quad y' = Ay.$$

Hence, $\psi(x', y') = \psi(Ax, Ay)$. But by the assumptions of the lemma $\psi(Ax, Ay) = \psi(x, y)$. Thus $\psi(x', y') = \psi(x, y)$. The lemma is proven.

Proof (Proof of Theorem 1). We rewrite the condition (3) as

$$\frac{\sum_{x, y \in X} L(x, y) f(Ax) f(Ay)}{\|f(Ax)\|^2} = \frac{\sum_{x, y \in X} L(x, y) f(x) f(y)}{\|f\|^2}. \tag{5}$$

The denominators in both parts of these expressions are equal. Hence, after the change of variables $u = Ax$, $v = Ay$ the equality (5) is equivalent to

$$\sum_{u, v \in X} L(A^{-1}u, A^{-1}v) f(u) f(v) = \sum_{u, v \in X} L(u, v) f(u) f(v). \tag{6}$$

By the assumptions of Theorem 1 this equality hold for any function f. Form this we immediately conclude that the function $L(A^{-1}u, A^{-1}v)$ and $L(u, v)$ coincide:

$$L(A^{-1}u, A^{-1}v) = L(u, v). \tag{7}$$

Since (7) holds for every isometry A, the theorem follows now from Lemma 1.

Theorem 1 allows us to rewrite expression (4) in the form

$$\mathcal{F}\{f(x)\} := \frac{\sum_{x,y \in X} L(\rho(x,y)) \, f(x) \, f(y)}{\|f\|^2}, \tag{8}$$

Using Theorem 1 one can show that the functional (8) is uniquely determined by the functional

$$\widetilde{\mathcal{F}}\{f(x)\} := \frac{\sum_{x,y \in X} L(\rho(x,y)) \left[f(x) - f(y) \right]^2}{\|f\|^2}, \tag{9}$$

by the formula

$$\widetilde{\mathcal{F}}\{f\} = 2 \left(C - \mathcal{F}\{f\} \right), \tag{10}$$

where C is a constant depending only on the kernel $L(\rho)$.

To define this constant we consider a function $S(\rho)$, whose value equals the number of points of symmetric space X that lie on the sphere $Sp_x(\rho)$ of radius ρ with center at an arbitrary point x. Since the space X is symmetric, $S(\rho)$ does not depend on x. The expression for constant C has the form

$$C = \sum_{\rho} L(\rho) \, S(\rho). \tag{11}$$

Indeed, opening the parentheses in formula (9), we obtain

$$\widetilde{\mathcal{F}}\{f(x)\} := 2 \frac{\sum_{x,y \in X} L(\rho(x,y)) f(y)^2}{\|f\|^2} - 2\mathcal{F}\{f(x)\}. \tag{12}$$

Let us first sum over x in (12), summing successively over spheres with radii $\rho = 0, \rho_1, \rho_2, \ldots$ with center in some fixed point y. Since the space X is symmetric, the number of points $S(\rho)$ on a sphere of radius ρ does not depend on the choice of the center y. For this reason

$$\sum_{x \in X} L(\rho(x,y)) = \sum_{\rho} L(\rho) \, S(\rho) = C$$

is independent of y. Summing now over y in (12) and using the definition of $\|f\|$ we obtain (10).

The notation of the quality functional in form (9) is convenient in the sense that it quite clearly reflects ones intuitive perceptions of a functions quality, as it directly includes the difference of the functions values at points x and y, located at a distance $\rho(x,y)$. In particular, if the kernel $L(\rho(x,y))$ is non-negative, then the functional takes the minimal (zero) value on constant functions. Meanwhile, since with the "worsening of function $f(x)$, the differences $[f(x)-f(y)]^2$ increase, broadly speaking, we can assume that with the growth of the value of functional (9) the function "worsens. Conversely, in accordance with formula (10) with a positive kernel $L(\rho(x,y))$, it is the decrease of functional (8) that corresponds to the "worsening of the function.

3   The Assignment of Classes for Functions of Equal Quality

In this section we will need some knowledge from the theory of representation of groups.

Let G be a group and let $\mathscr{L}$ be a linear space. A representation of G in $\mathscr{L}$ is an assignment of a linear transformation $T(A) : \mathscr{L} \to \mathscr{L}$ to each element $A \in G$ such that the product of elements A and B of a group G corresponds to the product of operators, i.e.

$$T(A \cdot B) = T(A) \circ T(B).$$

The *dimension* of a representation T is defined to be dimension of $\mathscr{L}$.

In this work $\mathscr{L}$ is the linear space of real-valued functions on X. If, as we assume here, the space X consists of a finite number N of points, then the dimension of space $\mathscr{L}$ is N.

Let us A be an isometric transformation of the space X into itself. To every function $f(x) \in \mathscr{L}$ we will assign a corresponding function $g(x) = f(A^{-1}x) \in \mathscr{L}$. This correspondence between functions $f(x)$ and $g(x)$ specifies the operator $T(A)$:

$$g = T(A)f.$$

The operator $T(A)$ depends, of course, on what kind of isometric transformation A is considered, but no matter how A is chosen, the corresponding operator $T(A)$ is linear. In fact, for any functions $f_1(x)$ and $f_2(x)$ and numbers λ_1, λ_2 we have

$$T(A)[\lambda_1 f_1(x) + \lambda_2 f_2(x)] \; = \; \lambda_1 f_1(Ax) + \lambda_2 f_2(Ax) \; = \; \lambda_1 T(A)f_1 + \lambda_2 T(A)f_2.$$

The correspondence $A \mapsto T(A)$ between the isometric transformations of space X and linear operators on space $\mathscr{L}$ of functions on X, is precisely the representation of group G of isometric transformations that interests us.

Since the dimension of $\mathscr{L}$ is finite, a choice of a basis $\phi_1(x), \ldots, \phi_N(x)$ in $\mathscr{L}$ allows us to represent the operator $T(A)$ by a square matrix $\|T_{ik}(A)\|$. Thus the representation $T(A)$ is given by the correspondence between isometric transformations and square $(N \times N)$-matrices.

If $\phi_1(x), \ldots, \phi_N(x)$ is an orthonormal bases then the matrix $\|T_{ik}(A)\|$ is orthogonal for any isometry A.

A representation T is called *irreducible*, if there are no non-trivial subspaces of L which are invariant for all operators $T(A)$. Otherwise the representation is called *reducible*.

It is a classical result of the representation theory of finite groups that a finite-dimensional linear space $\mathscr{L}$ on which a finite group G acts decomposes into a direct sum of irreducible representations. In particular,

$$\mathscr{L} \; = \; \mathscr{L}_0 \oplus \mathscr{L}_1 \oplus \cdots \oplus \mathscr{L}_m. \tag{13}$$

Here each $\mathscr{L}_i$ is an invariant subspace of $T(A)$ for all isometries $A : X \to X$. For $s = 0, \ldots, m$, we set $N_s := \dim \mathscr{L}_s$ and we denote by $T^s(A)$ the restriction of $T(A)$ to $\mathscr{L}_s$. Then each T^s is an irreducible representation of the group of isometries.

We note the representation T of the group of isometries of X in the space $\mathscr{L}$ of linear functions on X is always reducible. Indeed, the space $\mathscr{L}_0 \subset \mathscr{L}$ of constant functions is a non-trivial invariant subspace. Thus $\mathscr{L}$ is a direct sum of non-trivial irreducible representations. Every function $f(x) \in \mathscr{L}$ can be projected onto each subspace $\mathscr{L}_s$. We denote by f^s this projection. Then f can be written as a direct sum of its "irreducible" components

$$f(x) = \sum_{s=0}^{m} c_s \phi^s(x), \tag{14}$$

where $c_s := \|f^s\|$ and $\phi^s := f^s / \|f^s\|$ is the normalized projection onto $\mathscr{L}_s$.

The decomposition of the space $\mathscr{L}$ into orthogonal subspaces $\mathscr{L}_s$ has a direct relation to the question that interests us concerning the evaluation of the quality of functions. And indeed, the following theorem holds.

Theorem 2. *The functional* (8) *takes the same value $\mathscr{F}_s$ on all functions $f \in \mathscr{L}_s$ ($s = 0, \ldots, m$). The value $\mathscr{F}_s$ depends only on s and the choice of kernel $L(\rho(x,y))$.*

For the proof of Theorem 2 we will need the following lemma.

Lemma 2. *The representations $T_s(A)$ and $T_q(A)$ are not equivalent when $s \neq q$.*

Proof (Proof of Lemma 2). Let us assume the contrary, that the representations $T_s(A)$ and $T_q(A)$ are equivalent. Then one can choose the orthonormal bases $\phi_1, \ldots, \phi_l$ and $\psi_1, \ldots, \psi_l$ of the subspaces $\mathscr{L}_s$ and $\mathscr{L}_q$ respectively ($N_s = N_q = l$) such that the matrices $\|T_{ik}^s(A)\|$ and $\|T_{ik}^q(A)\|$ are orthogonal and coincide for all isometries A.

Consider the function

$$\Phi(x, y) := \sum_{i=1}^{l} \phi_i(x) \psi_i(y). \tag{15}$$

One immediately checks that for any isometric operator A the following identity holds:

$$\Phi(Ax, Ay) = \Phi(x, y). \tag{16}$$

By Lemma 1, any function satisfying (16) is a function of distance. Thus

$$\sum_{i=1}^{l} \phi_i(x) \psi_i(y) = \Phi(\rho(x, y)). \tag{17}$$

Since $\Phi(\rho(x,y)) = \Phi(\rho(y,x))$ it follows from (17) that

$$\sum_{i=1}^{l} \phi_i(x)\psi_i(y) \;=\; \sum_{i=1}^{l} \phi_i(y)\psi_i(x).$$

This equality is contradictory, since setting choosing $y = y^*$ such that at least one of the values $\phi_i(y^*)$, $\psi_i(y^*)$ $(i = 1,\ldots,l)$ is different from zero, we come to the conclusion, that the system of orthonormal functions $\phi_1,\ldots,\phi_l; \psi_1,\ldots,\psi_l$ is linearly dependent. The obtained contradiction refutes the assumption about the equivalence of the representations of $T_s(A)$ and $T_q(A)$. Lemma 2 is proven.

To each function $L(\rho(x,y))$ we associate a linear operator $\widehat{L} : \mathscr{L} \to \mathscr{L}$, defined by the formula

$$\widehat{L}f := \sum_{y \in X} L(\rho(x,y)) \cdot f(y). \tag{18}$$

One readily checks that $\widehat{L}$ commutes with all the operators $T(A)$:

$$T(A)\widehat{L} = \widehat{L}T(A). \tag{19}$$

Lemma 3. *For each $s = 0,\ldots,m$, all functions $f \in \mathscr{L}_s$ are eigenfunctions of $\widehat{L}$ with the same eigenvalue, i.e. there exists $\mu_s \in \mathbb{R}$ such that*

$$\widehat{L}f = \mu_s f, \quad \text{for all } f \in \mathscr{L}_s.$$

Proof (Proof of Lemma 3). Consider the set $L'_s := \widehat{L}(\mathscr{L}_s) \subset \mathscr{L}$. We will show that this subspace is invariant relative to each of the operators of the representation T(A), i.e.

$$T(A)f' \in \mathscr{L}_s, \quad \text{for all } f' \in \mathscr{L}_s. \tag{20}$$

Indeed, by (19), we have $T(A)f' = T(A)\widehat{L}f = \widehat{L}T(A)f$, and since $T(A)f \in \mathscr{L}_s$, we conclude that $T(A)f' \in \mathscr{L}_s$.

Let us now examine two possible cases:

(a) subspace $\mathscr{L}_s$ is a trivial subspace, containing only the null vector $f'(x) = 0$;

(b) in subspace $\mathscr{L}'_s$ there exists at least one nonzero vector.

In case (a) the lemmas assertion is obvious, since $\widehat{L}f = 0$ for all $f \in \mathscr{L}_s$ and, hence, all functions in $\mathscr{L}_s$ are eigenfunctions of $\widehat{L}$ with eigenvalue 0.

Therefore from now on we only consider the case (b). Representation $T^s(A)$ of G on L_s induces a representation $T'^s(A)$ of G on $\mathscr{L}'_s = \widehat{L}(\mathscr{L}_s)$. Since T^s is an irreducible representation, so is T'^s. Hence, $\mathscr{L}'_s$ coincides with one of L_q in the decomposition (13). We will now show that $\mathscr{L}'_s$ coincides with $\mathscr{L}_s$. Indeed, $\widehat{L} : \mathscr{L}_s \to \mathscr{L}'_s = \mathscr{L}_q$. By Shur's lemma this map is either 0 or an isomorphism of representations. As we assumed that $\mathscr{L}'_s \neq 0$ this map must be an isomorphism. The assertion follows now from Lemma 2.

Thus $\widehat{L} : \mathscr{L}_s \to \mathscr{L}_s = \mathscr{L}'_s$. Since the representation $\mathscr{L}_s$ is irreducible it follows form the Shur's lemma that the restriction of $\widehat{L}$ to $\mathscr{L}_s$ is a non-zero constant μ_s. Thus $\widehat{L}f = \mu_s f$ for all $s \in \mathscr{L}_s$. The lemma is proven.

Proof (Proof of Theorem 2). Let μ_s be as in Lemma 3. Notice that the functional $\mathcal{F}$ can be written as

$$\mathcal{F}\{f\} = \frac{\langle f, \widehat{L}f \rangle}{\langle f, f \rangle}.$$

Then for every $f \in \mathscr{L}_s$ we have

$$\mathcal{F}\{f\} = \frac{\langle f, \mu_s f \rangle}{\langle f, f \rangle} = \mu_s \equiv \mathcal{F}_s.$$

From Theorem 2 follows a simple formula for the value of the functional (4), if function f(x) is defined by the decomposition (14). Namely,

$$\mathcal{F}\{f\} = \frac{\sum_{s=0}^{m} c_s^2 \mathcal{F}_s}{\sum_{s=0}^{m} c_s^2}. \tag{21}$$

From (10) it follows that a similar formula holds for the functional $\widetilde{\mathcal{F}}$:

$$\widetilde{\mathcal{F}}\{f\} = \frac{\sum_{s=0}^{m} c_s^2 \widetilde{\mathcal{F}}_s}{\sum_{s=0}^{m} c_s^2}. \tag{22}$$

From (21) we see that the choice of the kernel $L(\rho(x, y))$ in functional (8) is reflected only in values of $\mathcal{F}_s$, evaluating the "quality of the irreducible component $\mathscr{L}_s$. Thus, for evaluating the complexity of an arbitrary function, one can define the numbers $\mathcal{F}_s$ instead of defining the kernel $L(\rho(x, y))$, see Sect. 4 below.

Theorem 2 allows us to identify the classes of equivalent (in terms of their quality) functions without specifying the concrete functional of quality. No matter how the intuitive conceptions concerning a functions quality may be formalized, the ordering of functions based on their quality is connected with the introduction of order relations between two functions f and g: $f \preceq g$. This notation is read as: *"function f is not worse than function g.* The relation $\preceq$ is transitive.

We will say: *"a function f is equivalent to a function g* and denote $f \sim g$, if $f \preceq g$ and $g \preceq f$. We now postulate the following properties of the introduced order relations:

$$f(Ax) \sim f(x); \tag{23}$$
$$\text{if } f \preceq g, \text{ then } \lambda f(x) + \mu g(x) \preceq g(x). \tag{24}$$

Condition (23), essentially, coincides with condition (3), while condition (24) is stronger than (2). Postulating condition (2) allows us to establish the equivalence of functions belonging to the same irreducible component, without using a quality functional.

Theorem 3. *If two non-zero functions $f(x)$ and $g(x)$ belong to the same irreducible component $\mathcal{L}_s$ then $f \sim g$.*

Proof (Proof of Theorem 3). Let $f, g \in \mathcal{L}_s$ be non-zero functions. Since $\mathcal{L}_s$ is an irreducible representation of G, g belongs to the linear span on functions $\{T(A)f : A \in G\}$. In other words, $g(x)$ can be written as

$$g(x) = \sum_{i=1}^{N_G} \lambda_i f(A_i x), \qquad (25)$$

where N_G is the rank of the group G. It follows from this equation, (23) and (24) that $g(x) \preceq f(x)$. Similarly, $f(x) \preceq g(x)$ and, hence, $f \sim g$.

Thus the equivalence classes of functions of "the same quality" are determined without specifying the quality functional. However, to compare two functions, which have nonzero projections on at least two different layers, we must use functional (8), i.e. to specify the kernel function $L(\rho)$. This immediately leads to the establishment of concrete "weights" $\mathcal{F}_s$, attributed to the irreducible components, and it is necessary that these weights correspond to intuitive conceptions about the complexity of functions from these components. One has to remember this when choosing the kernel $L(\rho)$. An example of such a choice will be given in the beginning of Sect. 5.

4   The Power Series Expansion of a Function of Distance

It was shown above, that the kernel of the quality functional (8) must be chosen to be a function of distance $\rho(x, y)$. In this section we examine the properties of a distance functions on symmetric spaces, related to their expansion into a series of some system of functions, also depending on the distance. The choice of this system of function is closely related to the decomposition of $\mathcal{L}$ into irreducible components, discussed in the previous section.

In each component $\mathcal{L}_s$ we choose an orthonormal basis ϕ_j^s ($j = 1, \ldots, N_s$, $s = 0, \ldots, m$). Clearly, $N = \sum N_s$ and the collection ϕ_j^s is a basis of $\mathcal{L}$.

For each $s - 0, \ldots, m$, define the function

$$K_s(x, y) := \sum_{j=1}^{N_s} \phi_j^s(x)\phi_j^s(y). \qquad (26)$$

Theorem 4. *The functions K_s are independent of the choice of the basis ϕ_j^s and are functions of distance*

$$K_s(x, y) = K_s\big(\rho(x, y)\big).$$

The system of functions $K_s(x, y)$ ($s = 0, \ldots, m$) is a complete system of function in the space of functions of ρ.

Proof (Sketch of the proof of Theorem 4). One readily checks that the operator
$\widehat{K}_s : \mathscr{L} \to \mathscr{L}$

$$\widehat{K}_s f(x) := \sum_{y \in X} K_s(x, y) f(y)$$

is the orthogonal projection onto $\mathscr{L}_s$. All the assertion of the theorem follow directly from this fact.

It follows from this theorem that any function of distance is a linear combination of the functions K_s. In case of the kernel function (8) one readily sees that

$$L\big(\rho(x, y)\big) = \sum_{s=0}^{m} \mathscr{F}_s K_s(x, y), \tag{27}$$

where $\mathscr{F}_s$ are the weights defined in Theorem 2.

We will now list some useful properties and relations connected with the functions $K_s(\rho)$.

First Property

For all ρ

$$|K_s(\rho)| \leq K_s(0) = \frac{N_s}{N} > 0. \tag{28}$$

Proof (Proof of the first property)

$$NK_s(0) = \sum_{x \in X} K_s(0) = \sum_{x \in X} \sum_{j=1}^{N_s} \phi_j^s(x)^2 = \sum_{j=1}^{N_s} \sum_{x \in X} \phi_j^s(x)^2 = \sum_{j=1}^{N_s} \|\phi_j^s\|^2 = N_s.$$

Hence, $K_s(0) = \frac{N_s}{N}$. This proves the second equality of (28). The first inequality of (28) follows from the Cauchy inequality.

Second Property

The functions K_s are orthogonal with weight $S(\rho)$:

$$\sum_{\rho} S(\rho) K_s(\rho) K_q(\rho) = \delta_{sq} K_s(0), \tag{29}$$

where $S(\rho)$ is the number of points in the sphere of radius ρ around any point $x \in X$ and δ_{sq} is the Kronecker symbol.

Using this property one can compute the coefficients μ_s by the formula

$$\mu_s = \frac{1}{K_s(0)} \sum_{\rho} L(\rho) K_s(\rho) S(\rho). \tag{30}$$

Proof (Sketch of the proof of the second property). Since K_s is the kernel of the orthogonal projection onto $\mathscr{L}_s$, K_s and K_q are orthogonal as functions on $X \times X$. Thus

$$\delta_{sq} K_s(x, x) = \sum_{x, y \in X} K_s(x, y) K_q(x, y) = \sum_{\rho} \sum_{\rho(x, u) = \rho} K_s(x, y) K_q(x, y)$$

$$= \sum_{\rho} S(\rho) K_s(\rho) K_q(\rho)$$

Third Property

The functions K_s satisfy the second orthogonality relation:

$$\sum_{s=0}^{m} \frac{K_s(\rho)K_s(\kappa)}{K_s(0)} = \frac{1}{S(\kappa)}\delta_{\rho\kappa}. \tag{31}$$

Proof (Proof of the third property). Fix κ and consider $\delta_{\rho\kappa}$ as a of ρ. Since $K_s(\rho)$ is a complete system of functions we can write

$$\delta_{\rho\kappa} = \sum_{s=0}^{m} \lambda_s(\kappa)K_s(\rho). \tag{32}$$

To compute the coefficients $\lambda_s(\kappa)$ we multiply (32) by $K_q(\rho)S(\rho)$ and sum over ρ. Then using (29) we have

$$\lambda_q(\kappa)K_q(0) = \sum_{\rho} \delta_{\rho\kappa}K_q(\rho)S(\rho) = K_q(\kappa)S(\kappa).$$

Hence,

$$\lambda_q(\kappa) = \frac{K_q(\kappa)S(\kappa)}{K_q(0)}. \tag{33}$$

Substituting this expression to (32) we obtain (31).

From Theorem 4 and the second property we obtain the following important result about symmetric spaces: *The number $m+1$ of irreducible components in decomposition* (13) *is equal to the number of different distances between the points of X.*

Notice now that the "complexity" of a function $K_s(\rho)$ also can be evaluated by the quality functional $\mathscr{F}_s$ or $\widetilde{\mathscr{F}}_s$. To make it precise, let us fix a point $x^* \in X$ and consider the function

$$g_s(y) := K_s\big(\rho(x^*, y)\big).$$

By (26) the function $g_s(y) \in \mathscr{L}_s$. It follows that $\mathscr{F}(g_s) = \mathscr{F}_s$ and $\widetilde{\mathscr{F}}\{g_s\} = \widetilde{\mathscr{F}}_s$. In this sense the "complexity" of K_s is the same as the "complexity" of the functions in $\mathscr{L}_s$. Using this result we can compute the "complexity" of any function $L(\rho)$ through its decomposition into a linear combination of K_s. Indeed if

$$L(\rho) = \sum_{s=0}^{m} \mu_s K_s(\rho),$$

then, using (21), we obtain

$$\mathscr{F}\{L(\rho(x^*, y))\} = \frac{\sum_{s=0}^{m} \mu_s^2 K_s(0)\mathscr{F}_s}{\sum_{s=0}^{m} \mu_s^2 K_s(0)}.$$

5   The Potential Function in a Symmetric Space

To explain how the above facts are used for the choice of a potential function in the method of potential functions, we need to specify a concrete form of the kernel function $L(\rho)$ in (8) and (9).

Let us define the kernel by the formula

$$L(\rho) = \frac{1}{4S(\rho_1)} \delta_{\rho\rho_1}, \tag{34}$$

where ρ_1 is the smallest distance between non-equal points of X. Then the value of the functional (9) is proportional to the sum of the squares of the differences of the values of f in the neighboring points. Thus the value of $\mathscr{F}\{f\}$ is bigger for the functions which we intuitively consider "worst".

One easily checks the following properties of the functional (9) with kernel (34):

(1) $0 \le \widetilde{\mathscr{F}}\{f\} \le 1$;
(2) if $f(x) \ge 0$ then $0 \le \widetilde{\mathscr{F}}\{f\} \le \frac{1}{2}$,

and as for any functional of this type

(3) $\widetilde{\mathscr{F}}(const) = 0$.

For the kernel (34) the weights $\mathscr{F}_s$ and $\widetilde{\mathscr{F}}_s$ are expressed in terms of the functions K_s as follows

$$\mathscr{F}_s = \frac{1}{4} \frac{K_s(\rho_1)}{K_s(0)}, \tag{35}$$

$$\widetilde{\mathscr{F}}_s = \frac{1}{2} \left(1 - \frac{K_s(\rho_1)}{K_s(0)} \right). \tag{36}$$

These formulas follow easily from (31).

From now on we enumerate the irreducible components in (13) by weights $\widetilde{\mathscr{F}}_s$ so that

$$0 = \widetilde{\mathscr{F}}_0 < \widetilde{\mathscr{F}}_1 < \cdots < \widetilde{\mathscr{F}}_m. \tag{37}$$

We now turn to the question of choosing the potential function $K(x, y)$ in the *method of potential functions*. In practice whenever this method is used the function $K(x, y)$ is chosen as a function of distance: $K(x, y) = K(\rho(x, y))$. This choice for function $K(x, y)$ is justified by the following reasons. In Sect. 4 we introduced the system of functions $K_s(\rho)$ and proved its completeness. For this reason, no matter how the function $K(x, y) = K(\rho(x, y))$ is chosen, it can be expressed as a sum

$$K(\rho(x, y)) = \sum_{s=0}^{m} \mu_s K_s(\rho(x, y)). \tag{38}$$

The method potential functions discussed in the previous sections of this book implies that the coefficients $\mu_s \geq 0$, which, in view of (28), that

$$K(0) > 0, \tag{39}$$

and for all ρ

$$|K(\rho)| < K(0). \tag{40}$$

In order to determine which further restrictions are reasonable, when choosing the potential function $K(\rho(x,y))$, i.e. for the assignment of non-negative numbers μ_s in decomposition (38), we will examine the machine realization of the method of potential functions (cf. Sect. 3 Chap. II of this book). On every n-th step the machine realization comes down to computing the sum

$$f^n(x) = \sum_{i=0}^{n-1} r^i K(x, x^{i+1}). \tag{41}$$

If we introduce the function

$$\pi^n(x) = \sum_{i=0}^{n-1} r^i \delta_{x,x^{i+1}},$$

then

$$f^n(x) = \sum_{y \in X} K\big(\rho(x,y)\big)\pi^n(y). \tag{42}$$

Function $\pi(x)$ is equal to zero everywhere, except at points x_i, used in the process of learning. The problem of learning makes sense only when the number of points used in the process of learning is much less than the total number of points in space X. For this reason the function $\pi(x)$ is different from zero only in points that are separated from each other and it intuitively becomes clear that it is very "tattered, "plateresque". This is evident also from the value of the functional $\widetilde{\mathscr{F}}\{\pi^n(x)\}$. Really, it is easy to calculate the value of this functional with the assumption that among the chosen points x_i there are no adjacent points (i.e. $\rho(x_i, x_j) > \rho_1$ when $i \neq j$). This value is equal to

$$\widetilde{\mathscr{F}}\{\pi^n(x)\} = \frac{1}{2}$$

and depend neither on the number n of featured points (as long as there are no adjacent ones among them), nor on the values r^i (i. e. on the concrete learning algorithm). The value of the functional, equal to $1/2$, corresponds to the very "tattered function (to which attests, for example, property 2) of functional $\mathscr{F}$). As for function $f^n(x)$, it should be smooth enough, since for large enough values of n (but still much less than than the total number of points in space X) it must approximate the function $f^*(x)$ that is being restored, which is assumed to be "smooth, not "tattered, i.e. possessing a high quality (c.f. Sect. 1.1 of this chapter). To such functions $f^n(x)$ must correspond the small value of the functional $\widetilde{\mathscr{F}}\{\pi^n(x)\}$.

In formula (42) the function $K(\rho(x,y))$ can be viewed as the kernel of the linear "integral" operator $\widehat{K}$, which transforms a function $\pi^n(x)$ into a function $f^n(x)$. It follows from the above discussion that the operator $\widehat{K}$ must map the a function of bad quality (with a large value of $\mathscr{F}$) to functions of good quality (with a small value of $\mathscr{F}$). Therefore it makes sense to introduce the following definition: operator $\widehat{K}$ with kernel $K(\rho(x,y))$ is called *bettering* (resp. *worsening*), if $\widetilde{\mathscr{F}}\{\widehat{K}f\} \leq \widetilde{\mathscr{F}}\{f\}$ (resp. $\widetilde{\mathscr{F}}\{\widehat{K}f\} \geq \widetilde{\mathscr{F}}\{f\}$ for any function $f(x)$.

Assume that the irreducible components $\mathscr{L}_s$, and hence also the coefficients μ_s, are numbered in accordance with (37). Then the following theorem holds.

Theorem 5. *Assume that the function $K(\rho)$, corresponding to operator $\widehat{K}$, is given by (38). Then the operator $\widehat{K}$ is bettering (worsening) only if and only if the sequence $|\mu_s|$, $s = 0, 1, \ldots, m$ is non-increasing (non-decreasing).*

Proof (Proof of Theorem 5). It is enough to proof the assertion of Theorem 5 for bettering operators.

(a) Proof of necessity. Suppose the requirement of the theorem are not satisfied, i.e. $|\mu_s| > |\mu_j|$ for some $k > j$. Consider the function

$$f(x) = \frac{1}{\sqrt{2}}\phi_j(x) + \frac{1}{\sqrt{2}}\phi_k(x),$$

where ϕ_i and ϕ_k are any functions from $\mathscr{L}_j$ and $\mathscr{L}_k$, respectively. By (22)

$$\widetilde{\mathscr{F}}(f) = \frac{1}{2}\left(\widetilde{\mathscr{F}}_i + \widetilde{\mathscr{F}}_j\right).$$

Since

$$\widehat{K}f = \frac{\mu_j}{\sqrt{2}}\phi_j + \frac{\mu_k}{\sqrt{2}}\phi_k(x),$$

we also have

$$\widetilde{\mathscr{F}}\{\widehat{K}f\} = \frac{\mu_j^2\widetilde{\mathscr{F}}_j + \mu_k^2\widetilde{\mathscr{F}}_k}{\mu_j^2 + \mu_k^2} = \widetilde{\mathscr{F}}_j + \frac{\mu_k^2}{\mu_j^2 + \mu_k^2}(\widetilde{\mathscr{F}}_k - \widetilde{\mathscr{F}}_j).$$

But, by our assumption, $\mu_k^2 > \mu_j^2$. Hence

$$\frac{\mu_k^2}{\mu_j^2 + \mu_k^2} > \frac{1}{2}.$$

Besides this, since the irreducible components are ordered in accordance with the values of the functional, $\widetilde{\mathscr{F}}_k - \widetilde{\mathscr{F}}_j > 0$. Thus

$$\widetilde{\mathscr{F}}\{\widehat{K}f\} > \widetilde{\mathscr{F}}_j + \frac{1}{2}(\widetilde{\mathscr{F}}_k - \widetilde{\mathscr{F}}_j) = \frac{1}{2}(\widetilde{\mathscr{F}}_k + \widetilde{\mathscr{F}}_j) = \widetilde{\mathscr{F}}\{f\}.$$

Hence, if the condition of the theorem is not satisfied, then the operator K is not a bettering operator.

(b) Proof of sufficiency. Let the conditions of the theorem be satisfied. Consider an arbitrary function $f(x) = \sum_{s=0}^{m} c_s \phi^s(x)$ and the number

$$\Delta := \widetilde{\mathscr{F}}\{f\} - \widetilde{\mathscr{F}}\{\widehat{K}f\}.$$

We need to show that $\Delta > 0$. Indeed, using (22) we obtain

$$\Delta = \widetilde{\mathscr{F}}\{f\} - \frac{\sum_{s=0}^{m} \mu_s^2 c_s^2 \widetilde{\mathscr{F}}_s}{\sum_{s=0}^{m} \mu_s^2 c_s^2} = \frac{\sum_{s=0}^{m} \mu_s^2 c_s^2 \left[\widetilde{\mathscr{F}}\{f\} - \widetilde{\mathscr{F}}_s\right]}{\sum_{s=0}^{m} \mu_s^2 c_s^2}.$$

Since the sequence $\widetilde{\mathscr{F}}_s$ is increasing and $0 = \widetilde{\mathscr{F}}_0 \leq \widetilde{\mathscr{F}}\{f\} \leq \widetilde{\mathscr{F}}_m$, there exists k such that

$$\widetilde{\mathscr{F}}\{f\} - \widetilde{\mathscr{F}}_{k-1} \geq 0, \quad \widetilde{\mathscr{F}}\{f\} - \widetilde{\mathscr{F}}_k \leq 0.$$

Hence, we can write

$$\Delta = \frac{1}{\sum_{s=0}^{m} \mu_s^2 c_s^2} \left[\sum_{s=0}^{k-1} \mu_s^2 c_s^2 \left(\widetilde{\mathscr{F}}\{f\} - \widetilde{\mathscr{F}}_s\right) - \sum_{s=k}^{m} \mu_s^2 c_s^2 \left(\widetilde{\mathscr{F}}\{f\} - \widetilde{\mathscr{F}}_s\right)\right], \quad (43)$$

where the summands in both sums are non-negative.

By our assumptions $\mu_s^2 \geq \mu_k^2$ for $s \leq k-1$ and $\mu_s^2 \leq \mu_k^2$ for $s \geq k$. Hence,

$$\sum_{s=0}^{k-1} \mu_s^2 c_s^2 \left(\widetilde{\mathscr{F}}\{f\} - \widetilde{\mathscr{F}}_s\right) \geq \mu_k^2 \sum_{s=0}^{k-1} c_s^2 \left(\widetilde{\mathscr{F}}\{f\} - \widetilde{\mathscr{F}}_s\right)$$

$$\sum_{s=k}^{m} \mu_s^2 c_s^2 \left(\widetilde{\mathscr{F}}\{f\} - \widetilde{\mathscr{F}}_s\right) \leq \mu_k^2 \sum_{s=k}^{m} c_s^2 \left(\widetilde{\mathscr{F}}\{f\} - \widetilde{\mathscr{F}}_s\right).$$

Substituting these inequalities into (43) we obtain

$$\Delta \geq \frac{\mu_k^2}{\sum_{s=0}^{m} c_s^2} \left[\sum_{s=0}^{k-1} c_s^2 \left(\widetilde{\mathscr{F}}\{f\} - \widetilde{\mathscr{F}}_s\right) - \sum_{s=k}^{m} c_s^2 \left(\widetilde{\mathscr{F}}\{f\} - \widetilde{\mathscr{F}}_s\right)\right]$$

$$= \mu_k^2 \frac{\sum_{s=0}^{m} c_s^2}{\sum_{s=0}^{m} \mu_s^2 c_s^2} \left(\widetilde{\mathscr{F}}\{f\} - \frac{\sum_{s=k}^{m} c_s^2 \widetilde{\mathscr{F}}_s}{\sum_{s=k}^{m} c_s^2}\right). \quad (44)$$

By (22) the last quantity in parentheses is equal to zero, and therefore $\Delta \geq 0$. This proves the sufficiency of the theorem. Theorem 5 is proven completely.

Theorem 5 establishes those additional restrictions on the choice of the potential function, which was discussed above. Namely, in connection with the fact that the operator $\widehat{K}$ must be bettering, the coefficients μ_s in decomposition (38) must not only be nonnegative, but also nonincreasing: $\mu_0 \geq \mu_1 \geq \cdots \geq \mu_m$.

In the conclusion of this point we will summarize those reasons which must be considered when choosing a potential function $K(x, y)$ in symmetric spaces:

i. It is expedient to choose the potential function $K(x, y)$ as a function $K(\rho(x, y))$, depending only on the distance $\rho(x, y)$.
ii. This function can be defined by the decomposition (38) into a series in the system of functions $K_s(\rho)$. The system of functions $K_s(\rho)$ is uniquely defined for a given space X.
iii. The coefficients μ_s in decomposition (38) must be positive.
iv. If we order the coefficients μ_s in accordance with (37), then the sequence $\mu_0, \mu_1, \ldots, \mu_m$ must be monotonically decreasing.

In the cases when the potential function $K(\rho)$ is initially defined in closed form (for example, with expressions $\widehat{K}(\rho) = e^{-\alpha\rho^2}$, $K(\rho) = 1/(1 + \alpha\rho^2)$ and so forth) to check properties (iii) and (iv) one can calculate the coefficients μ_s, making use of formula (30). Functions $K(\rho)$, for which properties (39) and (40) are not satisfied, are certainly not suitable as potential functions.

The practice of applying potential functions shows that the results of using the method depend little on how the coefficients μ_s are chosen, if the satisfy the above restrictions.

6   On the Choice of a Potential Function in the Space of Vertices of an m-dimensional Cube

Out of the various symmetric spaces which we encounter in practice, the most significant is the space of vertices of an m-dimensional cube. We encounter such a space, for example, in the recognition of black-and-white images. In the present paragraph we will show how the theory outlined above is applied in this concrete space. It will be convenient for us to consider that the coordinates $x_1, \ldots, x_m$ of the cubes vertices taking values ± 1, i.e. that the center of the cube is located at the origin, and the edge has euclidean length 2. As the distance between the vertices $x = (x_1, \ldots, x_m)$ and $y = (y_1, \ldots, y_m)$ we use the usual *Hamming distance*

$$\rho(x, y) = \frac{1}{4} \sum_{i=1}^{m} (x_i - y_i)^2 = \frac{1}{2} \left(m - \sum_{i=1}^{m} x_i y_i \right), \tag{45}$$

equal to the number of distinct places in the codes of the vertices. The space of vertices of an m-dimensional cube with metric (45) is from here on called a *Hamming space*. This space contains $N = 2^m$ points. A Hamming space is a symmetric space, as one can easily check.

Let us now find out, in what way the linear space $\mathscr{L}$ of functions on the Hamming space decomposes into irreducible components L_s.

Consider the system of functions in $\mathscr{L}$, consisting of the constant function $\phi_0 = 1/2^{m/2}$ and functions:

$$
\left.
\begin{array}{l}
\frac{1}{2^{m/2}}x_1, \quad \frac{1}{2^{m/2}}x_2, \dots, \quad \frac{1}{2^{m/2}}x_i, \dots, \quad \frac{1}{2^{m/2}}x_m; \\
\frac{1}{2^{m/2}}x_1x_2, \frac{1}{2^{m/2}}x_1x_3, \dots, \frac{1}{2^{m/2}}x_ix_j, \dots, \frac{1}{2^{m/2}}x_{m-1}x_m; \\
\cdots \quad\quad \cdots\cdots \quad\quad \cdots\cdots \quad\quad \cdots \\
\frac{1}{2^{m/2}}x_1x_2\cdots x_m
\end{array}
\right\}
\tag{46}
$$

so that the functions, written in the s-th line, have the form

$$
\phi_{i_1 i_2 \dots i_s} = \frac{1}{2^{m/2}} x_{i_1} x_{i_1} \cdots x_{i_s},
$$

and the indices $i_1, i_2, \dots, i_s$ take all possible values from 1 to m that satisfy the condition $i_1 < i_2 < \cdots < i_s$. Thus the s-th line contains exactly $\binom{m}{s}$ functions, and their total number (including the constant function) equals 2^m, i.e. equals the number N of points of the Hamming space. Bearing in mind, that the above system of functions is orthonormal, we conclude, that it is the orthonormal basis in linear space $\mathscr{L}$ of functions on the Hamming space.

One can easily check that the functions written in the s-th line (46) belong to one irreducible component $\mathscr{L}_s$ and make up its basis. It follows that the number of irreducible components (including the component $\mathscr{L}_0$ consistent of constant functions) is equal to $m + 1$, i.e. in accordance with the general theory, to the number of distinct distances in the examined symmetric space. The dimension of the s-th component $\mathscr{L}_s$ equals $N_s = \binom{m}{s}$.

Next we calculate the functions of $K_s(\rho)$ for a Hamming space.

For the component $\mathscr{L}_0$

$$
K_0(\rho) = \frac{1}{2^{m/2}} \cdot \frac{1}{2^{m/2}} = \frac{1}{2^m}.
$$

For component $\mathscr{L}_s$ we have

$$
K_s\big(\rho(x,y)\big) = \frac{1}{2^m} \sum_{i_1 < \cdots < i_s} x_{i_1} \cdots x_{i_s} \cdot y_{i_1} \cdots y_{i_s}.
$$

Set $z_i := x_i y_i$ $(i = 1, \dots, m)$. Then

$$
K_s\big(\rho(x,y)\big) = \frac{1}{2^m} \sum_{i_1 < \cdots < i_s} z_{i_1} \cdots z_{i_s}.
\tag{47}
$$

If the distance between x and y equals ρ, then in accordance with (45) there will be ρ negative (-1) and $m - \rho$ positive $(+1)$ values among z_i. The summands in (47) can be broken into groups with j positive and $s - j$ negative z_i's. In both cases $0 \le j \le \min\{s, \rho\}$. Every such summand equals $(-1)^j$. Their number equals $\binom{\rho}{j}\binom{m-\rho}{s-j}$. Now, summing over j, we will get

$$
K_s(\rho) = \frac{1}{2^m} \sum_{j=0}^{\min\{s,\rho\}} \binom{\rho}{j}\binom{m-\rho}{s-j}(-1)^j.
\tag{48}
$$

With this formula one can calculate, in particular,

$$K_1(\rho) = \frac{1}{2^m}(m - 2\rho),$$

$$K_2(\rho) = \frac{1}{2^m}\frac{(m - 2\rho)^2 - m}{2},$$

etc. Also

$$K_s(1) = \frac{1}{2^m}\sum_{j=0}^{1}\binom{\rho}{j}\binom{m-\rho}{s-j}(-1)^j = \left(\binom{m-1}{s} - \binom{m-1}{s-1}\right)\frac{1}{2^m}$$

$$K_s(0) = \frac{1}{2^m}\binom{m}{s}.$$

Thus

$$\frac{K_s(1)}{K_s(0)} = 1 - \frac{2s}{m}.$$

and, consequently, by (36) the value $\widetilde{\mathscr{F}}_s$ of the quality functional $\widetilde{\mathscr{F}}$ with kernel (34) on functions $f \in \mathscr{L}_s$ equals

$$\widetilde{\mathscr{F}}_s = \frac{s}{m}. \tag{49}$$

Formula (49) shows that the values of the functional increase with the growth of s. This is in full compliance with our intuitive understanding of the change of complexity of functions (46) when moving from the top to the bottom lines in (49). Indeed, one can show that each function, in the s-th line, possesses the following property: among the m vertices of the cube that are located at the minimal distance $\rho = 1$ from any fixed vertex x^*, there are exactly s vertices in which the values of the function differ by a sign from its value in x^*; in the remaining $m - s$ adjacent vertices the values of the function coincide with the value in x^*. The modules of the values of all the functions in (49) are the same in all vertices and equal $1/2^{m/2}$.

In accordance with the note at the end of Sect. 4, the values (49) of the quality functional characterize the complexity of the functions $K_s(\rho)$. With the growth of the number s the function $K_s(\rho)$ becomes more complex. In this case function $K_s(\rho)$ is a polynomial of order s in ρ, and, correspondingly, with the growth of s grows the number of its sign changes, extrema and other intuitive indicators of complexity.

Let us now turn to the question of the decomposition a function of distance (in particular, a potential function), defined in a Hamming space into a linear combination of functions $K_s(\rho)$. Here, of course, we can use formula (30), where the function $S(rho)$ (the number of points on a sphere of radius ρ) in this case, as can be easily shown, has the form

$$S(\rho) = \binom{m}{\rho}.$$

However, the practical use of formula (30) leads to difficult calculations. In a number of cases it is possible to calculate (precisely or approximately) the decomposition coefficients μ_s without resorting to a direct calculation by formula (30), but using the following identity:

$$\left(\frac{1-u}{2}\right)^\rho \cdot \left(\frac{1+u}{2}\right)^{m-\rho} = \sum_{s=0}^{m} K_s(\rho) u^s, \tag{50}$$

whose proof is left to the reader (see page 134 of the Russian edition for a detailed proof).

As an example of a precise calculation of coefficients μ_s we use this identity for the decomposition of function $K(\rho) = e - \alpha\rho$ (where α is some constant). With this goal we choose the value of u in (50) so that

$$u = \frac{e^\alpha - 1}{e^\alpha + 1} \quad \Leftrightarrow \quad \ln\frac{1+u}{1-u} = \alpha.$$

Substituting this value of U in (50) we obtain

$$e^{-\alpha\rho} = \sum_{s=0}^{m} (1 - e^{-\alpha})^s (1 + e^{-\alpha})^{m-s} K_s(\rho).$$

Thus for $K(\rho) = e^{-\alpha\rho}$ the decomposition coefficients μ_s are given by

$$\mu_s = (1 - e^{-\alpha})^s (1 + e^{-\alpha})^{m-s}. \tag{51}$$

This, in particular, shows that (in the Hamming space) the function $e - \alpha\rho$ can be used as a potential function.

We will now show, how formula (50) can be used for an asymptotic (for $m \to \infty$) evaluation of the decomposition coefficients of one quite broad class of functions of distance. Namely, we consider functions of the form

$$K(\rho) = f(\rho/m),$$

where $f(z)$ is a sufficiently smooth function, defined on the segment $0 \le z \le 1$. To calculate the decomposition coefficients μ_s of function $K(\rho)$ we will multiply both sides of formula (50) by

$$K(\rho)S(\rho) = f(\rho/m)\binom{m}{\rho}.$$

Summing over ρ between 0 and m, recalling (30) and setting $z = (1-u)/2$, we get:

$$\sum_{\rho=0}^{m} \binom{m}{\rho} z^\rho (1-z)^{m-\rho} f(\rho/m) = \sum_{s=0}^{m} (-2)^s \left(z - \frac{1}{2}\right)^s \binom{m}{s} \frac{\mu_s}{2^m}. \tag{52}$$

The left hand side of this expression is a polynomial of S.N. Bernstein of function $f(z)$. It is known that this polynomial approximates function $f(z)$ for $m \to \infty$ uniformly on the segment $0 \le z \le 1$ and therefore

$$f(z) \sim \sum_{s=0}^{m} (-2)^s \left(z - \frac{1}{2} \right)^s \binom{m}{s} \frac{\mu_s}{2^m}, \tag{53}$$

and the error diminishes with the growth of m uniformly in z, for example, as $1\sqrt{m}$, if we demand only the continuity of $f(z)$, and as $1/m$, if $f(z)$ is twice differentiable. Suppose now that $f(z)$ has a Taylor series expansion in a neighborhood of $z = 1/2$. Then, comparing this Taylor series with the right hand side of (52), we get the expression for the coefficients μ_s:

$$2^{-m} \binom{m}{s} \mu_s \sim \frac{1}{s!(-2)^s} f^{(s)} \left(\frac{1}{2} \right), \tag{54}$$

which is asymptotically accurate for $m \to \infty$. However, for a finite m it makes sense to use formula (54) only for relatively small values of s, since with the growth of s the right side of (54) becomes comparable with the error of this formula. We can easily assess the errors for the use of formula (54), if we use the well-known results concerning the evaluation of approximation precision by the polynomials of S.N. Bernstein.

Formula (54) allows us to check, whether or not the function $K(\rho) = f(\rho/m)$ can serve as a potential function in a Hamming space.

Causality Modeling and Statistical Generative Mechanisms

Igor Mandel[(✉)]

Telmar Inc., New York, NY 10015, USA
igor.mandel@gmail.com

Abstract. Causality notion lies at the heart of science, but when statistics tries to address this issue some profound questions remain unanswered. How statistical inference in probabilistic terms is linked with causality? What modern causality models offer that is substantially different from the traditional dependency models like regression or decision trees, and if yes, do they deliver these promises? How causality models are related to statistical and machine learning techniques? What is the relationship between causality modeling, statistical inference, and machine learning on one side – and operations research and optimization on the other? Or, more generally: if the causal picture of the world is a commonly accepted goal of any science, could the non-causal statistical models be of any use? If yes – in what sense? If not – why are they so widely used? The insufficient level of detail in discussions of these and similar problems creates a lot of confusion, especially now, when lauded terms like Data Mining, Big Data, Deep Learning and others appear even in the non-professional media. This paper inspects the underlying logic of different approaches, directly or indirectly, related with causality. It shows that even established methods are vulnerable to small deviations from the ideal setting; that the leading approaches to statistical causality, Structural Equations Modeling (SEM), Directed Acyclic Graphs (DAG) and Potential Outcomes (PO) theories do not provide a coherent causality theory, and argues that this theory is impossible on pure statistical grounds. It also discusses a new approach in which the concept of causality is replaced by the concept of dependent variable generation. Separation of the variables generating the outcome from others just correlated with it (which often separates also causal from non-causal variables) is proposed.

Keywords: Dependency modeling · Statistical inference · Causality modeling
Counterfactual statements · Statistical learning · Intrinsic probability
Generative statistical mechanisms

1 Introduction

As is very often in science, definitions of important concepts are not commonly accepted. Notion of causality is not an exception (see Sect. 3.1). Any causality model, broadly speaking, belongs to the wider class of "dependency modeling", which is also, in turn, not defined well. Vapnik (2006), for example, uses this term in the title of his book, yet making it practically interchangeable with "inference" and "statistical learning"; in (Bontempi and Flauder 2015) "dependency" is countered to "causality" in

© Springer Nature Switzerland AG 2018
L. Rozonoer et al. (Eds.): Braverman Readings in Machine Learning, LNAI 11100, pp. 148–186, 2018.
https://doi.org/10.1007/978-3-319-99492-5_7

(Efron and Hastie 2016) statistical learning and inference are tightly intermixed, and so on. Dependency modeling include such approaches as regression analysis of any kind, discriminant analysis, mixed models, decision trees, Bayesian models of different types, pattern recognition, neural networks and other large groups of methods. It is often, under a different angle, could be identified with supervised learning, yet such fields like structural equations modeling (SEM) and causal modeling, which are not usually considered a part of statistical learning, also go into this category.

The importance of the distinction between causal and not is quite obvious. When one does anything with data – classifies, makes convenient groups, create hypotheses about "factors" and "clusters" and so on – they basically answer the question *what* (is going on). But when dependent variables step into play, a different (causal) question is invoked - *why* (something happened). It's enough to write simple linear equation with Weight (Y) as a function of Height (X) with OLS estimates of the linking parameter – the next question immediately is something like: how strongly X affects Y; what else affects Y; is X the "real cause" of Y or are they just mutually correlated; can one use this equation to reliably forecast Y and so on. In that sense the difference between non-causal and causal statistics is approximately the same as the difference between collecting forensic data and making the right conclusion (based on it) about the criminal. Whatever good the forensic job has been done, without spotting the criminal it is useless. More accurately – data has great potential but until this potential has materialized in the final action of the "establishing strong dependence," it is just in its dreaming "latent" phase.

The most serious objection to this could be that: if dependency model has no causal character, but still works very well (like if system recognizes handwritten characters but has no idea why it does it, like in neural network) – what is bad with that? The answer is – nothing, it's great. But not all problems are like that. Too often causality is the only needed solution. Doctor cannot give a medicine, if doesn't understand the causal mechanism between drug and the illness. There are many studies which cannot be reproduced (Open Science Collaboration 2014, 2015), the whole "crisis of repro-ducibility" (Wasserstein et al. 2016; Johnson, et al. 2017) pointing to some deep problems in statistical not just the causal methodology.

Why do "traditional statisticians", regularly citing the mystical mantra that "cor-relation is not causation," still pursue the paradigm of dependency modeling without any hues to causality in countless books and papers? If these models are not causal – what are they? Do they mean something which is no less important than causal, just have no special name? If the dependent variable is perfectly approximated (or "ex-plained") by other variables but these explanations have no proven causal nature – is that a legitimate model, besides just predictive value, if any, in the best case?

Why statisticians, publishing within causal modeling framework, do not openly deny the traditional dependency modeling techniques and conclusions as non-adequate? Are causal models going to replace, say, traditional regression, due to their superiority or should they co-exist due to…what? Is it because they answer different questions and these questions are "equally important"? Or is it because they answer the same questions but in a different way and each answer has its own merit? Which one?

How do both, traditional methods and causality modeling, correspond with the general scientific super-quest for causes and effects of any phenomena? Do they follow its (un)written requirements or establish their own paradigm? Surprisingly, this question is rarely asked and even more rarely answered in a transparent manner.

How is the classical theory of statistical significance (or wider: statistical inference) related to causality? If something is "significant" – can one say it is "causal"? If yes – why is a new special causality theory needed? If no – what sense does that significance have? Or is there perhaps some middle ground? If so, what is it?

In my view, a clear answer to these and other questions is very important, especially now, when some feel "identity crisis" (http://simplystatistics.org/2015/10/29/the-statistics-identity-crisis-am-i-really-a-data-scientist/) for statistics in the era of Big Data, Deep Leaning, and other remarkable changes we all face. It would help to formulate the correct questions, shift focus to the right directions, and ultimately make statistics closer to its final goal, understanding of the complex uncertain situations for making decisions.

I cannot evenly consider all these problems for understandable reasons – it would require a volume of a large size. The main ***purpose of the paper*** is to attract attention to many conflicting ideas permeating the field of causal statistical literature and propose some new approach to it. In particular, a lot of space is designated to demonstrate the troubles of the dominating causality approaches, especially DAG theory, to show their limitations and underlying assumptions, which too often are ignored by both theorists and practitioners.

The proposed generative approach to dependency modeling somehow smooths ungrounded claims of the current causal modeling, but also goes beyond the approximations of the traditional dependency modeling. It makes an attempt to distinguish between causal and non-causal variables in some specific sense, which looks like a very important problem by itself. It still does not cover the real causal mechanisms, but at least allows us to be closer to their understanding. It may play a role of intermediator between useful causal simulation models and statistical data.

My general conclusion about the possibility to create coherent and non-contradictive theory of causation within just statistical framework is very pessimistic. The viable alternative is switching to agent based or other simulation models supported by machine learning – a topic, which is just outlined but not developed in this paper.

The paper skips technical details and discusses concepts lying in the foundations of the respective theories. Some important methodological questions about relations between statistical inference and causality, statistical inference and data analysis (in its direct sense, as expressed in Mirkin 2011) and others relevant to the topic are covered in (Mandel 2017a) and not touched in this paper. The remaining part organized as follows. Part 1 describes some motivating numerical examples with actual and artificial data, helping to clarify the main questions raised in the paper. Part 2 discusses the problem of causality from philosophical and statistical perspectives and critically considers the most popular approaches. Part 3 describes the author's approach to dependent modeling, combining causal and generative concepts. The conclusion summarizes the findings.

2   Lessons from the Ideal Causal Models

The *ideal causal statistical model* describes the process, which is clear and transparent from causal perspective, in a sense that practically no one would argue a nature of causal relations between its components. It allows clarifying some rather controversial concepts about statistics and causality. Energy provided by food is a good example. It is known, that it could be obtained from fats, proteins and carbohydrates, not from fibers or water – so, these ingredients are causes of energy. For illustration I used one dataset from British site http://www.weightlossresources.co.uk/ with 87 products and https://www.csun.edu/science/ref/spreadsheets/xls/nutrition.xls, an American site with 873 products (after some data cleaning). Then I generated an artificial data set (10,000 observations) simulating food-like relationships and made some experiments on it as well. Here, I would describe findings from all the three data sets (Food_S, for small food dataset; Food_L, for large food dataset and A_L for the artificial large dataset).

Complete and Incomplete Models; Determination and p-Values. In all datasets, the models were extremely simple: the linear regressions of the calories by volume of ingredients without the intercept. Table 1 summarizes results. Complete model yields the coefficients (9 for fats, 4 for carbohydrates and proteins), fully complying with those published in literature. But incomplete models need comments.

1. Fat's coefficients are more or less stable (with differences around 10–12% of the correct value) for any model. Carb's coefficients vary more significantly (15–20%); protein doesn't stand the test at all, its coefficients vary up to almost 3 times in any direction from 1.6 to 11.
2. Quality of models, in general, is very good by standard criteria. All coefficients but one, in column 3 for protein, have extremely low p-value – so, typically, it would be considered as a sign of "significance" of the coefficient and good quality of the model. Coefficient of determination in column 3 (82%) is very high; many statisticians would accept the p-value slightly exceeding the magic 0.05 level to say that it is a good model anyway; even more researchers would say that individual model

Table 1. Complete and incomplete models of energy produced by different products (Food_S; all linear regression models are calculated without intercept)

	Complete model	Fat - Carb	Fat - Protein	Carb - Protein	Individual models	
	1	2	3	4	5	6
R squared	100%	97%	83%	46%	Coefficients	R^2
Fat	8.95	9.94	9.42			9.84	82%
Carbohydrates	3.97	3.49		4.61	3.29	13%
Protein	4.02		1.64	11.13	8.84	22%
p-values for coefficients of regression						
Fat	0.0000	0.0000	0.0000		0.0000	
Carbohydrates	0.0000	0.0000		0.0000	0.0006	
Protein	0.0000		0.0857	0.0000	0.0000	

for fat with $R^2 = 82\%$ is very good, too. Model in column 4, with very low p-values and $R^2 = 46\%$ would be by many considered nice as well, because 46% is still much more than zero. So, by all standard criteria, there is a bunch of "good models" there, what p-values and R squared confidently justified.

3. Models like these are routinely made in practice. If, say, one doesn't know about the existence of carbs and has only data about fat and protein – her choice will be limited by models in column 3 or two individual models. And it is unclear what would be preferred – either pairwise model with doubtful (and far from true) protein coefficient or individual model for fat. In any case, the result will inflate contribution of fat. It means that even when relation between dependent and independent variables is surely linear and exhaustive, any *deviation from the complete model is immediately punished* by distortion of the correct values, while wrong models are strongly justified by both very low p-value and high enough determination.

4. These correct values have nothing to do with the "parameter" of certain theoretical distribution, which is allegedly the main target for estimating procedures in mathematical statistics. They are estimated from the physics of the food consumption, not from statistical theory, and could be measured directly with a scale and a calorimeter, just to be later confirmed by regression.

5. Data set Food_L allows to look at the ideal model with more confidence due to 10-fold increase in the sample size. The coefficients values and p-values were actually the same as before, with small differences (explained by measurement errors), but some other features turned out to be very interesting.

Figure 1 depicts the scatter plot of carbohydrates and energy for different products (similar charts are for fats and proteins). The amazing triangular form tells us that "predictive power" of each X variable is very good when X is large and horribly bad when X is small. This effect is called *heteroscedasticity* and is considered a serious impediment for statistical modeling but it doesn't actually alter the fact that relations between Y and X are causal and coefficients do not change in different domains of food ingredients' values.

Clearly, this weird shape was formed because two other ingredients play their role in the outcome. But it illustrates very vividly, that specific "influence" of each factor, considered separately, *even when we know for sure that it is causal*, may be interpreted misleadingly. Heteroscedasticity in partial models goes unnoticed in a complete model, it doesn't spoil anything there.

Sample Size, Sample Character and Measurement Errors. We may rightfully assume that causal relationship holds onto any subset of the data (contingent to measurement errors). To see how it works, I sorted the dataset Food_L by content of fat in ascending order and calculated the values of the regression coefficients in a cumulative style: first regression was calculated on the first 4 data points, started from the beginning (the product with minimal fat content); second – on five points, and so on until the end. Results are presented in Fig. 2 in two versions: solid lines show the coefficients for cumulative data; dotted lines show the same coefficients but calculated starting from the data point at which fat is not equal to zero, i.e. from the point 240. Interesting conclusions could be drawn from this chart.

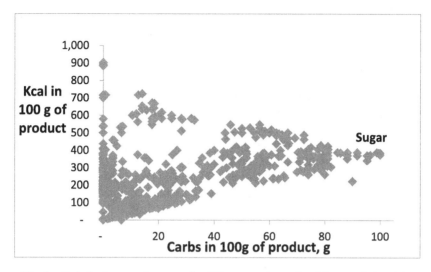

Fig. 1. Carbohydrates and energy for different products; Food_L; correlation 0.30

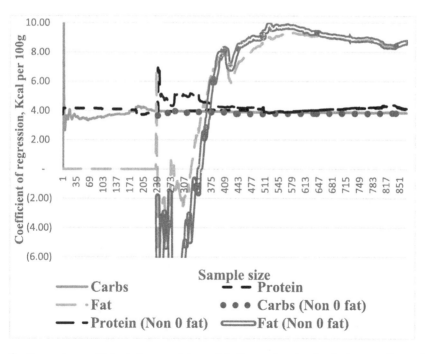

Fig. 2. Regression coefficients for complete model after sorting data by fat in ascending order; Food_L

1. The negative value of the fat's coefficients (the Y axe was cut off for simplicity sake) reaches −11; it becomes positive after only 90 points with non-zero fat values. It's not a small sample size by any standard. The convergence to its proper value (gray solid line) happened just on the end of the data. Importantly, one cannot attribute this high variability to the presence of many zeroes: the dotted line shows practically the same pattern and converges at the same point.
2. In absence of fat (a very good situation for any causal modeling algorithm which likes to fix values as much as possible – see Sect. 3.3) two other coefficients on the left part of the chart demonstrate different level of convergence – carbs have small but constant bias, protein shows deviation about 17% of 70 observations. Protein, however, diverges later to the very end up to 8%. Protein, measured in non-zero subset (dotted brown line), does not converge for the long time, with differences up to 26%.

In general, those results give a rather sad picture: even with "ideal complete model" coefficients strongly depend on sample size but not in a meaning of traditional sample theory (where sample 100 usually considered as fine) but on a very different scale. The same procedure after sorting by protein shows similar things:

- the coefficient for protein converges very slowly to the correct value, at about 740 observations, where first 113 are zeroes;
- coefficients for protein with or without zeroes converge from different directions and even with 550 observations still significantly differ from both the correct value and themselves;
- if one makes incorrect model with the intercept and all three variables, the protein's coefficient converges with about 600 sample size only!

Clearly the food data have different (yet small) inconsistencies and, undoubtedly, some measurement errors, which contributed into effects described above. In my mind, all these results are very telling nevertheless. But let's see how it works on artificial data, where the same logic is applied in fully controlled conditions.

The artificial dataset A_L contains 10,000 observations and three uncorrelated uniformly distributed positive random variables X with different average values. Each variable was spoiled by adding the ***measurement error (ME)***. To each individual value of Xi a random value was added. The random value was a normally distributed variable with zero mean and standard deviation, equal to particular fraction of Xi, varied from zero to 5% of Xi. Then variable Y was formed as a sum of products of original (before errors) X values and given "coefficients of influence" (4, 5 and 9). It means, that the real value of Y is a direct function of the proper values of X, exactly like in the food example above (calories would depend on the real fat content, regardless of errors with which this content is measured). But in turn, Y variable was also disturbed by adding its own errors in the same way as for X. The goal was to estimate the causal coefficients of influence using only disturbed data, both for Y and X, as it happens in real life. Figure 3 shows some results.

In this chart, each line shows how coefficient was changing when data were sorted by the respective variable (in the same manner as in Fig. 2). Only the coefficients for

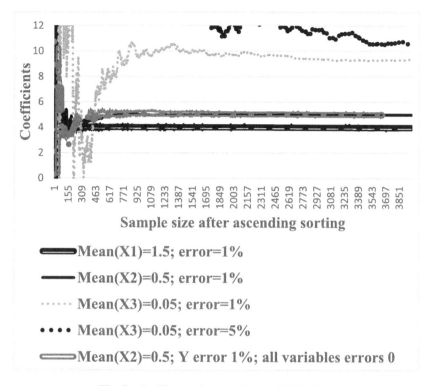

Fig. 3. Coefficients for complete model for A_L

variables used for sorting are depicted, i.e., the line for X1 shows the coefficients after sorting by X1, line for X2 – after sorting for X2 and so on. The results are interesting.

Three lines belong to variables with very low error, just 1% (a minimal level, taking place, I suppose, in any real data). Coefficients for X1 with maximal average value 1.5, are estimated correctly very fast. X2, with an average value of 0.5, varies much more and stabilizes only with 700 (!) observations.

If errors are a bit higher (5%), which is also a very realistic estimate for many observational data, then the divergence from the correct values is so large that coefficient for X3 does not reach it even with whole sample: 9.75 instead of 9 for 10,000 data points.

Possibly, the more surprising is the thick broken line. It shows the coefficients for X2, when all X variables have no errors at all and Y has just 1% error. After sorting for X2 the coefficient is stabilizing just after 550–570 observations, despite the fact that X2 has 10 times higher average value than X1.

Correlation in Regression Context. After simple manipulation with dataset A_L with zero error level for all variables (a lower - left chart in Fig. 4), I selected a subset of 620 data points with the following correlation matrix:

	X1	X2	X3
X2	0.56	1.00	
X3	**−0.95**	−0.67	1.00
Y	0.97	0.74	**−0.96**

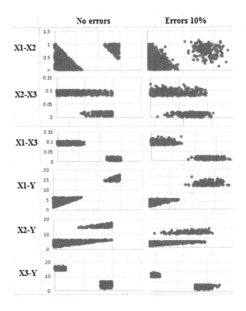

Fig. 4. Scatterplots of two datasets

By all classical criteria, data with such unusual features should disturb any model. Yet the *regression yields absolutely the same coefficients*, as were used in the data generations: 4, 5, and 9. It's remarkable for two reasons: high multicollinearity should have distorted coefficients, but it didn't; very strong negative correlation X3-Y should have led to a negative regression coefficient, but it didn't.

The instinctive reaction of any statistician would be to reduce the number of variables by taking off one of those highly correlated. The results are: if X3 is taken away, C1 = 3.81, C2 = 6.13, i.e. significantly different from the correct values, as was observed all the time with the food example. If X1 is excluded, then C2 = 14.9 and C3 = −6.6, i.e. values again are way off the truth. So, only a complete model works here again, i.e. it is best not to struggle with multicollinearity, but accept it as "neutral" in this situation. Intuition and common sense tell us that opposite signs of regression and correlation coefficients are not a good thing, and indeed, special methods are developed to avoid it (Lipovetsky and Conklin 2001, 2014). Yet intuition does not work here.

Regression and Clusters. The data in Fig. 4 demonstrate a clear cluster structure. If a researcher would see such sharply visible two clusters – it would be considered as a very interesting finding. The left and right panels of the chart are not substantively different. Most likely, some statisticians would incline to do models separately in each

Table 2. Regression coefficients for models in clusters and in the whole sample

	Correct model, no intercept			Models with intercept			
	x1	x2	x3	x1	x2	x3	Intercept
Data without errors							
Models in all data and in clusters	4.0	5.0	9.0	4.0	5.0	9.0	0.0
Data with 10% error							
All data	4.2	5.2	5.7	2.8	3.9	−39.2	5.2
Cluster 1	3.2	4.2	14.0	3.2	4.1	2.8	1.1
Cluster 2	3.9	5.5	62.0	0.7	2.0	−6.1	12.2

cluster with a hope to find different regularities and compare them. Some others would make all models – for the whole data and for each cluster; some would make mixed models. Table 2 summarizes results.

As one can see, the clustering did not affect the estimates in the ideal situation without any errors. Adding the errors, however, even as small as 10%, dramatically changes the results, yielding even negative values in the case with intercept. How may such sensitivity to small changes be acceptable in standard practice? Everyone would agree, that some kind of errors is always present in the data – so, an effect like that may appear in any concrete situation.

What, finally, the ideal model experiments tell us? The main findings from the consideration of the ideal models could be summarized in several statements.

1. The point of the "ideal model" was to give an example of the data where the real process behind the observed values has a proven physical (i.e. causal) understanding.
2. In this situation, the whole machinery of the linear regression, the most developed, probably, part of statistics from the times of F. Galton, is supposed to work ideally in any sense, and yet expected features were violated to a larger or smaller extent. Namely, *presence of any condition* below leads to a certain error:

- adding intercept, where it should not be;
- reducing the number of variables, even if high correlation leads to that;
- measurement errors, even as small as 1%;
- changing the sample size, especially making a model on a subset, started from minimal values of one X variable after sorting by this variable in ascending order;
- combining measurement errors and clustering.
  On the other hand, some *violations of the typical requirements* for a good regression do not make any harm – but only in an ideal situation of the complete model:
- an extremely high negative correlation between a variable and Y does not prevent a complete model to be correct, but makes the signs of regression and correlation coefficients different;

- multicollinearity does not necessarily prevent correct estimates, contrary to the common wisdom;
- the recommended elimination of one of the correlated variables distorts the entire model;
- heteroscedasticity or piecewise separation do not make the complete model wrong, but distorts incomplete or imprecise models.

3. Very small p-values signal the "non-random" results even if coefficients are very far from the correct values.
4. Freezing one variable on the same level (zero in Fig. 2) - an approach, lying in a basis of say, do-operators (see Sect. 3.2), instead of helping may make estimates worse.

Some of these and other aspects of regression were considered and analyzed in literature (see Berk 2004); each of them may find a mathematical explanation. For example: extensive literature on measurement errors in statistical modeling (Cheng and Van Ness 1999; Wansbeek and Meijer 2000; Viswanathan 2005; Carroll et al. 2006; Buonaccorsi 2010; Masiuk et al. 2017) covers the topic comprehensively; in (Lipovetsky and Conklin 2001, 2014) the problem of "matching" the signs of the coefficients of regression and correlation is carefully considered in combination with multicollinearity effects; importance of the "small p-values", which was a beacon of the traditional statistical thinking and practice for decades, was debunked by many specialists (Wasserstein and Lazar 2016; discussion and references in Mandel 2017a); there is a big literature about combining clustering and regression, or any other predictive modeling techniques (Hastie et al. 2009 provides references). But many effects mentioned here, in my view, have very strong combined sobering power – it is clear, that all these techniques may easily fail, even if the real relationship between Y and X is firmly established. In fact, *only combination "complete and correct model – ideal data" guarantees perfect results*, and bad thing is, that any deviation from that has unpredictable consequences, not just "everything is approximately correct". No, it's not.

Regression models lie at the heart of practically any approach to causal modeling. For that simple reason any unpleasant features of the elementary regression should be just multiplied in much more complicated causal models.

3   Causal Models in Statistics

3.1   General Ideas and Definitions

Causality as a philosophical notion could be traced back to the earliest writings available; at least in Bible, the "cause" has about 170 appearances. An update overview of the field, with different levels of bias towards one or another concept, and important references could be found in many books (Kistler 2006; Pearl 2009; Groff 2008; Bunge 2009; Berzuini 2012; Illari and Russo 2014; Morgan 2014; Morgan and Winship 2014; VanderWeele 2015; Imbens and Rubin 2015; Hoover 2016 and others) and very detailed periodically updated papers published in Stanford Encyclopedia of Philosophy (Dowe 2007; Menzies 2014; Hitchcock 2010 and others). I focus primarily only on main concepts of statistical causality, which represent the mainstream in certain sense and would go to wider philosophical issues only when it is clearly necessary.

These main approaches are: *simultaneous structural equations (SEM)* (historically the first, derived from path-analysis by S. Wright in 20-s (Kline 2011)); *potential outcomes (PO)*, proposed methodologically by J. Neyman in 20-s and technically developed by D. Rubin and others from 70-s (Rubin 2006; Imbens and Rubin 2015); and the concept of *Directed Acyclic Graphs (DAG)* developed by J. Pearl and others (Pearl 2009; Spirtes et al. 2001) from 80-s and tightly related to Bayesian networks (popular terms "DAG" and "DAG theory" are used below interchangeably). This last theory was presented recently in a compact form in the book (Pearl et al. 2016) which I would actively quote below under the abbreviation PGJ. Similar classification of approaches (without PO) was given, for example, in (Hoover 2016) for economics; some different, yet affiliated with mentioned branches were developed also by Robins (see Geenland et al. 1999; Dawid 1979; Berzuini et al. 2010), Mansky and other researchers; a little bit aside stays "Granger's causality" theory; recent developments try to merge causal theories with statistical learning (Scholkopf 2012). There is an opinion, defended prominently by J. Pearl, that almost all of these approaches in fact talk about the same things, but use different languages. This thesis was neither supported by others, nor found any strong opposition (see Sect. 3.3).

Statistical literature usually either does not give a *clear definition of the cause* or have obvious difficulties when it is trying to; I considered this issue in details in (Mandel 2017a). Maybe, the vagueness of the notion "causality" is one of the main reasons why some prominent scientists do not really like it. There is a famous saying by Bernard Russell, more than one hundred years ago: *"The law of causality, I believe, like much that passes muster among philosophers, is a relic of a bygone age, surviving, like the monarchy, only because it is erroneously supposed to do no harm"* (quoted from Dowe 2007). L. Zadeh once noted that he does not see any way to define causality strongly, and for that reason, cannot see a possible theory for causality (Zadeh 2001). In particular, he brings the following example: *a friend of mine calls me up on the telephone and asks me to drive over and visit him. While driving over, I ignore a stop sign and drive through an intersection. Another driver hits me, and I die. Who caused my death?* Out of all four chained events which caused the death – *friend's call, car riding, ignoring a stop sign, and being hit by a car* - the argument goes, one cannot formally separate "real causes" from "fictitious" ones, and so no theory is possible.

Just to illustrate how differently the notion of causality can be understood, I list several concepts in Table 3. It's easy to see that items there are not completely compatible with each other: some are more comprehensive than the others. The main divisions among concepts are the following.

1. *The actuality*: causes as something really happened vs causes conditioned on something which has not happened, i.e. hypothetical. The first is defended strongly by concepts in rows 3, 7, 8, 9; the second – in rows 4 and partly 10 (with regard to information transfer). Concepts in rows 1, 2, 3, 5, 6 are rather neutral over this issue, for they do not specify it and they could be applied either to real or logical processes.

2. *The singularity*: is theory applicable to each single event or to many events of certain types. The first is presented by all rows, while the second – by rows 3 and 5,

Table 3. Some definitions and concepts of causality

	Title	Content	References
1	General definition	Causality is the agency... that connects one process (the cause) with another process or state (the effect), where the first is understood to be partly responsible for the second	https://en.wikipedia.org/wiki/Causality
2	Efficient cause	Something what brings a thing into being (parents are efficient cause of children)	Aristotle
3	Regularity theory	A is cause of B if B typically follows A, what is proven by our experience	D. Hume http://www.iep.utm.edu/hume-cau/
4	Counterfactual theory	An event E causally depends on C iff (i) if C had occurred, then E would have occurred, and (ii) if C had not occurred, then E would not have occurred	G. Galileo; D. Hume; D. Lewis (Lewis 1973; Bunge 2009; Menzies 2014)
5	Probabilistic theory	A probabilistically causes B, if the information that A occurred increases the likelihood of B occurrence	H, Reichenbach; P. Suppes (Good 1983; Hitchcock 2010)
6	Manipulation theory	X causes Y only if one can change X to affect Y	R. Collingwood https://en.wikipedia.org/wiki/Causality
7	Process (Transmission) theory	Causal process is something changing the world; pseudo-process (like moving shadow) does not	B. Russell; W. Salmon (Dowe 2007)
8	Physical connections (i.e. conserved quantity theory)	Interactions of the "world lines" assuming some physical connections. Different things are exchanged in the process: properties; tropes; structures; conserved quantities	J. Aronson; D. Fair; Dowe; D. Ehring; H. Castaneda (Dowe 2007)
9	Connectivity via forces theory	Forces are "species of causal relations"; causation cannot be materialized without them	J. Bigelow, B. Ellis, R. Pargetter (Bigelow et al. 1988)
10	Causality mechanisms	Causal mechanisms are "ultimately unobservable physical, social, or psychological processes through which agents ... operate ... to transfer energy, information, or matter to other entities"	A. Bennet (Bennet 2008)

with the disclaimer that probability is not to be attached to each individual event; it is disputable though in Bayesian and other approaches.

3. **The substance**: is it something "tangible" vs "anything". The first is supported in rows 7–10, the second – in rows 2–6.

The further discussion will be limited by the following:

(a) It will be considering only three mainstream lines of statistical modeling mentioned above (SEM, DAG and PO). I will consider SEM and DAG theories together for one important reason – it was convincingly shown, that, on the one hand, there are no contradictions in applying DAG theory to historically much older SEM in order to make it "more causal" (see Sect. 3.3) and, on the other hand, there is nothing specifically causal in classic SEM that doesn't exist in DAG theory (Pearl 2012).

(b) I will touch only basic premises of the theories, without technical analysis, assuming that if there is a problem in the basic stuff – it affects the whole theory;

(c) I will go primarily to causal problems over observational data, considering experimental data only when needed.

The main feature which makes DAG and PO approaches at least similar to each other is their deep orientation to the **counterfactual theory of causation**. In PO it is the core of the entire approach; in DAG – one of the cornerstones, together with others (like do-calculus). For that reason, discussion of this concept would be applicable to both and it makes sense to start with this important topic. Here I just briefly summarize main findings of my previous analysis of the topic (Mandel 2017a, Sect. 3.2; see also skow 2013; Dawid 2009 and others):

1. The science successfully worked **without counterfactual assumptions** for centuries and made as big successes as it did so far – so, no need to overturn this path.

2. Some counterfactual statements (CS) one can imagine comparatively easy (what would happen if I take the next turn on the road instead of the one I took in reality), but others one **cannot** (what would have been my reaction to an assault had I been a woman, while I'm a man). The boundary between the two classes is rather fuzzy, if any.

3. A CS claim can be applied not only to statistically "similar" situations, but literally to the very **individual cases**, to the "...*very juncture of history*" (PGJ, p. 91). In my mind, it is self-denouncing – I cannot imagine, what would happen with such an exercise, although this kind of statements is often used in daily talks.

4. CS of certain types **cannot be independent** on others CS. Why the gender CS above in 2 faces this intuitive repulsion? Because, presumably, women react differently than men to too many things, and making CS about gender means to make many other CS about other aspects - this CS is organically connected with others (not in a clear way).

5. Counterfactual statements **exclude time** from consideration. Say, in reality I hit the glass by hand, thus it has fallen and broken. CS: if I didn't hit the glass, it would not be broken. The first 2 min (after what moment?) glass is where it was – so, CS is true. But if 1,000 years passed, the table becomes rotten and glass will be broken anyway.

6. CS are not substantially different from **alternative history** narratives, which could be whatever but science, a commonly accepted view. Point 4 above is here especially relevant – so many other "alternative" CS should be produced to make any conclusion about original CS plausible (imagine the chain of events to assume that Hitler won the WWII in order to recover a "possible future").

7. It is not exactly clear, how CS are applied to **numerical variables:** what is a CS alternative to the normal body temperature 36.6 C? 40? 34? 34.2?).

8. There are many more theoretical issues, which are hard to express briefly here, but among those I would stress just one important thing. Counterfactual logic found an unexpected support from the general philosophical perspective. In J. Pearl's words *"This quasi-deterministic functional model mirrors Laplace's conception of nature..., according to which all nature's laws are deterministic, and randomness surfaces merely due to our ignorance of the underlying boundary conditions. (The structural equation models used in economics, biology, and stochastic control are typical examples of Laplacian models.) Dawid detests this conception."* http:// bayes.cs.ucla.edu/r269-web.html. Well, not only he (see Dawid 2009). The view that *"randomness surfaces merely due to our ignorance"* (see also Pearl 2009, p. 26) is weird to hear in our days (see Mandel 2011). However, this view is further advanced in the whole DAG theory.

3.2   Some Problems and Drawbacks of the DAG Theory

A theory assumes that relationship between many causally related variables are to be presented as the graph, where each node represents the variable and arrow – the direction of the causal relation. The graph should be acyclic, i.e. do not have loops. Each arrow has the assigned value of the causal coefficient. What differentiates DAG from SEM (also having graph like that), it seems, is the following:

- new ways to explore a **structure of the causal graph** (to find chains of causally related variables);
- new ways to intervene the graph precisely with the purpose to reveal its causal structure; it is assumed, that SEM themselves are not "causal enough". These new operations are *"surgery", "do-operators"*, and **counterfactual** values applications.

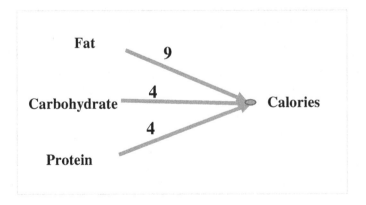

Fig. 5. A simple DAG chart for food

I would not consider here the exploration of the causal graph structure (one of the most, if not the most important topics of DAG theory) for the very simple reason: I'm interested just in simplest graphs with one node and many arrows (like one in Fig. 5 on a food example). If something does not work in a simple case – I don't see a reason to go to more complicated things.

In the DAG context, **Surgery** means taking off some arrows from the graph (and, perhaps, replacing them by others) in order to test some hypothesis about the model. Surgery by itself is a legitimate procedure, which just means, that researcher tests different models with different variables and different graph's topology. It was widely practiced in SEM and in statistics in general. Applying to the simplest structure with one target node, surgery would mean the elimination of some input variables. I showed already what difference it makes for food data in part 1 – coefficients could be seriously changed. DAG theory gives no tools to address this simple situation, however.

The third type of intervention, **counterfactual assumption**, was briefly considered in Sect. 3.1. Let's stop now on do-operators, the key element of the DAG theory.

Do-operators and Index Numbers. *Do-operator* is, perhaps, the most innovative element in DAG theory. "*Using do-expressions and graph surgery, we can begin to untangle the causal relationships from the correlative*" (PGJ, p. 55). It is defined as follows (PGJ, p. 55): "*$P(Y = y|do(X = x))$ is the probability that $Y = y$ when we intervene to make $X = x$. ... $P(Y = y|do(X = x))$ represents the population distribution of Y if everyone in the population had their X value fixed at x.*" Then it is illustrated on the example on p. 57, referring to the data about the Simpson paradox in Table 1.1 on p. 2. The table shows all relevant statistics for men and women regarding the drug and demonstrates that while the recovery rate for both men and women is higher for drug takers than for others, for the whole population it is not. To "untangle" this paradox, consider two quantities, one for $X = 1$ (drug was taken) and another for $X = 0$:

$$P^1 = P(Y = 1 \mid do(X = 1)) = R_M^1 \alpha + R_F^1(1 - a) \qquad (1)$$

$$P^0 = P(Y = 1 \mid do(X = 0)) = R_M^0 \alpha + R_F^0(1 - a), \qquad (2)$$

where P^1 is the probability of recovery in the whole population assuming that everyone would have a recovery rate of the drug-takers; P^0 – the same probability, assuming that everyone has a rate of non-drug-takers; the left part is a do-operator for condition $Y = 1$, i.e. for these recovered; R_M^0, R_F^0 - recovery rates for males and females if they didn't take drugs (with upper index 1 – for those who did); a - percent of Males in the total sample.

If $P^1 - P^0 > 0$ (as in the example) – it means, that taking drug is advantageous for the entire population and one should ignore the actual recovery rate for the sample. The paradox is considered to be untangled by causal means, i.e. by enforced (*doing* instead of conditioning or *seeing*) assignment of specific values to the data: one rate was forced to be everywhere in (1) and another – everywhere in (2).

From these do-formulas easily follows:

$$\Delta P = P^1 - P^0 = \alpha(R_M^1 - R_M^0) + (1-a)(R_F^1 - R_F^0) = \alpha\Delta_M + (1-a)\Delta_F, \quad (3)$$

where $\Delta_M = R_M^1 - R_M^0; \Delta_F = R_F^1 - R_F^0$ are the differences in rates for drug takers and non-takers for men and women respectively, i.e. (3) is the weighted average of the two differences for the whole population (the weights are proportions of men and women).

Let's look closer at the source of the "paradox". Average recovery rate for takers $\overline{R}^1 = \beta R_M^1 + (1-\beta)R_F^1$, and for non-takers $\overline{R}^0 = \lambda R_M^0 + (1-\lambda)R_F^0$, where β and λ are the fraction of males in groups of drug-takers and non-takers, respectively. Subtracting the second quantity from the first and regrouping gives the average rates difference for the entire population:

$$\Delta\overline{R} = \overline{R}^1 - \overline{R}^0 = \Delta_F + \beta R_M^1 - \lambda R_M^0 - \beta R_F^1 + \lambda R_F^0 \quad (4)$$

Assuming that $\beta = \lambda$, we obtain

$$\Delta\overline{R} = \beta\Delta_M + (1-\beta)\Delta_F, \quad (5)$$

and assuming further that $\beta = \alpha$, (5) would become equal to (3), i.e. $\Delta P = \Delta\overline{R}$. These calculations help us clarify the meaning of the do-operators.

(a) Substantially, do-operator changes values of some variable (the rate in this example) and keeps other variables intact (the structure of the population, or subgroup weights). It is done in order to contrast two artificially created indicators (what if all have this rate vs what if all have another rate), none of which was actually observed As a methodological tool it is one of the oldest in statistics and lies in the basis of the *index numbers* theory.

The whole idea of index number is to make some weights (classically, a volume of production) fixed on one level while letting the variable of interest change over time or space. For instance, Laspeyres price index is:

$$P_L = \frac{\sum\limits_{i=1,N} p_i^1 q_i^0}{\sum\limits_{i=1,N} p_i^0 q_i^0}, \quad (6)$$

where individual prices p in the numerator are from the later period 1 and quantities q are from period 0, while in the denominator both values are from period 0. The idea is to see, what happens if old quantities in period 1 remain the same as in period 0, but only prices are changed. This strong assumption, by the way, is followed by the next one, that the old structure of goods also remains in period 1 – which is, in life, also unreal. So, the numerator here is some imaginary quantity not only because actual quantities are different in two periods, but also because the structure of goods is different. Yet the index is a very useful tool to say, what "on average" happened to prices.

These techniques are known at least from 1823 (J. Lowes index) and routinely used everywhere (Consumer Price 2004; Ralph et al. 2015 and references therein). So, methodologically, the comparison of the two do-operators does the same thing – fix one value and change another and look for the difference. It is not bad by itself, I just don't see why it should be called causal. It seems, that relation between do-operators and index numbers is not perceived as such in DAG theory (my long email discussion with a moderator of a J. Pearl's causality blog A. Forney showed it clearly), although, perhaps, it could be very helpful.

(b) It is clear from (3) and (5) that the final conclusion about drug efficiency (the real causal inquiry of interest) is a function of the coefficient α, with given rate difference. The results vary drastically depending on the gender distribution in a sample. If $\Delta_M = -2\%, \Delta_F = 5\%$, then if $\alpha = 42\%$, the total effect $\Delta P = 1.64\%$, but if $\alpha = 75\%$, $\Delta P = -1.00\%$, i.e. the sign changes. What would be recommended to the doctors by the statistician, equipped with this magical do-operator and having these results from two different studies, the only difference between which is in the different gender proportions?

The Simpson paradox, by definition, means the following:

$$sign(\Delta_M) = sign(\Delta_F) \neq sign(\Delta R) \tag{7}$$

Only in the special case of the full equality of all weights in the whole sample and in subgroups Eq. (4) becomes (5) and Simpson paradox becomes impossible (and do-operators lose any meaning). Otherwise, if one wants to test analytically, in which combination of the values the requirement (7) holds, she would have to check numerous possibilities in (4) to make (7) true. There is nothing causal in it. Ironically, also without any reference to the causality, the "paradox" was resolved long time ago, and again with the index numbers (Lipovetsky and Conclin 2006); see more in (Mandel 2017a).

To summarize: the ***do-operators could be considered as a special class of index number-like procedures***, to measure the changes of certain imaginary quantities, but not directed at estimation of the actual forces causing the changes. Similar interpretation can be applied to potential outcomes, too (see Zagar et al. (2017); I. Lipkovich, private correspondence).

Coming back to ***counterfactual interventions*** used in DAG, they are similar to do-operators except the type of values used: do-operators replace the observed values by imaginary ones, whereas do-operators in counterfactual setting do the same for values which cannot be observed. In both cases, the resulting quantity is as imaginary, as the input is, and real causes escape from capture. The idea, that cause is something which could be felt only as a contrast between the real and imaginable is interesting – however, it is not what the cause is (see part 3).

3.3 DAG and PO (Potential Outcomes) Theories – Similarities and Differences

A detailed analyses of differences and similarities of two leading causal statistical theories is given in (Mandel 2017a). Here I just briefly sketch main points without going into technical details.

1. The two approaches, in fact, address *different problems*. The DAG is very general, it assumes the presence of complicated chains of causally related variables, clearly within the old paradigm of path analysis and SEM. Its main results are in area of finding the chains of effective influences in a graph (*d-separation, back door criteria* and so on), which in certain sense also a topic for graph theory (Bang-Jensen and Gutin 2009). PO has nothing to do with that. Its main purpose is to improve an experimental design, which is already causal, to test the hypothesis that treatment was conducted randomly indeed. If an ideal experimental design will be ever possible, PO would have nothing to do with that. There is a completely different situation with DAG: even with this ideal situation, its task would not disappear, for one should anyway test how one cause affects another one, how the causes "collide", and so on.

2. PO approach is based *entirely* on a counterfactual idea, while DAG works both with CS and realistic statements. DAG assumes manipulation over the graph's nodes and arrows. In PO the whole idea is to find a proper estimation of the counterfactual values, which are treated as a specific missing values problem.

3. The ways of estimating counterfactual values in the two approaches, are strikingly different. In DAG it is a simple manipulation with the existing graph. Do-calculus simulates the situation of intervention when SEM are recalculated under an assumption of constancy of the variable of interest, $X = x$: "*...using the same operation in a slightly different context, we can use SEMs to define what counterfactuals stand for*" (PGJ, p. 91).

   The completely different situation is in PO approach. It doesn't use any graphs to begin with, and relies only on the existing data to estimate "missing" counterfactual values. Either it is propensity scores or different weighting balancing techniques – it is much more "statistical" in the traditional meaning of the word.

4. Counterfactual values in DAG are determined by the values of its other variables and the model, while in PO it is a function of a complex statistical procedure, like propensity scores, matching and grouping, which involves many aspects, personal decisions, settings parameters for the models, etc. It would be very interesting to see the results of *comparison* of the two ways to estimate the counterfactual values on the same artificially controlled data – but, I could never find one. It is rather strange, considering the huge volume of literature on the subject.

5. PO treats CS as a very natural way to make a causal statement and *uses CS in model parameters estimation*. DAG treats CS as a possible version of the event and doesn't use in estimation of the model's parameters, which are considered given. CS is used here only for estimation of the "causal effects". It easily considers graphs with both actual and counterfactual values – "everything is possible", it says, let's just play a game. Even a nice Lewis' term "*cross worlds probability*" is used as if no problems with it existed at all (PGJ, p. 98). Indeed, in DAG the main orientation is to use CS as a tool for "what if" scenario analytics. It easily takes questions like "*what would be if this and that has changed*" and recalculates the outcome, i.e. the entire graph, accordingly. In PO, on the other hand, CS are made once and forever; they are assumed as unknown values to be estimated via the model. In DAG, CS is a tool for model's *manipulation*. This, I guess, explains attractiveness of DAG theory for social scientists –the easiness with which graphs

are manipulated is a golden key for answering many "what if" questions the policymakers are accustomed to ask.

6. These differences between DAG and PO, possibly, were the reason, why PO is much more oriented toward **traditional statistical practice** (theory of experiments and so on), while DAG delegitimizes it in a certain way, trying to give a different meaning to statistics. The virtual lack of a statistical type of thinking and tools in all DAG theory related materials has far-reaching consequences (see 7–9 below).

7. Regression as a tool for **parameter estimation** has many drawbacks – multi-collinearity, nonlinearity, heteroscedasticity, confounding variables, heterogeneity, outliers, and so on (see part 1 and references therein). But for the DAG theory all that doesn't exist. This issue typically is not discussed; wrong coefficients would be just taken and manipulated with as they were the correct ones.

8. This creates a paradoxical situation of **non-distinction between causal and non-causal relationships**. Regression itself is not a causal tool per se, this is commonly accepted. But the same regression estimates in SEM framework, after application of DAG theory, should be miraculously transformed into correct causal estimates. How may it happen? That represents an interesting mental gap when causality is considered as a feature of only complex systems, but it disappears in very "simple" ones, with just one effect of interest. This view is completely opposite to PO and to mainstream statistics. Example: my modest statement that regression coefficient is equal to causal coefficient in a setting when all X variables are exogenous, was discussed on ASA blog (Mandel 2017b), where, expectedly, this statement was just confirmed. Yet the very same statement was not allowed to appear on the causality blog http://causality.cs.ucla.edu/blog/. Bayesian software (Conrady and Jouffe 2015), a commercial product using DAG theory, makes things at least more transparent: when it comes to a graph like one on Fig. 5, it yields the simple regression as a "causal results".

Another, even more important aspect of this non-distinction was clearly expressed in my direct discussion with Dr. J. Pearl. To my question *"Can a DAG distinguish variables with zero causal effects (on Y) from those having non-zero effects"* the answer was *"Of course not, no method in the world can do that without further assumption"* causality.cs.ucla.edu/blog/index.php/2018/01/24/can-dags-do-the-un-doable/. As follows from his further comments there, it is assumed that the causal structure is given (what implies the automatic validity of the relevant regression estimates, etc.). But in my mind this problem – **what is causal and what is not** – is a key issue of the entire theory, and bypassing it means throwing the baby out with the bathwater (see more in Sect. 4).

9. All structural equations are treated in DAG at the face value as if they are just usual equations from the textbook of linear algebra; no statistical features are required. Two X variables with different coefficients of correlation with Y will be treated identically, when their values are made fixed for the do-operators.

10. **Propensity scores matching,** a most popular tool for estimating potential outcome coefficients, has one intrinsic problem (Mandel 2011, 2013, 2017b). For calculation one should build a logistic regression, where the dependent variable is a treatment, and independent ones – confounded covariates. The probabilities of each unit to

belong to the treated group - propensity scores - then matched among those who received treatment and those who did not. Let us assume, that R^2 (or any other goodness-of-fit statistic) of the logistic regression is close to 1. It means that covariates predict propensity scores ideally, i.e. treatment is completely determined by those covariates. Respectively, the units would have the same propensity scores, and matching will be almost ideal. But it also means that covariates determine the treatment mechanism – i.e. it is the *worst* scenario possible, because treatment should be, by design, completely random. If, on the other hand, $R^2 = 0$, the matching will be completely random (i.e. useless), but treatment – the *best* possible. So, in one case we sacrifice the matching, in the other – the treatment. The question is: what are we trying to obtain with this process, where optimum criteria could be neither max nor min?

11. Both theories do not seriously consider one important question, *"what type of "no effect" data is needed to make a causal statement"*, or what is the "control group"? This is one of the very big issues in statistics in general, in both observational and experimental studies: a *huge asymmetry* between "target" and "non-target outcomes". In marketing, for example, it is never clear what to put into the "control sample" against, say, *"Mercedes buyers"*. All rich? All urban residents? Whole population? But depending on the answer the whole model will be completely different.

12. *Causal mechanisms* – seemingly the core of the DAG approach, the graphs themselves – are not the real mechanisms. More precisely – the DAG theory does not know exactly what they are. One thing is to "test the hypothesis" about causal inference, another – to draw the real mechanism structure. I consider this problem deeper in Sect. 4.1.

13. The two considered approaches (and others, for this matter) are not seriously *computationally tested*, in contrast to statistical learning algorithms which obtained tremendous success for that reason in the last decades. In (Craycroft 2016) author reports positive results about propensity scores, but in (King and Nielsen 2016) the results are negative; (Kaplan and Chen 2011), using special Bayesian version of the propensity scores, make some mixed conclusions. In (Li et al. 2016) results of the testing of some aspects of the DAG approach are also unclear, with several negative conclusions. Under the good testing I understand either some kind of direct comparison of the results on datasets with known features, following strong testing methodology (like the one proposed in Imai and Tingley (2012)), or making artificial model of the data generation with known parameters and then recovering these parameters using PO and DAG theories, respectively (similarly to what I do in Sect. 4.2). It would resolve, hopefully, long-standing issues not only about the comparative validity of PO and DAG theories but much more general questions about the plausibility of counterfactuals, interventions, predictions and so on. Until it is done, it's hard to consider the whole concept of causality modeling as a useful practical tool. Comparison of two most influential approaches was and remains a debatable issue both theoretically and empirically. The latest book (Peters et al. 2017), presenting new promising approach to causality, also carefully leaves this question open.

4   Causality and the Concept of the Effect Generation

I propose a concept which is not causality theory per se, but rather a specific view on the nature of relationships between different aspects of the universe in a process of its modeling, which has causality aspects in it, but yet denies the possibility of the universal causality statistical theory. It may, however, help in creating a real causal explanation in some cases. It was described in general in Mandel (2013); technically developed in Lipovetsky and Mandel (2015a, b), applied to marketing problems in Mandel (2016), Dodson and Mandel (2015); and advanced in Mandel (2017a).

4.1   What the Actual Causes Are?

What are the relations, which could be called causal? I'll try to answer this question, orienting to several non-counterfactual theories of causation listed in Sect. 3.1.

1. The only thing, capable of changing anything in this world, is the *force*. Definitions of force in physics as *"any interaction that, when unopposed, will change the motion of an object"* or *"a power that causes an object to move or that changes movement"* have some mechanistic overtones but capture the essence. I have no need to list different specific forces in physics. It's enough to postulate a **hypothesis, that all actual changes in the physical world are ultimately caused by some forces** applied to certain objects. In the very detailed account of the topic forces are called *"species of the causal relation itself, and as such have a different ontological status from the sorts of entities normally considered to be related as causes to effects"* (Bigelow et al. 1988). I agree with that and just add that the claim also is applicable to forces in **biology, psychology, and social life**. Are they also "forces", to begin with, or something else? In my mind – yes, I don't see any principal difference between physical and social forces in respect to the causality. The very terminology used in expressions like *"economic forces swept over the traditional habits"*, *"the law enforced the court ruling"* and in countless others show that this understanding is very natural.

2. A shift from the causes to the forces has a serious advantage – force, at least in physics, is a well-developed notion. Sociophysics, econophysics and mediaphysics (Kuznetsov and Mandel 2007) show how classical physical concepts like forces, fields, energy, etc. can be organically applied in biology, sociology, and economics. *"Science requires causation... Forces are among the causal relations discovered by science. (We leave it an open question whether there are other species of causal relations in addition to forces, either within science or outside it.)"* – was said 30 years ago (Bigelow et al. 1988). Maybe, it is still an open theoretical question, but for my pragmatic purposes I didn't find any other "causation species".

3. The fact that only certain forces can finally cause changes in the world is helpful in another important aspect: it **eliminates the need for the counterfactual approach** to causality. If one wants to make any theory of causation – she should demonstrate how to measure actual forces and how they influence the effect. This problem, in general, is of very high complexity. The huge success in physics, chemistry and especially engineering was achieved only because all working forces were

investigated in isolation, in very simple situations where few (usually 1 to 3) forces worked simultaneously. Any serious deviation from these simple situations almost inevitably changes results, usually in unpredictable way.

4. Usually, people are interested in causes for two types of things: *events* and *processes*. Some examples of events (car accident, shattered bottles and glasses) were considered earlier. A typical example of the process was a food one; similarly, it could be "causes of economic decline", "causes of poor performance" and so on. Very often statistical data have static, not dynamic nature, like income measured in many states for one particular year, and still, the causal question is asked, like: "why states have very different income". Yet it doesn't change the point: the process took place earlier, but was not, perhaps, measured; it produced just "cumulative result", which is also a subject to causal questioning. In fact, it's hard to draw an accurate borderline between these two types of things (each event, ultimately, is a certain point of some process), but I still leave this convenient distinction.

Events are associated with *categorical* (usually binary) *variables*, they either happened or not. Behind each event, there is some causal tree, like in Zadeh's example in Sect. 3.1. In a data set, respectively, one may assume that the number of trees is equal to the number of the observations, and their equality to each other is by no means guaranteed (people in road accidents die for different causes with different causal trees). It means that statistical casual theory should ideally work with a set of the trees in order to cover the whole story. But the radical simplification may help if the concept of the only *one last force* will be applied. It means, the causal tree is cut just before the event, leaving just one "last cause" for consideration. Physically, it is not inappropriate: only last force applied makes any event to happen. In the case of "death in car accident" the very moment of death was the hitting of some vital body's organ by a piece of metal or plastic. This kind of things is usually stated after an autopsy. Using the one last cause would eliminate a need in anything like DAGs, because the primary goal of this hypothetical theory would be to find and measure only last forces for specific events. If one needs to go somewhat deeper – of course, she can go further on the tree and find next level of the forces, etc. The net of causal events would be reconstructed as a chain of the last forces for each one.

5. Processes, unlike events, typically are measured with *numerical variables*. Consumption of food and associated calories cannot be deconstructed into a chain of events with the "last cause" – it is a process, and all ingredients, at least for practical purposes, are consumed simultaneously. If in the events there is one last cause – here all causes are distinguishable not by the time of making an effect, but only by their relative contributions. Let us assume these processes *additive*. Additivity does not necessary mean linearity or even monotonicity; it just means that effect has cumulative nature, depending somehow on all causing factors working at the same time. These causes are, however, also "last" but in a different sense: we consider protein as a cause of energy, but do not look into the next "why" layer – why this amount of protein is in this piece of meat. The same eternal chain of causes as for events can be built for additive processes.

6. The relevant data set for analyzing events may be just cases with different X variables, representing the situations, which may trigger or not certain events of a

given type. If, for example, one is looking for the causes of death on the roads (cases when Y = 1) and considers car accidents as one of the causes, then respective X is "hitting a car". If X = 1 – the car was hit, meaning, it may (but not necessarily) trigger the death, i.e. Y could be 0 or 1. It creates a special data set, which is different from the usual setting only in one respect: if X = 0, by definition, Y = 0 (no trigger – no possibility of Y = 1). This representation is equivalent to the description of the world, in which events of certain type happen for one and only one reason – but the reasons may be different. Or, in other words, it reflects the principle: *one effect may be obtained by different causes*, but if at least one of them has worked – other causes are irrelevant, with exact concordance to the last event principle.

7. *Additive* processes need a different approach. For example: 100 people dig the trench with shovels, working the different number of hours each with different individual intensity per hour. The total effect is a sum of individual contributions, which may or may not be approximated by a simple regression model as in part 1 - it depends on the difference in individual productivity. This example is very typical for all economic situations and is applicable to countries, regions, etc. Processes like that are often reduced to simple *multiplicative* function, like "*Total work = Number of work hours × Work per hour*", similarly to the food example.

8. The combination of events and additive processes is a very common phenomenon, too. Any established threshold transforms a certain process into an event: if say, we can talk about "causes of economic growth" meaning the process, the question about "causes of the economic growth exceeding 10%" may be considered as the event and treated respectively. If one stands on the position, that only triggered cause is considered in a model, it simplifies the required data representation significantly.

If one wants to consider causation seriously – one should assume an underlying process, at least in terms of the simplest classification above. Common practice that any model could be approximated by several equations of the additive type, where components have different nature and have nothing to do with the actual causal mechanism – the way how SEM and DAGs are organized via regression models – *departs from causality* much farther than it is usually perceived.

9. *Causal mechanism* of any process may work on individual and on the group's levels. I agree with A. Bennet, who defines "*...causal mechanisms as ultimately unobservable physical, social, or psychological processes through which agents with causal capacities operate in specific contexts to transfer energy, information, or matter to other entities.*" (Bennet 2008, p. 207). It merges the "force definition" of the causality accepted here with other "transformation theories" from the Table 3. What is important, it refers also to "*ultimately unobservable*" things, which, possibly, needs additional comment. Some of the processes are, of course, observable, and could be, at least in theory, measured, like in the food example (see also *causalisation* procedure in Sect. 4.3). But the typical situation is: "*...the scientific practice is to seek explanation at a lower level than the explanandum. If we want to understand the pathology of the liver, we look to cellular biology for explanation. ...To explain is to provide a causal mechanism, to open up the black*

box and show the nuts and bolts. ... they enable us to go from the larger to the smaller: from molecules to atoms, from societies to individuals.." (J. Ester; quoted from Groff (2008), p. 209). Much more about mechanisms and their examples could be found in (Mandel, 2017a).

4.2   Statistical Estimations of the Generated Outcomes

What is following is an approach, intended to help in solving "*quasi-causal*", or "approximately causal" problem, without pretense that the real causal mechanisms are captured. I just consider a common situation of the one target variable, causes of which are of interest.

We may not know the concrete causes of the target values appearing, but we can safely say that each particular situation in the given part of the universe generates values of this variable. This "particular situation" is described by specific variables which we consider relevant. The **generation mechanism** for values of $Y = y$, in general, is unknown, and revealing it is our focus. This mechanism is defined only on population, not on the individual level. It is just a way to say: if we observe $X = x$, there is a certain probability that we will observe $Y = y$, and the goal is to estimate this probability. Solving this problem, we at least approach to something, less ambitious than causality in the full meaning, but much more than just correlation, as will be shown. If I may say, that having a car, regardless of anything else, forces 15% of all car owners to visit car mechanics at least once annually, I may say, that having a car causes the visits to the mechanic, in that very accurate sense. If all car owners visit a mechanic – that would mean we found a **sufficient** cause. If only car owners make this visit - we found a **necessary** cause. In reality, of course, neither one is usually found, but different frequencies work. Since generation of causal outcome differs for events and processes, two approaches are proposed.

Intrinsic Values and Probabilities of the Events. Let us say, we are interested in some outcome, expressed as a binary variable – say, we want to understand, who buys Mercedes ($Y = 1$) against those who do not ($Y = 0$), a typical problem for both statistical learning and causal modeling. We have no exact causal mechanisms with measured components, but have instead a lot of information about different variables, which supposedly affect the purchasing behavior, $X1, X2...$; let us assume they all are also binary and that only situations $X = 1$ may generate $Y = 1$; $X = 0$ always yields $Y = 0$.

We can say that each X variable generates $Y = 1$ values with certain **intrinsic probability** p. It is intrinsic, because it is not equal to the marginal frequency of Y over X, and cannot be measured directly. It could be interpreted in the following way: there are some features in being a male, that force them to buy Mercedes more often that female; there are also some features in being wealthy, which force to buy stronger than being poor, etc. All these intrinsic forces, interlinked, may produce ultimately just one result – either buy or not. The "exact cause" of it is impossible to point out for an individual unless we make a deep dive into the individual psyche and make her "causal tree". It's enough to make the group's estimation of these causes interplay, i.e. to recover the parameters of the generation mechanism. Some outcomes $Y = 1$ may also appear just **randomly** with probability r – they depend only on the human desire to buy Mercedes, regardless of any specific features X we managed to measure.

This type of thinking matches trigger mechanisms for event discussed in Sect. 3.1, where just one "final force" was applied. With this concept the outcome Y = 1 occurs as a union of the independent events coming from two different sources – measured variables X and random occurrence. The same value could appear when any one or several causes make the impact, so the total probability of the outcome follows the rule of sum of probabilities because the actual and random causes are independent of each other. Thus, the basic equation of the process can be defined as follows:

$$S = S_{causal} + S_{random} - S_{causal}S_{random},\qquad(8)$$

where S is an observed frequency of Y = 1 occurrence, S_{causal} is the frequency of the events Y = 1, generated by "pseudo-causal" X variables; S_{random}, the frequency due to random causes. Using p, r notation, for one X variable we can rewrite (8):

$$S = p_1 + r - p_1 r\qquad(9)$$

From the data, we can estimate proportions of Y = 1 within the cases X = 0 and X = 1 as S0 and S1, respectively. Then the problem is how to estimate the parameters p_1 and r. The relation (9) can be represented in a symmetric form:

$$1 - S_i = (1 - r)(1 - p_i)^{x_i},\qquad(10)$$

where i = 1, 2 identifies two cells for X1 = 0 and X2 = 1.

For two binary variables X1 and X2 we observe four proportions: S1 for the case X1 = 0, X2 = 0; S2 for the case X1 = 0, X2 = 1; S3 for the case X1 = 1, and X2 = 0; and S4 for the case X1 = 1 and X2 = 1, i.e., the system of four equations with three unknown values $p1$, $p2$, r. In a general case of K variables the model can be constructed as an extension of the system (10) and presented in the generalized form:

$$S_i = 1 - (1 - r)\prod_{k=1}^{K}(1 - p_k)^{x_{ik}},\qquad(11)$$

where k = 1, 2, ..., K is a number of variable xik identifying the k-th parameter of probability in i-th cell. The values of xik are 1 or 0 when the variable is present or not, respectively. So, for an i-th value Si, with values xik = 1 the term $1 - p_k$ enters the product in (10), and for values xik = 0 the term $1 - p_k$ is absent from i-th row of the data. The cells as the new units are denoted by the current index i = 1, 2, ..., N.

The relations (11) show that if any probability p_k is close to 1, the total probability of the event occurrence Si is close to 1 as well. Regrouping and taking logarithm of each Eq. (11), and using notations

$$y_i = \ln(1 - S_i),\quad b_0 = \ln(1 - r),\quad b_k = \ln(1 - p_k),\qquad(12)$$

we represent (11) as a linear system:

$$y_i = b_0 + b_1 x_{i1} + \ldots + b_K x_{iK} \tag{13}$$

So the problem of the estimation of the parameters is reduced to the linear regression. We search for a solution which can approximate the Eq. (13), so let us rewrite them in the matrix form:

$$y = Xb + e, \tag{14}$$

where y is a vector of N-th order, X is the design matrix of X_{ik} values (completed by the additional column of all 1s which corresponds to the intercept b0 in the model), b is the vector of all $K + 1$ parameters in (12), and e is the vector of deviations of N-th order. With enough observations (cells) $N > K$ it can be solved in the regular ordinary least squares (OLS) approach by minimizing squared deviations in the LS objective. With the found parameters b, each original unknown probability can be derived from the relations (12):

$$r = 1 - \exp(b_0), \quad p_k = 1 - \exp(b_k), \tag{15}$$

There are generalizations associated with ridge regression in case of multi-collinearity (Lipovetsky and Mandel 2015b), but they are not considered here. The same paper shows, how to calculate the contributions of each X into the global outcome, which is very convenient for the practical purposes. It also presents the results of many experiments (see also Dodson and Mandel 2015).

The model above describes a situation with binary output and categorical (binary) input. Let me just outline possible techniques for other types of data, obeying the same events related mechanisms. Concrete algorithms are to be a subject of the separate work.

Numerical Outcome – Categorical Input. Interpretation of the input remains the same as before, but the outcome is numerical – say, the duration of survival in medical trials or an individual income of the people with different socio-demographic status, etc. In this case, each value $X = 1$ generates not just probability of $Y = 1$ appearance, but the certain spectrum of values for Y. Quite reasonable to assume, that it would be normally distributed random variables $Y(x1), Y(x2), \ldots$, each of which has two intrinsic values (parameters of distribution): average and standard deviation. The observed value of Y appears if it could be generated by any of those distributions. The union principle of (8) remains as it is, but the final distribution would be the convolution of the original ones (for respective $X = 1$ and random term in any given combination) adjusted by the product of these distributions.

Categorical Outcome – Numerical Input. The probability of outcome $Y = 1$ changes when a value of each independent variable changes, what makes the model, at the first glance, identical to the logit or multinomial logit models. Yet, as in previous cases, each variable should contribute into the result individually: for particular observation value $Y = 1$ may happen either when X1 generated that, or X2, and so on. The closest are the models of regression with random coefficients (Hildreth and Houck

1968; Demidenko and Mandel 2005; Lipovetsky 2007), where each observation gets its own regression coefficient for each independent variable. There could be considered at least two types of modifications: first, the coefficient should be limited in value from 0 to 1 (because it is the intrinsic probability), and second, observations should be somehow grouped for each X to build a small zone, where the same frequency of Y = 1 is to be applied. It makes a problem of estimation more complex, linking it with a problem of breaking domain of each variable into several parts, but doesn't change the logic of the triggered events.

4.3 Additive Models: Detection of the Generative Variables

Causal additive models, like those considered in part 1, are the favorite topic of statistics and needs no additional comments, besides the one that they are rarely called "causal". Regression analysis allows estimating parameters for any imaginable additive model. However, the concept of the causal-like outcome generation makes traditional regression estimates inadequate, because regression by itself doesn't distinguish variables *correlated* with the outcome and those which *generate* this outcome. The key issues are as follows.

Let assume, that values of dependent variable Y are completely determined (up to measurement errors) by the set of independent variables $Z\sim$. In fact, there is always some random component R in Y, which even in theory cannot be explained by any $Z\sim$, but let's skip this for a while. In the case of food example, this set is known. In real studies, only some variables from it may be included in a model; let's call them X variables; the remaining Z variables are not known and not observed, i.e. $Z\sim$ is a union of Z and X. There are other variables S included into the model, although they do not belong to $Z\sim$. So, the actual model aims to establish relationship Y = F(X, S), while the correct model should estimate the proper relationship $Y = F(Z\sim)$. What could one do in this situation?

In traditional modeling, the only answer is to increase the cardinality of the sets X and S, without distinguishing between them, because adding any new variable into model increases the goodness of fit statistics like R^2, and keep eye on the model overfitting at the same time. I showed in Sect. 1 the difficulties related to missing variables.

The causal considerations may help to address the issue of how **to *distinguish* between variables used and not used in Y generation**, i.e. between X and S. It seems very important by itself.

To illustrate the issue and show possible approaches to its solution let's return to generation data examples from part 1 with some modifications. The following dataset was created: 10,000 observations; 3 X variables, 4 S variables and 3 Z variables; all were generated as uniformly distributed random values, with normally distributed measurement errors. Y was a function of the X (observed) and Z (unobserved) variables; both X and Y have 5% measurement errors. Estimation of the parameters was done in several versions. Short data characteristics and results of estimation are presented in Table 4. There are no serious multicollinearity issues in data. Here are some comments.

Table 4. Estimation of the coefficients used in data generation

	Regression coefficients					T - statistics			
	Values in data generation	X and S model	X only	S only	X, S and Z	X and S model	X only	S only	X, S and Z
	1	2	3	4	5	6	7	8	9
X1	**1.0**	0.38	1.07		1.01	5.58	8.91		38.80
X2	**2.0**	0.58	1.96		1.96	27.19	58.83		199.24
X3	**3.0**	0.87	3.11		2.97	9.62	9.92		84.07
S1		0.29		0.32	0.01	81.63		89.03	3.22
S2		−0.01		0.01	−0.00	−13.07		−13.37	−0.22
S3		0.09		0.10	0.00	44.37		47.43	1.58
S4		0.01		0.01	0.00	24.79		26.23	−0.23
R^2, %		76.7	28.3	74.8	96.7				

1. The models were made with intercept, although Y was generated without. It was done simply because if the real model is unknown, intercept stands for normalization of the coefficients. I also made models without the intercept – they are, generally, worse, unless model coincides exactly with a way data were generated.
2. Coefficients for all observed 7 variables (column 2) are far from these used in data generation, despite high t-statistics and extremely low p-values (not shown here).
3. But if only correct variables used in a model (column 3), the coefficients are very close to the right values, although determination is low.
4. Irrelevant S variables give much better R^2 than relevant X variables (75% vs 28%). In a normal situation, when the difference between X and S is unknown, either this model or the full model in column 2 would be preferred – although both have nothing to do with real values of the coefficients.
5. Adding 3 unobserved Z variables, naturally, makes all coefficients for X almost ideal, and S show finally low t-statistics and could be eliminated. Yet, even in this grotesque situation t for S1, 3.22, is higher than 3 (a commonly accepted threshold) – so, it stubbornly nudges to the wrong direction no matter what! Moreover, taking off even one Z variable out of three immediately makes X coefficients sharply distorted (0.5, 0.9, 1.3, respectively). Yet the standard recommendations would advise eliminating variables based on t-statistics or similar criteria in any situation, not only in extremely rare and non-recognizable situations like one in column 5, the only one where S makes no harm.

From this example is clear, how important is to select only specific X variables and nothing else. It is not guaranteed in general, that coefficients for them will be always estimated correctly, as in this case, but the probability of that will increase.

In (Mandel 2017a) was convincingly demonstrated, that shifting from the goodness of fit and statistical inference statistics to the *statistical learning* type of errors (making model on training set and applying it to testing set) doesn't change the conclusion. The more is departure for the exactly specified model – the worse are distortions of the existing ingredients. In one example was shown, for instance, that the *wrong model only with S variables is much superior to the correct one only with X variables*.

The traditional approaches do not help to separate S and X variables. The proposed solution is based on the following idea. Since Z is independent on X, it should be relatively more disturbing for Y recovery via X, if X is small, because variance of Z should overwhelm X contribution to Y (bX). However, this effect should be much smaller for S variables. Variation of the Y estimates based on S should also vary a lot for small S, but estimates should not converge to the "right values" for the simple reason that there are no right values. It should continue to vary for any values of S, regardless on S-Y correlation, because this correlation is not supported by the real contribution of S to Y. Those assumptions were confirmed by the experiment.

1. Datasets were generated with varying sample size, number of X and S variables, level of Z, correlations between X, S and Y, different R^2 for *complete* model (Y as a function of X and S variables) and *correct* model (which uses only X variables).
2. Each dataset was standardized to make all variables having mean 0 and standard deviation 1 and sorted in ascending order sequentially by each of X and S variables. Each time the linear regression of Y by the respective one variable (X or S), used for sorting, was estimated in cumulative style as described in part 1 – the first regression was calculated for the first three data points, second – for the first four points and so on.
3. These regression coefficients were normalized within the interval 0–1, which makes comparisons easier. The different statistics based on those coefficients were applied to separate S and X variables with different types of errors.

A typical chart for one of the dataset is depicted in Fig. 6. As one can see, the difference between behavior of the coefficients for X (solid lines) and S (broken lines) variables is astonishing. What is especially amazing – all S variables had very high correlations with Y, much higher than X variables, what in any other situation would create immediate temptation to use those variables in a model – but still they are visibly separated from the really generating the outcome X variables.

The concrete shape of the charts is not necessarily like on Fig. 6, some S variables may be raising, not declining – but the striking visual difference between two types of variables practically always (in my experience) emerges. Two features could be especially useful to create a formal criterion for X and S separation, besides immediate visual separation, as here.

First is the presence of "trend" for S variable in about the second half of the data, showing that convergence, with very rare exceptions, never happens, unlike X, where it happens usually very fast (does not mean it converges to the correct value, but it is a different issue). Second, S variable has much higher variance in any given interval than X, especially, again, in a second half of the data, where sample size is higher.

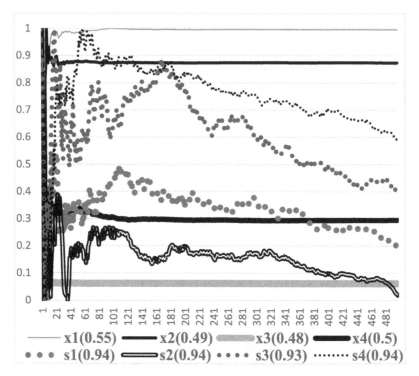

Fig. 6. Moving regression univariate coefficients (normalized from 0 to 1) after sorting for each X and S variable, 500 observations (value in brackets in variable's title is a correlation of this variable and Y)

The relevant reflection of those features could be obtained by the following indicators, which would provide different values for X and S variables:

L1 – standard deviation of the second half of the data;
L2 – trend of the second half of the data;
L3 – standard deviation of the residuals after approximation of the coefficients by the trend line.

All three indicators supposedly should be higher for S variables than for X. L1 and L3 are usually correlated and one of those could be redundant, but it is not clear yet.

The procedure of separation using L3 is as follows. For particular dataset the values of L3 are calculated and sorted in ascending order. Those values are presented in Table 5 for six different types of datasets.

It is immediately clear from the table, that for each dataset values of L3 (mainly between 0.01 and 0.14) are much lower for X variables than for S (mainly between 3 and 28), as expected. But it is still not sufficient to make a correct separation. Ideally,

Table 5. Indicator L3 for six different datasets with 8 variables and 500 observations

##	Types of datasets/Correlations with Y	X variables after sorting				S variables after sorting			
		x	x	x	x	s	s	s	s
		1	2	3	4	5	6	7	8
1	One S variable have high correlation; all other X, S - low	0.01	0.04	0.05	0.06	4.0	5.4	8.6	12.5
2	All X and S correlations are high	0.00	0.01	0.02	0.05	4.0	4.5	4.7	8.3
3	All X and S correlations are low	0.03	0.07	0.09	0.55	1.0	4.4	12.8	21.4
4	All X and S correlations are low	0.02	0.02	0.05	0.07	9.1	9.6	19.9	25.7
5	All X and S correlations are low	0.01	0.02	0.02	0.03	13.6	16.6	25.5	27.9
6	One X has high correlation, all other X, S - low	0.01	0.05	0.06	0.14	1.4	3.6	5.7	6.5

transition from X to S variables should be seen by certain spike of the indicator on a border between the two. But to measure that quantitatively presents a challenge. Two questions should be answered.

(a) What *statistics* in a row of sorted data should be used to capture the expected spike among two classes of variables? Closest candidates are: the original values; chain absolute (or relative) increments between the neighboring values (between values in columns 2 and 1; 3 and 2, etc.); chain increments between average value on the left and the given value (for value in column 5: average in columns 1–4 minus value in column 5); chain increments between cumulated averaged values on the left and right of the given value (average in columns 1–4 minus average in columns 5–8) and so on.

(b) What *definition of the spike* should be used, for any statistics? It could be, for example, the first large (how large?) value among previously small ones; the maximal value out of all observed; second maximal value (following logic of robustness and trimming the data) and so on. Another dimension for that is – the spike as a probabilistic/statistical meaning (higher than some theoretically pre-scribed value), or in data analytical meaning (above certain threshold, or just certain rank statistics).

These difficulties could be seen based on example in Fig. 7, where shown the chain increments between neighboring indicators from Table 5 (0.04 − 0.01 = 0.03; 0.05 − 0.04 = 0.01 from the row 1, etc.). Five datasets demonstrate very visible spikes between X and S, but set 3 (broken line) does not. It means, that broken vertical line can separate correctly X from S in five cases out of six, i.e. this separation criterion generates certain error. Other combinations of criteria and spikes definitions would generate errors as well.

In fact, these problems of strong separation are typical statistical learning problems and could be solved respectively by very different methods, besides the ones described. I'm working on the larger experiment but it seems the main idea works very well. Table 5 speaks for itself – with very different situations variables are perfectly

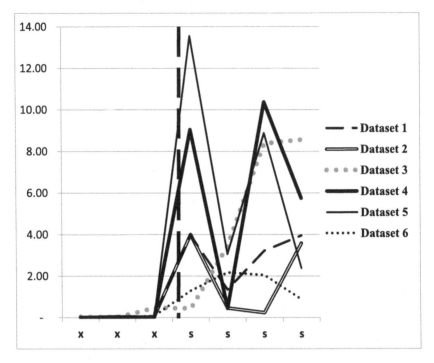

Fig. 7. Chain absolute increments of L3 by datasets from Table 5 and separation of S and X variables (vertical broken line)

separated, even when S are correlated with Y on a level of 0.92–0.96 (row 2). Tree complicated cases with very low correlations (rows 3–5) are successfully resolved, too – yet in this case erroneous decision is not that important, because anyway influence of X is very small.

For practical purposes, even in absence of the strong indicator of separation, it could be recommended that one runs bootstrap or some other random sampling technique and calculates L1–L3 for each variable. If certain variables show the sharp and stable separation for majority of the bootstrap runs in a line with illustration above – they are the best candidates for elimination from the additive model.

The proposed approach is just a first attempt to solve the separation problem. It needs to be further explored, both theoretically and experimentally; some new indicators may be found as well. The most important is not the concrete form of the separation criteria, but the ***recognition of the importance of the separation problem***. Creation of the model based on only relevant variables, not on those of S-type, is the main message.

In (Hofmann et al. 2008; Scholkopf 2012; Scholkopf et al. 2013; Peters et al. 2017) authors propose some effective methods to detect the correct direction of the causality in paired variables (to distinguish "age affects health" vs "health affects age"). It could be very interesting to compare this type of criteria with proposed one, although they address not exactly the same (yet similar) issues.

The separation between X and S makes sense only if the real generative process between input and output took place. Traditional interpretation of the regression coefficients as conditional change of one variable (Y) when another variable (X) is changing on one unit remains valid, of course, here as well, but it is not what is of interest. Only the presence of the *transmissible quantity*, aX, the actual contribution of X to Y with coefficient a, is going to be captured by those criteria of separation. Conditional values will be found every time and everywhere, and this is a topic for countless methods of the "number of variables reduction" regression literature (Leightner and inoue 2012, etc.). It has nothing to do with the real additive models. But if one actually finds those contributive "fractions of the Y by X", it means that generative process works, i.e. some variables do the real job.

Recalling the data for quasi-causal events (3.1), one can ask: why that X – S separation problem did not appear there? But in fact it did, it was just solved differently. If the intrinsic probability is zero, a variable belongs to S class. This simple procedure is possible for one fundamental reason: effects of each variable for events were accounted separately from the others, unlike situation with additive models. For both data types *model should not rely on non-generative variables*.

The notion of data generation is tightly related with a concept of *variable's causalization*: transformation of the original X variable to another variable X', which is closer to the real causal mechanism or represents it directly. In a food example all variables have been physically causal. Now imagine, that instead of measuring carbs, protein and fats one just measures price of one pound of food bought in different sections of the supermarket. The higher price – the more calories this given pound would yield, in average. But price is definitely not the causal variable, despite of this correlation. The real (physical) causes are still carbs, etc.; what happened - higher prices are usually assigned to fatty foods, which create more calories. So, the switch from price to fat would be a causalization of the price variable. Or take the well-known fact that air temperature on mountains is lower that on a sea level, which is considered as example of causal relation in (Peter et al. 2017). But altitude itself is not a physical cause of the low temperature (on high erupting volcanos temperature will be high, for instance); the air becomes thinner with altitude, and according to Gay-Lussac law the temperature goes down. For casualization one should use not altitude, but the pressure of the air – and that model would be "more causal" indeed. Or: the more money has been spent on advertising during campaign, the more new customers a company has acquired. But customers do not care about money: only actually seen ad may affect their behavior. Respectively, X variable "money" should be somehow converted into X' variable "impressions" to become more causal. Transition "impressions – new customers" is causal because one can see a mechanism, but "money – customers" is not. After this type of data modification the generative mechanisms will be much better revealed and estimated, and speculations about "contributions" above will be clearer.

The original intention of many researchers to untangle the conundrum "correlation – causation" is very well intended. But the way it is typically addressed in causal literature, through adding one more layer of relationship (the graph of many variables) is not a relevant replacement of the real problem – what is the essence of any single relationship (as was clearly emphasized, in particular, in Peters et al. (2017)). *The causality lies not above, but under the surface of the correlation.*

5 Conclusion

The one hundred years' history of the development of the causal approaches in statistics has not resulted in one clear and commonly accepted theory. Different versions are based on numerous assumptions often without appropriate philosophical and scientific foundations. Counterfactual logic, never actually used in natural sciences, has become a basic premise for the leading theories.

The sharp contradictions between statistical inference and causal explanations were not explained and not even properly addressed. Universal "market-like" approach to statistics facilitates a coexistence of hundreds of methods and algorithms, some of which are short lived, but some, regardless of their merits, become very popular due to internal and external forces acting in such complex systems as a scientific community. The speed of life and chasing of specific (often material) interests do not stimulate a tedious but accurate testing of statistical conclusions in practice, rather than within the very artificial subsets of data used. All these and other factors derail attention from the fundamental questions of validity of statistical conclusions, reproducibility of the results, and, finally, the place of statistics in a scientific picture of the world.

The tremendous success of data science and machine learning in the last years adds a new layer of uncertainty to these problems, because it promotes completely new paradigms, having nothing to do with causality, and, almost always, with traditional statistical inference. It actually sweeps under the rug all inconvenient questions and replaces them with one unbeatable argument that if the error on the testing sets or on the new objects is low, the procedure is excellent, no matter on what it is based. ***Deep learning*** is the best and most triumphant bearer of such ideology. There is no thinnest hue to causality in neural nets, just their ability to learn from the experience - which is, of course, a great deal. And, indeed, the number of problems to be solved with those techniques is huge, but not unlimited. The danger of overestimating the power of the conclusions of this type when applied to the new areas, besides the domain in which the model was built, is growing. The uncontrollable internal logic of the neural networks in absence of causal components creates a new type of uncertainty – perhaps, not appreciated by statisticians yet. Yet causal lows in life will be never cancelled – the problem is how to deal with them. The further we go, the clearer that all-embracing theory of causality in statistics is not built and, most likely, not possible.

The alternative to statistical approach – agent based and other ***simulation models*** (Squazzoni 2012, etc.) – assume active human participation at least during the initial stage (creating the rules for agents), which makes them at the same time more subjective and more open to causal representation. It is both very natural and promising. I see the future as a wise combination of these two lines of thinking: ***merging machine learning***, ***simulation*** and ***decision-making*** models in organic combination, where each part plays its own role. It seems, we are at the very first steps in this direction now.

The proposed *generative approach to the data analysis* is trying to soften unrealistic claims from the causality theorists, to admit the fact that axiomatic theory of causality cannot be built, and to replace it by a more pragmatic view. Not only "correlation is not causation", but also "generation is not causation" – yet the generative approach at least moves us a bit closer to the understanding of what are the actual causal mechanisms – the final goal of any scientific inquiry.

Acknowledgements. The study of causality was supported by Telmar Inc. and some of the results were incorporated in its software. Author sincerely thanks I. Lipkovich and S. Lipovetsky for the numerous fruitful discussions and B. Mirkin for very meaningful comments and suggestions.

References

Bang-Jensen, J., Gutin, G.: Digraphs: Theory, Algorithms and Applications. Springer, Heidelberg (2009). https://doi.org/10.1007/978-1-84800-998-1

Bennett, A.: The mother of all "isms": organizing political science around causal mechanisms. In: Groff, R. (ed.) Revitalizing Causality: Realism About Causality in Philosophy and Social Science, pp. 205–219. Routledge (2008)

Berk, R.: Regression Analysis: A Constructive Critique. Sage Publications, Newbury Park (2004)

Berzuini, C., Dawid, P., Bernardinelli, L. (eds.): Causality: Statistical Perspectives and Applications. Wiley, Chichester (2012)

Bigelow, J., Ellis, B., Pargetter, R.: Forces. Philos. Sci. **55**, 614–630 (1988)

Bontempi, G., Flauder, M.: From dependency to causality: a machine learning approach. J. Mach. Learn. Res. **16**, 2437–2457 (2015)

Bunge, M.: Causality and Modern Science. Transaction Publishers, New Brunswick (2009)

Buonaccorsi, J.P.: Measurement Error: Models, Methods, and Applications. Chapman and Hall, Boca Raton (2010)

Carroll, R., et al.: Measurement Error in Nonlinear Models: A Modern Perspective. Chapman and Hall, New York (2006)

Cheng, C.L., Van Ness, J.W.: Statistical Regression with Measurement Error. Arnold Publishers, London (1999)

Conrady, S., Jouffe, L.: Bayesian Networks & BayesiaLab: A Practical Introduction for Researchers. Bayesia USA, Franklin (2015)

Consumer Price Index Manual: Theory and Practice. International Monetary Fund (2004)

Craycroft, J.: Propensity score methods: a simulation and case study involving breast cancer patients. Paper 2460 (2016). https://doi.org/10.18297/etd/2460

Dawid, P.: Conditional independence in statistical theory. J. R. Stat. Soc. B **41**, 1–31 (1979)

Dawid, P.: Beware of the DAG! In: JMLR: Workshop and Conference Proceedings, vol. 6, pp. 59–86 (2009)

Dowe, P.: Causal processes. In: Stanford Encyclopedia of Philosophy (2007). http://seop.illc.uva.nl/entries/causation-process/

Demidenko, E., Mandel, I.: Yield analysis and mixed model. In: Proceedings of Joint Statistical Meeting. ASA, Alexandria, VA (2005)

Dodson, D., Mandel, I.: Causal Analytics for Media Planning (2015). https://et220.etelmar.net/index.aspx

Efron, B., Hastie, T.: Computer Age Statistical Inference Algorithms, Evidence, and Data Science. Cambridge University Press, New York (2016)

Good, I.J.: Good Thinking: The Foundations of Probability and Its Applications. The University of Minnesota, Minneapolis (1983)

Greenland, S., Robins, J.M., Pearl, J.: Confounding and collapsibility in causal inference. Stat. Sci. 14(1), 29–46 (1999)

Groff, R. (ed.): Revitalizing Causality: Realism about Causality in Philosophy and Social Science. Taylor and Francis Group, London (2008)

Hastie,T., Tibshirani, R., Friedman, J.: The Elements of Statistical Learning: Data Mining, Inference, and Prediction, Springer (2009)

Hildreth, C., Houck, J.P.: Some estimators for a linear model with random coefficients. J. Am. Stat. Assoc. 63, 584–595 (1968)

Hitchcock, C.: Probabilistic causation. In: Stanford Encyclopedia of Philosophy (2010). http://plato.stanford.edu/entries/causation-probabilistic/

Hofmann, T., Scholkopf, B., Smola, A.J.: Kernel methods in machine learning. Ann. Stat. 36(3), 1171–1220 (2008)

Hoover, K.D.: Causality in economics and econometrics. In: The New Palgrave Dictionary of Economics. Springer, Heidelberg (2016). https://doi.org/10.1057/978-1-349-95121-5_2227-1

Illari, P., Russo, F.: Causality: Philosophical Theory meets Scientific Practice. Oxford University Press, London (2014)

Imai, K., Tingley, D.: A statistical method for empirical testing of competing theories. Am. J. Polit. Sci. 56(1), 218–236 (2012)

Imbens, G., Rubin, D.: Causal Inference for Statistics, Social, and Biomedical Sciences: An Introduction. Cambridge University Press, New York (2015)

Johnson, V., Payne, R., Wang, T., Asher, A., Mandal, S.: On the reproducibility of psychological science. J. Am. Stat. Assoc. 112, 517 (2017)

Kaplan, D., Chen, C.: Bayesian Propensity Score Analysis: Simulation and Case Study (2011). https://www.sree.org/conferences/2011/program/downloads/slides/20.pdf

King, G., Nielsen, R.: Why Propensity Scores Should Not Be Used for Matching (2016). https://pdfs.semanticscholar.org/8ed9/88fa9e9ed4b7569faaab920639953c881b27.pdf

Kistler, M.: Causation and Laws of Nature. Routledge, London (2006)

Kline, R.: Principles and Practice of Structural Equation Modeling. The Guilford Press, New York (2011)

Kuznetsov, D., Mandel, I.: Statistical physics of media processes: mediaphysics. Phys. A 377, 253–268 (2007)

Leightner, J., Inoue, T.: Solving the omitted variables problem of regression analysis using the relative vertical position of observations. Adv. Decis. Sci. 2012 (2012). Paper ID 728980

Lewis, D.: Counterfactuals. Harvard University Press, Cambridge (1973)

Li, H., Yuan, Z., Su, P., Wang, T., Yu, Y., Sun, X., Xue, F.: A simulation study on matched case-control designs in the perspective of causal diagrams. BMC Med. Res. Methodol. BMC Ser. 16, 102 (2016)

Lipovetsky, S., Conklin, M.: Analysis of regression in game theory approach. Appl. Stochastic Models Bus. Ind. 17, 319–330 (2001)

Lipovetsky, S., Conklin, M.: Data aggregation and Simpson_s paradox gauged by index numbers. Eur. J. Oper. Res. 172, 334–351 (2006)

Lipovetsky, S.: Iteratively re-weighted random-coefficient models and Shapley value regression. Model Assist. Stat. Appl. 2, 201–212 (2007)

Lipovetsky, S., Conklin, M.: Predictor relative importance and matching regression parameters. J. Appl. Stat. (2014)

Lipovetsky, S., Mandel, I.: Review on: handbook of causal analysis in social research, Springer, 2015. Technometrics **57**(2), 298–300 (2015a)

Lipovetsky, S., Mandel, I.: Modeling probability of causal and random impacts. J. Mod. Appl. Stat. Methods **14**(1), 180–195 (2015b)

Mandel, I.: Sociosystemics, statistics, decisions. Model Assist. Stat. Appl. **6**, 163–217 (2011)

Mandel, I.: Fusion and causal analysis in big marketing data sets. In: Proceedings of JSM. ASA, Alexandria, VA, pp. 1719–1732 (2013)

Mandel, I.: Causal models in estimation of the advertising ROI. In: Proceedings of JSM. ASA, Alexandria, VA, pp. 1720–1725 (2016)

Mandel, I.: Troublesome Dependency Modeling: Causality, Inference, Statistical Learning (2017a). https://papers.ssrn.com/sol3/papers.cfm?abstract_id=2984045

Mandel, I.: Regression coefficients vs causal coefficients. Post in ASA blog, 19 July 2017 (2017b). http://community.amstat.org

Masiuk, S., Kukush, A., Shklyar, S., Chepurny, M., Likhtarov, I.: Radiation Risk Estimation: Based on Measurement Error Models. Walter de Gruyter, Boston (2017)

Menzies, P.: Counterfactual theories of causation. In: Stanford Encyclopedia of Philosophy (2014). http://seop.illc.uva.nl/entries/causation-counterfactual/

Mirkin, B.: Core Concepts in Data Analysis: Summarization, Correlation and Visualization. Springer, Heidelberg (2011). https://doi.org/10.1007/978-0-85729-287-2

Morgan, S.L. (ed.): Handbook of Causal Analysis in Social Research. Springer, Heidelberg (2014). https://doi.org/10.1007/978-94-007-6094-3

Morgan, S.L., Winship, C.: Counterfactuals and Causal Inference: Methods and Principles for Social Research. Cambridge University Press, Cambridge (2014)

Open Science Collaboration: Investigating variation in replicability: a "Many Labs" replication project. Soc. Psychol. **45**, 142–152 (2014)

Open Science Collaboration: Estimating the reproducibility of psychological science. Science **349**(6251) (2015)

Pearl, J.: Causality. Cambridge University Press, Cambridge (2009)

Pearl, J.: The Causal Foundations of Structural Equation Modeling. Technical report R-370 (2012). http://ftp.cs.ucla.edu/pub/stat_ser/r370.pdf

Pearl, J., Glymour, M., Jewell, N.: Causal Inference in Statistics: A Primer. Wiley, Chichester (2016)

Peters, J., Janzing, D., Schölkopf, B.: Elements of Causal Inference: Foundations and Learning Algorithms. The MIT Press, Cambridge (2017)

Ralph, J., O'Neill, R., Winton, J.: A Practical Introduction to Index Numbers. Wiley (2015)

Rubin, D.: Matched Samples for Causal Effect. Cambridge University Press, New York (2006)

Scholkopf, B.: Causal Inference and Statistical Learning (2012). http://ml.dcs.shef.ac.uk/masamb/schoelkopf.pdf.   http://machinelearningmastery.com/machine-learning-statistical-causal-methods/

Scholkopf, B., Janzing, D., Peters, J., Sgouritsa, E., Zhang, K., Mooij, J.: Semi-supervised learning in causal and anticausal settings. In: Schölkopf, B., Luo, Z., Vovk, V. (eds.) Empirical Inference, pp. 129–141. Springer, Heidelberg (2013). https://doi.org/10.1007/978-3-642-41136-6_13

Skow, B.: An Argument Against Woodward's Theory of Causal Explanation (2013). http://web.mit.edu/bskow/www/research/manipulationism.pdf

Spirtes, P., Glymour, C., Scheines, R.: Causation, Prediction, and Search. The MIT Press, Cambridge (2001)

Squazzoni, F.: Agent-Based Computational Sociology. Wiley, Chichester (2012)

VanderWeele, T.: Explanation in Causal Inference: Methods for Mediation and Interaction. Oxford University Press, New York (2015)

Vapnik, V.: Estimation of Dependences Based on Empirical Data: Empirical Inference Science. Springer, Heidelberg (2006). https://doi.org/10.1007/0-387-34239-7

Viswanathan, M.: Measurement Error and Research Design. SAGE Publications, Thousand Oaks (2005)

Wansbeek, T., Meijer, E.: Measurement Error and Latent Variables in Econometrics. Elsevier, Amsterdam (2000)

Wasserstein, R., Lazar, N.: The ASA's statement on p-values: context, process, and purpose. Am. Stat. **70**(2), 129–133 (2016)

Zagar, A., Kadziola, Z., Lipkovich, I., Faries, D.: Evaluating different strategies for estimating treatment effects in observational studies. J. Biopharm. Stat. **27**(3), 535–553 (2017)

Zadeh, L.: Causality is Undefinable. Toward a Theory of Hierarchical Definability (2001). http://link.springer.com/chapter/10.1007/3-540-45813-1_2#page-1

Novel Developments

One-Class Semi-supervised Learning

Evgeny Bauman[1(✉)] and Konstantin Bauman[2]

[1] Markov Processes Inc., Summit, USA
ebauman@markovprocesses.com
[2] Fox School of Business, Temple University, Philadelphia, USA
kbauman@temple.edu

Abstract. One-class classification problem aims to identify elements of a specific class among all other elements. This problem has been extensively studied in the last decade and the developed methods were applied to a large number of different problems, such as outlier detection, natural language processing, fraud detection, and many others. In this work, we developed a new semi-supervised one-class classification algorithm which assumes that the class is linearly separable from other elements. We proved theoretically that the class is linearly separable if and only if it is maximal by probability within the sets of elements with the same mean. Furthermore, we constructed an algorithm for identifying such linearly separable class based on linear programming. We considered three application cases including an assumption of linear separability of the class, Gaussian distribution, and the case of linear separability in the transformed space of kernel functions. Finally, we examined the work of the proposed algorithm on the USPS dataset and analyzed the relationship of its performance and the size of the initially labeled sample.

1 Introduction

In machine learning, one-class classification problem aims to identify elements of a specific class among all other elements. This problem has been extensively studied in the last decade [8] and the developed methods were applied to a large variety of problems, such as detecting outliers [15], natural language processing [6], fraud detection [17], and many others [8].

The traditional supervised approach to the classification problems infers a decision function based on labeled data. In contrast, semi-supervised learning deals with the situation where relatively few labeled training elements are available, but a large number of unlabeled elements are given. This approach is suitable for many practical problems where it is relatively expensive to produce labeled data, such as automatic text classification. Semi-supervised learning may refer to either *inductive* learning or *transductive* learning [19]. The goal of *inductive* learning is to learn a decision rule that would predict correct labels for the new unlabeled data, whereas *transductive* learning aims to infer the correct labels for the given unlabeled data only. In this paper we address the semi-supervised one-class transductive learning problem, i.e. the main goal of the proposed algorithm is to estimate the labels for the given unlabeled data.

© Springer Nature Switzerland AG 2018
L. Rozonoer et al. (Eds.): Braverman Readings in Machine Learning, LNAI 11100, pp. 189–200, 2018.
https://doi.org/10.1007/978-3-319-99492-5_8

In order to make semi-supervised learning work, one has to assume some structure in the underlying distribution of data. The most popular such assumptions include smoothness assumption, cluster assumption, or manifold assumption [4]. In the proposed algorithm we make an assumption of linear separability of the class. Therefore, we study the question of what information is needed to estimate the linearly separable class. As a result, we prove that linearly separable class can be detected based on its mean.

In this paper we made the following contributions:

- Proposed a new semi-supervised approach to one-class classification problem.
- Proved that the class is linearly separable if and only if it's maximal by probability within all sets with the same mean.
- Presented an exact algorithm for detecting the linearly separable class by its mean, utilizing linear programming.
- Described three application cases including an assumption of linear separability of the class, Gaussian distribution, and the case of linear separability in the transformed space of kernel functions.
- Demonstrated the efficacy of the proposed algorithm on the USPS dataset and analyzed the relationship between the performance of the algorithm and the size of the labeled sample.

2 Related Work

There are two main blocks of the related work. The first block consists of works pertaining to one-class classification problem. The most popular supervised support vector machine approach to the one-class problem (OC-SVM) was independently introduced by Schölkopf et al. in [16] and Tax and Duin in [18]. This studies extended the SVM methodology to handle training sets consisting of labels of only one class. In particular, [16] proposed an algorithm which computes a binary function that captures the regions in the input space where the probability density lives (its support), i.e. a function such that most of the data will live in the region where the function is nonzero. Later, OC-SVM was successfully applied to the anomaly and outliers detection [5,9,11,20], and the density estimation problem [12]. This approach was effective in remote-sensing classification in the situations when users are only interested in classifying one specific land-cover type, without considering other classes [10]. Authors of [20] applied OC-SVM approach to detect anomalous registry behavior by training on a dataset of normal registry accesses. In [5] authors use OC-SVM as a means of identifying abnormal cases in the domain of melanoma prognosis. More applications of OC-SVM can be found in [8].

The second block of the related works consists of semi-supervised approaches to the classification problems. Book [4] provides an extensive review of this field and [14] presents a survey describing the current state-of-the-art approaches. There are several works that study one-class semi-supervised learning [2,7,10,13]. In particular, authors of [7] evaluate the suitability of semi-supervised one-class classification algorithms as a solution to the low default portfolio problem. In [13] authors implemented modifications of OC-SVM in the context of information retrieval. All these works utilize various modifications of OC-SVM approach. Note, that in this cases the optimization problem utilizes quadratic programming.

In contrast to these prior works, we consider a one-class semi-supervised transductive learning problem. It means that initially within a large unlabeled set we have a small labeled subset of elements pertaining to class A. Our proposed algorithm addresses the problem of identifying elements of class A among the large unlabeled set. In particular, we make an assumption of linear separability of the class, and, therefore, we study the question of what information is needed to estimate the linearly separable class. As a result, we prove theoretically that linearly separable class can be detected based on its mean. We use the notion of fuzzy sets to assign elements on the boundary to the class with a certain weight. This approach helps us to go beyond the finite set of discrete means. Finally, we also presented an exact algorithm for detecting the linearly separable class by its mean, utilizing linear programming.

3   Algorithm

We consider an arbitrary probability space $(\mathbb{R}^n, \Sigma, P)$, where $\mathbb{R}^n$ is Euclidean space, Σ is a Borel σ-algebra over $\mathbb{R}^n$ (it contains all half-spaces of $\mathbb{R}^n$), and P is a probability measure in $\mathbb{R}^n$. We assume that $(\mathbb{R}^n, \Sigma, P)$ satisfies the following conditions:

1. There exists the finite second moment $\int x^T x dP(x)$
2. The probability of any hyperplane in $\mathbb{R}^n$ is not equal to 1.

An arbitrary set $A \in \Sigma$ is defined by its indicator function

$$h_A(x) = \begin{cases} 1, x \in A \\ 0, x \notin A \end{cases}$$

Further, in the paper instead of the sets, we work with the indicator functions defining them and, therefore, we also define concepts of "linear separability" and "maximum by probability" for indicator functions.

Based on Condition 1, indicator functions $h_A \in L^2(\mathbb{R}^n, \Sigma, P)$. We denote by $\mathcal{H} = \{h_A | A \in \Sigma\}$ – the set of indicator functions for all measurable sets. The convex hull of $\mathcal{H}$ would be the set of functions

$$\hat{\mathcal{H}} = Conv(\mathcal{H}) = \{h \in L^2(\mathbb{R}^n, \Sigma, P) | \forall x : \ 0 \le h(x) \le 1\}.$$

By construction $\mathcal{H} \subseteq \hat{\mathcal{H}}$. Elements of $\hat{\mathcal{H}}$ usually interpreted as fuzzy sets, saying that set defined by h contains element x with the weight $h(x)$, where $0 \le h(x) \le 1$.

According to Condition 1 for each indicator function h defining measurable fuzzy set with the probability greater that 0, we can construct its mean (i.e., first normalized moment of the set defined by h):

$$\mu(h) = \frac{M(h)}{P(h)} = \frac{\int xh(x)dP(x)}{\int h(x)dP(x)}, \tag{1}$$

where $P(h) = \int h(x)dP(x)$ – scalar equal to the probability of h, and $M(h) = \int xh(x)dP(x)$ – n-dimensional vector, equal to first un-normed moment of h.

Further, we construct linear mapping of $\mathcal{H}$ to $(n+1)$-dimensional Euclidean space:

$$\varphi(h) = (M(h), P(h)). \tag{2}$$

$\mathcal{M} = \varphi(\mathcal{H})$ is the image of the set $\mathcal{H}$ with mapping φ to $\mathbb{R}^{n+1}$. Since φ is linear, then $\hat{\mathcal{M}} = Conv(\mathcal{M}) = \varphi(\hat{\mathcal{H}})$.

Based on the fact that $\hat{\mathcal{H}}$ is convex and closed set, and Conditions 1 and 2 we conclude that $\hat{\mathcal{M}}$ is also convex, closed and bounded. Moreover, $\hat{\mathcal{M}}$ does not belong to any hyperplane in $\mathbb{R}^{n+1}$.

3.1   Linear Separability and Maximum by Probability

Definition 1. *Indicator function $h \in \hat{\mathcal{H}}$ is linearly separable with vector $\bar{b} = (\bar{c}, d)$, where $\bar{c} \in \mathbb{R}^n$ ($\bar{c}$ is non-trivial) and d is a constant, if the following conditions are satisfied almost everywhere:*

- *if $c^T x + d > 0$, then $h(x) = 1$*
- *if $c^T x + d < 0$, then $h(x) = 0$.*

Note, that Definition 1 implies that if $h \in \hat{\mathcal{H}}$ is linearly separable then $P(\{x \in \mathbb{R}^n | c^T x + d \neq 0\} \cap \{x \in \mathbb{R}^n | 0 < h(x) < 1\}) = 0$. It means that fuzziness of h concentrates only on the hyperplane defined by equation $c^T x + d = 0$.

Lemma 1. *Image $y^* = \varphi(h^*)$ of function $h^* \in \hat{\mathcal{H}}$ (with $P(h^*) \neq 0$), is a boundary point of $\hat{\mathcal{M}}$ if and only if there exists a non-trivial vector $\bar{b}$, such that h^* is linearly separable with $\bar{b}$.*

Proof. Based on the Supporting hyperplane theorem [3], y^* is a boundary point of convex closed set $\hat{\mathcal{M}}$ if and only if there exists a non-trivial vector $\bar{b} = (\bar{c}, d)$, where $\bar{c} \in \mathbb{R}^n$ and d is a constant, such that $b^T y^* \geq b^T y$, $\forall y \in \hat{\mathcal{M}}$, which is equivalent to $b^T \varphi(h^*) \geq b^T \varphi(h)$, $\forall h \in \hat{\mathcal{H}}$. If we denote by $\Delta_b(h_1, h_2) = b^T \varphi(h_1) - b^T \varphi(h_2)$, then the inequality transforms to

$$\Delta_b(h^*, h) \geq 0, \ \forall h \in \hat{\mathcal{H}}. \tag{3}$$

Sufficiency. If h^* is linearly separable with $\bar{b} = (\bar{c}, d)$, then for arbitrary $h \in \hat{\mathcal{H}}$ we consider the difference:

$$\Delta_b(h^*, h) = b^T \varphi(h^*) - b^T \varphi(h) = \left(c^T M(h^*) + dP(h^*)\right) - \left(c^T M(h) + dP(h)\right)$$

that can be transformed to

$$\Delta_b(h^*, h) = \int (c^T x + d)(h^*(x) - h(x))dP(x) = \Delta_b^+(h^*, h) + \Delta_b^0(h^*, h) + \Delta_b^-(h^*, h), \text{ where}$$

$\Delta_b^+(h^*, h) = \int_{c^T x + d > 0}(c^T x + d)(1 - h(x))dP(x),$
$\Delta_b^0(h^*, h) = \int_{c^T x + d = 0}(c^T x + d)(1 - h(x))dP(x) = \int_{c^T x + d = 0} 0 \cdot (1 - h(x))dP(x) = 0,$
$\Delta_b^-(h^*, h) = \int_{c^T x + d < 0}(c^T x + d)(0 - h(x))dP(x).$

Since $0 \leq h(x) \leq 1 \; \forall x$, then $\Delta_b^+(h^*, h) \geq 0$ and $\Delta_b^-(h^*, h) \geq 0$, and, therefore, $\Delta_b(h^*, h) \geq 0 \; \forall h \in \hat{\mathcal{H}}$. It shows that there exists $\bar{b}$ that satisfies inequality (3) and implies that $\varphi(h^*)$ is a boundary point of $\hat{\mathcal{M}}$.

Necessity. If $y^* = \varphi(h^*)$ is a boundary point of $\hat{\mathcal{M}}$, then there exists $\bar{b}$ satisfying inequality (3). Vector $\bar{b} = (\bar{c}, d)$ defines the half-space in $\mathbb{R}^n$ with the following indicator function:

$$h_b(x) = \begin{cases} 1, & \text{if } c^T x + d \geq 0, \\ 0, & \text{if } c^T x + d < 0. \end{cases}$$

Based on the same calculation as in proof of Sufficiency: $\Delta_b(h_b, h) \geq 0 \; \forall h \in \hat{\mathcal{H}}$, and thus $\Delta_b(h_b, h^*) \geq 0$. However, inequality (3) implies that $\Delta_b(h_b, h^*) \leq 0$, which means that $\Delta_b(h_b, h^*) = 0$. Therefore, $\Delta_b^+(h_b, h^*) = 0$, and $\Delta_b^-(h_b, h^*) = 0$, that can be expressed in

$$\int_{c^T x + d > 0} (1 - h^*(x)) dP(x) = \int_{c^T x + d < 0} (0 - h^*(x)) dP(x) = 0.$$

This equality implies that the following conditions satisfied almost everywhere:

– if $c^T x + d > 0$, then $h^*(x) = 1$
– if $c^T x + d < 0$, then $h^*(x) = 0$,

and, therefore, h^* is linearly separable with $\bar{b} = (\bar{c}, d)$. $\square$

Definition 2. *Let's denote by W_α a set of all indicator functions that define sets in $\mathbb{R}^n$ with the means in α, i.e. $W_\alpha = \{h \in \hat{\mathcal{H}} : \mu(h) = \alpha\}$. Indicator function $h^* \in W_\alpha$ is called maximal by probability with mean α if $\forall h \in W_\alpha$: $P(h) \leq P(h^*)$.*

Lemma 2. *$\varphi(W_\alpha)$ is a finite interval in $\mathbb{R}^{n+1}$, i.e. for any $h^* \in W_\alpha$ if $\varphi(h^*) = (M(h^*), P(h^*)) = y^*$ then $\varphi(W_\alpha) = \{y = \lambda \cdot y^* | 0 < \lambda \leq \lambda_{max}^*\}$, where $\lambda_{max}^* \geq 1$.*

Proof. For an arbitrary $h^* \in W_\alpha$ with $\varphi(h^*) = (m^*, p^*) = y^*$, we construct a ray $D_{h^*} = \{y = \lambda \cdot y^* | \lambda > 0\}$ in $\mathbb{R}^{n+1}$. Since $\hat{\mathcal{M}}$ is convex, closed and bounded, then intersection of D_{h^*} and $\hat{\mathcal{M}}$ is a finite interval $I_{h^*} = D_{h^*} \cap \hat{\mathcal{M}} = \{y = \lambda \cdot y^* | 0 < \lambda \leq \lambda_{max}^*\}$. Since $y^* \in I_{h^*}$, then $\lambda_{max}^* \geq 1$.

The statement $h \in W_\alpha$ means that $\mu(h) = \frac{M(h)}{P(h)} = \alpha = \frac{M(h^*)}{P(h^*)} = \mu(h^*)$. If we denote $\frac{P(h)}{P(h^*)} = \lambda$, then $M(h) = \lambda \cdot M(h^*)$. Which is equivalent to the fact that there is exists a certain $0 < \lambda \leq \lambda_{max}^*$, such that $\varphi(h) = (m, p) = (\lambda \cdot m^*, \lambda \cdot p^*) = \lambda \cdot y^*$. This means that $I_{h^*} = \varphi(W_\alpha)$. $\square$

Theorem 1. *Function $h^* \in W_\alpha$ is maximal by probability, if and only if h^* is linearly separable.*

Proof. **Necessity.** Based on Lemma 2, $\varphi(W_\alpha) = \{y = \lambda \cdot y' | 0 < \lambda \leq \lambda'_{max}\}$ for an arbitrary $h' \in W_\alpha$. If $h^* \in W_\alpha$ is maximal by probability, then $P(h^*) = \lambda^* \cdot P(h') = \lambda_{max} \cdot P(h')$. Therefore, $\varphi(h^*) = y^* = \lambda'_{max} \cdot y'$, which means that $\varphi(h^*)$ is a boundary point of set $\mathcal{M}$. Finally, based on Lemma 1, h^* is linearly separable.

Sufficiency. If h^* is linearly separable, then based on Lemma 1, $\varphi(h^*) = y^*$ is a boundary point of set $\mathcal{M}$. Based on Lemma 2, $\varphi(W_\alpha) = \{y = \lambda \cdot y^* | 0 < \lambda \leq \lambda^*_{max}\}$. Since y^* is a boundary, then $\lambda^*_{max} = 1$. Therefore, $\forall h \in W_\alpha : P(h) = \lambda \cdot P(h^*)$ where $0 < \lambda \leq 1$. This means that h^* is a maximal by probability with mean in α. $\qquad\square$

3.2   Detecting Linearly Separable Class by Its Mean

Let $x_1, \ldots, x_N$ be i.i.d. random variables in $\mathbb{R}^n$ with distribution P. The special case where P is the empirical distribution $P_N(A) = \frac{1}{N} \sum_{i=1}^{N} h_A(x_i)$.

Based on Theorem 1, if class A is linearly separable it is also maximal by probability within all sets with the same mean $\alpha = \mu(A)$. Therefore, the problem of detecting class A would be equivalent to the problem of identifying the maximal by probability indicator function $h \in \hat{\mathcal{H}}$ with its mean in a given point $\alpha \in \mathbb{R}^n$, where the mean and the probability of h are calculated in the following way: $M(h) = \frac{1}{N} \sum_{i=1}^{N} x_i h(x_i)$, $P(h) = \frac{1}{N} \sum_{i=1}^{N} h(x_i)$. In other words, in this case, we are searching for the set of elements with the specified mean α and containing the maximum number of elements. More formally we solve the following problem:

$$\begin{cases} P(h) = \frac{1}{N} \sum_{i=1}^{N} h(x_i) \xrightarrow[h(x_1),\ldots,h(x_N)]{} max; \\ \mu(h) = \frac{\sum_{i=1}^{N} x_i \cdot h(x_i)}{\sum_{i=1}^{N} h(x_i)} = \alpha; \\ 0 \leq h(x_i) \leq 1, \ i = 1, \ldots, N. \end{cases} \Leftrightarrow \begin{cases} \sum_{i=1}^{N} h(x_i) \xrightarrow[h(x_1),\ldots,h(x_N)]{} max; \\ \sum_{i=1}^{N} (x_i - \alpha) \cdot h(x_i) = 0; \\ 0 \leq h(x_i) \leq 1, \ i = 1, \ldots, N. \end{cases} \quad (4)$$

This problem can be solved using linear programming optimizing by vector $(h(x_1), \ldots, h(x_N))$. If there is exists at least one set with a given center α, then Problem (4) would have feasible solution.

In conclusion, we proved the equivalence of the concepts of "linear separability" and "maximal by probability with a given mean", and proposed the exact algorithm of detecting linearly separable class based on its mean using linear programming.

3.3   Algorithms of Semi-supervised Transductive Learning for One-Class Classification

Problem. Set $X = \{x_1, \ldots, x_N\} \subset \mathbb{R}^n$ consists of two parts: sample $X_A = \{x_1, \ldots, x_l\} \subset X \cap A$ containing labeled elements for which we know that they belong to class A, and the rest of the set $X_{unlabeled} = \{x_{l+1}, \ldots, x_N\} \subset X$ containing unlabeled elements for which we do not know if they belong to class A or not. The problem is to identify all elements in X that belong to class A based on this information.

Linearly Separable Class. In real practical problems we usually do not have the exact value of $\mu(A)$ and, therefore, we have to estimate it. The proposed approach assumes that X_A is generated with the same distribution law as A and class A is *linearly separable* in $\mathbb{R}^n$. Based on the first assumption, mean of A is estimated by the mean of X_A calculated as follows: $\mu(X_A) = \frac{1}{l} \sum_{j=1}^{l} x_j$. Based on the assumption of linear separability of A it can be estimated by solving linear programming Problem (4) with $\alpha = \mu(X_A)$. More formally, we solve the following problem:

$$
\begin{cases}
\sum_{i=1}^{N} h(x_i) \xrightarrow[h(x_1),\ldots,h(x_N)]{} max; \\
\sum_{i=1}^{N} \left(x_i - \frac{\sum_{j=1}^{l} x_j}{l} \right) h(x_i) = 0; \\
0 \le h(x_i) \le 1, \ i = 1,\ldots,N.
\end{cases}
\tag{5}
$$

Since $\mu(X_A)$ is an estimation of the $\mu(A)$, in calculations we replace equalities $f(h) = 0$ by two inequalities $-\varepsilon \ge f(h) \ge \varepsilon$ for a certain small value of $\varepsilon > 0$.

Note that the performance of such estimation depends on two following factors:

- The real linear separability of class A in space $\mathbb{R}^n$.
- The difference between the mean estimated by $\mu(X_A)$ and the real mean $\mu(A)$.

In order to show how the proposed method works in practice, we constructed a synthetic experiment presented in Fig. 1 (left). In particular, it shows 300 points belonging to class A (red crosses, yellow triangles, and blue pluses) and 750 points which do not belong to A (purple crosses and gray circles). Class A is linearly separable in $\mathbb{R}^2$ by construction. Initially we have 100 labeled points in set X_A (red crosses on Fig. 1 (left)) and we calculate its mean $\mu(X_A)$. Using the proposed method we construct the estimation $\hat{A} \in W_\alpha$ of class A (yellow triangles and purple crosses Fig. 1 (left)). As a result just 1% of elements from class A were classified incorrectly (*False Negative*), making *Recall* = 99%. Moreover, there are no elements in *False Positive* group, which makes the *Precision* = 100%.

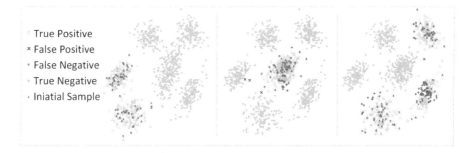

Fig. 1. Synthetic examples of (left) linearly separable class; (middle) normally distributed class; (right) linearly separable in transformed kernel space class. (Color figure online)

Class with Gaussian Distribution. In a certain class of problems, we can assume that elements of class A have a Gaussian distribution. This distribution is defined by its mean and covariation matrix, and, therefore, class A would fit an ellipsoid in $\mathbb{R}^n$ where most of the elements from A are inside the ellipsoid and most of the others are outside. We can estimate such class A by fixing not only the mean of A but also its covariation matrix, i.e. every second moment of the class. We do it by enriching the space $\mathbb{R}^n$ with all pairwise products between the n initial coordinates (features). In this case Problem (4) transforms to the following form:

$$
\begin{cases}
\sum_{i=1}^{N} h(x_i) \xrightarrow[h(x_1),\ldots,h(x_N)]{} max; \\
\sum_{i=1}^{N} \left(x_i - \frac{\sum_{j=1}^{l} x_j}{l} \right) h(x_i) = 0; \\
\sum_{i=1}^{N} \left(x_i^{(t)} \cdot x_i^{(r)} - \frac{\sum_{j=1}^{l} x_j^{(t)} \cdot x_j^{(r)}}{l} \right) h(x_i) = 0; \ t = 1,\ldots,n; \ r = 1,\ldots,n; \\
0 \le h(x_i) \le 1, \ i = 1,\ldots,N.
\end{cases}
\tag{6}
$$

In this case, the resulting class would be separable in $\mathbb{R}^n$ with a surface of the second order.

Figure 1 (middle) illustrates the work of the proposed method in the case of normally distributed class. All colors have the same meaning as in Fig. 1 (left) described in Sect. 3.3. There are 300 elements that belong to class A and 750 elements that do not belong to A. In this example class, A has a Gaussian distribution. Based on the initial sample X_A consisting of 100 elements we calculate the estimation of the mean and the covariation matrix of A and construct the estimation of A by solving the Problem (6) using linear programming. The results of this estimation are presented in Fig. 1 (middle). In this case we got $Precision = 96\%$ and $Recall = 98\%$.

Linearly Separable Class in the Transformed Space of Kernel Functions. One of the ways of identifying the space where the class is linearly separable is introducing kernel function $K(x_i, x_r)$ [1]. The linear programming Problem 5 for identifying the linearly separable set in the transformed space states as follows:

$$
\begin{cases}
\sum_{i=1}^{N} h(x_i) \xrightarrow[h(x_1),\ldots,h(x_N)]{} max; \\
\sum_{i=1}^{N} \left(K(x_i, x_r) - \frac{\sum_{j=1}^{l} K(y_j, x_r)}{l} \right) h(x_i) = 0; \ r = 1,\ldots,N; \\
0 \le h(x_i) \le 1, \ i = 1,\ldots,N.
\end{cases}
\tag{7}
$$

In order to illustrate the use of the proposed method in the transformed space of kernel functions, we construct an experiment where 450 elements belong to class A and 600 elements do not (Fig. 1 (right)). We optimize the indicator function h in the Problem (7) within the sets having the same mean in the transformed space as of the initial sample X_A consisting of 150 elements from A. As a result of this experiment we got $Precision = 94.5\%$ and $Recall = 94.6\%$.

4   Experiments

In order to show how the proposed algorithm works in practice, we applied it to the problem of classification of the handwritten digits in USPS dataset. This dataset was used previously in many studies including [15]. It contains 9298 digit images of size $16 \times 16 = 256$ pixels. In our experiment, we tried to identify the class of digit "0" within the rest of the digits. We denote the class of "0" by A. There are 1553 elements of class A in the dataset.

First, we ran linear SVM algorithm in the original 256-dimensional space and it had an error in the classification of only one element, which means that class A is almost linearly separable in this space. Secondly, we applied the proposed algorithm. We identified the real mean $\mu(A)$ of class A and used it to find the maximal by probability set with the same mean, as defined in Problem (4). Based on the Theorem 1 this estimation $\bar{A}$ is linearly separable. The results show that $\bar{A}$ differs from A in only one element. It means that our approach works well if we use the real mean of class A for the calculations.

However, in practice we can only estimate the real mean of class A based on a small sample of elements $X_A \subset A$ and solve Problem (5) as described in Sect. 3.3. Therefore, in the next experiment we study the relationship between the size of initial sample X_A an the performance of the constructed classification in terms of *precision* and *recall* measures. In particular, we run 100 times the following experiment:

1. Divide the dataset randomly to T_{3100} and S_{6198} with 3100 and 6198 of digit images correspondingly. The average number of elements from class A in S_{6198} is equal to 1046.
2. Within $T_{3100} \cap A$ create 20 random samples $\{X_A^{25}, X_A^{50}, X_A^{75}, \ldots, X_A^{500}\}$ with sizes from 25 to 500 with a step of 25.
3. For each sample X_A^* calculate its mean $\mu(X_A^*)$ and solve the linear programming Problem (5) for the set S_{6198} and mean $\mu(X_A^*)$.
4. Calculate the *precision* and *recall* measures of classification performance of elements in S_{6198}.

Note, that in this experiment we used separate set X_A^* that is not included into S_{6200}. We did it in order to eliminate the growth of the performance with the size of X_A^* based on the larger number of known labels and not on the better estimation of the mean of class A.

Finally, for each size of sample X_A we obtained 100 random experiments with corresponding values of *precision* and *recall*. Figure 2 shows the plots of the average *precision* and *recall* as functions from the initial sample size with 10% and 90% quantiles. In particular, Fig. 2 shows that the results are stable since the corridor between the 10% and 90% quantiles is narrow. The average value of *precision* does not grow significantly, and the standard deviation decreases with the growth of the sample size. The average value of *recall* grows with the sample size and its standard deviation decreases. Note that, *precision* reaches 90% level within 100 elements in the labeled set, and *recall* reaches this level within only 300 elements in the set X_A. It means, that in this experiment the high error in

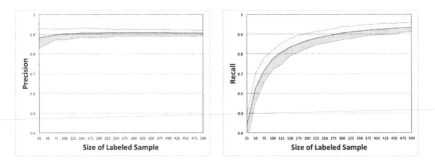

Fig. 2. Precision (left) and Recall (right) as functions from the size of the training sample X_A.

the estimating mean of the class A leads to identifying only a subset of A as a maximal by probability set.

In [15] authors used the same dataset and they also tried to separate class of "0" from other digits. However, [15] work in transformed space of kernel functions, whereas in our experiments we work in the original 256-dimensional space. The training set in their case consisted of all "0" images from 7291 images, which means that they had about 1200 images of "0". The test set contained 2007 images. They got $precision = 91\%$ and $recall = 93\%$. Figure 2 shows that our approach reaches the same level of performance in terms of $precision$ and $recall$ with the initially labeled sample of 500 elements.

5   Conclusion

In this paper, we presented a novel semi-supervised one-class classification method which assumes that class is linearly separable from other elements. We proved theoretically that class A is linearly separable if and only if it is maximal by probability within the sets with the same mean. Furthermore, we presented an algorithm for identifying such linearly separable class utilizing linear programming. We described three application cases including the assumption of linear separability of the class, Gaussian distribution, and the case of linear separability in the transformed space of kernel functions. Finally, we demonstrated the work of the proposed algorithm on the USPS dataset and analyzed the relationship of the performance of the algorithm and the size of the initially labeled sample.

As a future research, we plan to examine the work of the proposed algorithm in the real industrial settings.

References

1. Aizerman, M.A., Braverman, E.A., Rozonoer, L.: Theoretical foundations of the potential function method in pattern recognition learning. In: Automation and Remote Control, Number 25 in Automation and Remote Control, pp. 821–837 (1964)
2. Amer, M., Goldstein, M., Abdennadher, S.: Enhancing one-class support vector machines for unsupervised anomaly detection. In: Proceedings of the ACM SIGKDD Workshop on Outlier Detection and Description, ODD 2013, pp. 8–15 (2013)
3. Boyd, S., Vandenberghe, L.: Convex Optimization. Cambridge University Press, New York (2004)
4. Chapelle, O., Schölkopf, B., Zien, A.: Semi-Supervised Learning. MIT Press, Cambridge (2010)
5. Dreiseitl, S., Osl, M., Scheibböck, C., Binder, M.: Outlier detection with one-class SVMS: an application to melanoma prognosis. In: AMIA Annual Symposium Proceedings/AMIA Symposium, vol. 2010, pp. 172–176. AMIA Symposium (2010)
6. Joffe, E., Pettigrew, E.J., Herskovic, J.R., Bearden, C.F., Bernstam, E.V.: Expert guided natural language processing using one-class classification. J. Am. Med. Inform. Assoc. 22(5), 962–966 (2015)
7. Kennedy, K., Namee, B.M., Delany, S.J.: Using semi-supervised classifiers for credit scoring. J. Oper. Res. Soc. 64(4), 513–529 (2013)
8. Khan, S.S., Madden, M.G.: A survey of recent trends in one class classification. In: Coyle, L., Freyne, J. (eds.) AICS 2009. LNCS, vol. 6206, pp. 188–197. Springer, Heidelberg (2010). https://doi.org/10.1007/978-3-642-17080-5_21
9. Lee, G., Scott, C.D.: The one class support vector machine solution path. In: 2007 IEEE International Conference on Acoustics, Speech and Signal Processing - ICASSP 2007, vol. 2, pp. II-521–II-524 (2007)
10. Li, W., Guo, Q., Elkan, C.: A positive and unlabeled learning algorithm for one-class classification of remote-sensing data. IEEE Trans. Geosci. Remote Sens. 49(2), 717–725 (2011)
11. Manevitz, L.M., Yousef, M.: One-class SVMS for document classification. J. Mach. Learn. Res. 2, 139–154 (2002)
12. Muñoz, A., Moguerza, J.M.: One-class support vector machines and density estimation: the precise relation. In: Sanfeliu, A., Martínez Trinidad, J.F., Carrasco Ochoa, J.A. (eds.) CIARP 2004. LNCS, vol. 3287, pp. 216–223. Springer, Heidelberg (2004). https://doi.org/10.1007/978-3-540-30463-0_27
13. Munoz-Mari, J., Bovolo, F., Gomez-Chova, L., Bruzzone, L., Camp-Valls, G.: Semisupervised one-class support vector machines for classification of remote sensing data. IEEE Trans. Geosci. Remote Sens. 48(8), 3188–3197 (2010)
14. Prakash, V.J., Nithya, L.M.: A survey on semi-supervised learning techniques. Int. J. Comput. Trends Technol. (IJCTT) 8(1), 25–29 (2014)
15. Schölkopf, B., Platt, J.C., Shawe-Taylor, J.C., Smola, A.J., Williamson, R.C.: Estimating the support of a high-dimensional distribution. Neural Comput. 13(7), 1443–1471 (2001)
16. Schölkopf, B., Williamson, R.C., Smola, A.J., Shawe-Taylor, J., Platt, J.C.: Support vector method for novelty detection. In: Solla, S.A., Leen, T.K., Müller, K. (eds.) Advances in Neural Information Processing Systems, vol. 12, pp. 582–588. MIT Press (2000)

17. Sundarkumar, G.G., Ravi, V., Siddeshwar, V.: One-class support vector machine based undersampling: application to churn prediction and insurance fraud detection. In: IEEE International Conference on Computational Intelligence and Computing Research (ICCIC), pp. 1–7 (2015)

18. Tax, D.M., Duin, R.P.: Support vector data description. Mach. Learn. **54**(1), 45–66 (2004)

19. Vapnik, V.: Transductive inference and semi-supervised learning. In: Chapelle, O., Schölkopf, B., Zien, A. (eds.) Semi-Supervised Learning, Chap. 24, pp. 453–472. MIT Press (2006)

20. Zhang, R., Zhang, S., Muthuraman, S., Jiang, J.: One class support vector machine for anomaly detection in the communication network performance data. In: Conference on Applied Electromagnetics, Wireless and Optical Communications, pp. 31–37 (2007)

Prediction of Drug Efficiency by Transferring Gene Expression Data from Cell Lines to Cancer Patients

Nicolas Borisov[1,2]([✉]), Victor Tkachev[2], Anton Buzdin[1,2], and Ilya Muchnik[3]

[1] Institute for Personalized Medicine,
I.M. Sechenov First Moscow State Medical University, 8/2, Trubetskaya Street, Moscow 119991, Russian Federation
borisov@oncobox.com
[2] Department of R&D, OmicsWay Corp., 40S Lemon Ave, Walnut, CA 91789, USA
[3] Rutgers University, Hill Center, Busch Campus, Piscataway, NJ 08855, USA

Abstract. The paper represents a novel approach for individual medical treatment in oncology, based on machine learning with transferring gene expression data, obtained on cell lines, onto individual cancer patients for drug efficiency prediction. We give a detailed analysis how to build drug response classifiers, on the example of three experimental pairs of data "*kind of cancer/ chosen drug for treatment*". The main hardness of the problem was the meager size of patient training data: it is many many hundred times smaller than a dimensionality of original feature space.

The core feature of our transfer technique is to avoid extrapolation in the feature space when make any predictions of the clinical outcome of the treatment for a patient using gene expression data for cell lines. We can assure that there is no extrapolation by special selection of dimensions of the feature space, which provide sufficient number, say M, of cell line points both below and above any point that correspond to a patient. Additionally, in a manner that is a little similar to the *k nearest neighbor* (kNN) method, after the selection of feature subspace, we take into account only K cell line points that are closer to a patient's point in the selected subspace. Having varied different feasible values of K and M, we showed that the predictor's accuracy considered AUC, for all three cases of cancer-like diseases are equal or higher than 0.7.

1 Introduction

Let us describe a special medical application for data analysis, when there are two training data sets in the same feature space, associated each with its specific problem to build a predictor. Imagine also that the analyst wants to find a way, how to incorporate

Electronic supplementary material The online version of this chapter (https://doi.org/10.1007/978-3-319-99492-5_9) contains supplementary material, which is available to authorized users.

L. Rozonoer et al. (Eds.): Braverman Readings in Machine Learning, LNAI 11100, pp. 201–212, 2018.
https://doi.org/10.1007/978-3-319-99492-5_9

a solution for one problem to improve a solution for the second problem. In the considered case, the training data for one problem is a large enough, but another one is very small. Moreover, the problems are really different: their labels have different meaning.

It is natural to interpret the case as an example from area of machine learning, for which Vapnik gave name "knowledge transferring" [1]. However, it can't be included in scenarios, which he emphasized in papers [2–4]. It is a novel case, as we will show below. At the same time, our main attention will be focused on a very narrow special medical application example for two reasons.

Primarily, our approach is heuristic, based on a common geometrical sense, and we required to achieve a strong demonstration that it works. Secondarily, a foundation of our heuristics becomes more clearly observable, when it involves real content. Its items are the following: (1) when constructed predictor analyzes a new sample to produce associated prediction, it has to be adapted to that new observation, re-learn; (2) the re-learned predictor has to be built within a new specific subspace and uses a reconstructed training data. Now let us move to a basic story, our medical application.

Individual medicine implies that distinct treatment methods are prescribed to individual patients according several features that may be obtained from, e.g., gene expression profile. It is well known, that machine learning methods suffer from the deficiency of preceding samples, i.e. the gene expression data on patients combined with the confirmed outcome of known treatment methods. At the same time, there exist thousands of various cell lines that were treated with hundreds of anti-cancer drugs in order to check the ability of these drugs to stop the cell proliferation, and all these cell line cultures were profiled in terms of their gene expression.

Here we present a new method to predict efficiency of anti-cancer drugs for individual patients by transferring features obtained from the expression-based data from cell lines. The method was applied on three datasets for cancer-like diseases (chronic myeloid leukemia, as well as lung adenocarcinoma and renal carcinoma) treated with targeted drugs – kinase inhibitors, such as imatinib or sorafenib.

We describe the method rather narrowly, with a lot of details and pre-processing procedures, focused on very particular specifics of analyzed applications. Even that, as was describe above, we hope, that the main idea can be generalized into a broader-sense case, when one has several prediction problems in the same space, and wants to take solutions from some of them to improve solution for a particular one[1].

The sophisticated nature of intracellular processes and events that lead to cell proliferation and cancer progression gives a hint to combine prediction of drug efficiency of certain drugs and treatment methods for individual patients [6]. This approach, sometimes termed *drug scoring* and/or *personalized (individual) medicine*, may therefore involve the analysis of expression-based features and values to distinguish the patient who will respond to suggested treatment method. Two principal approaches may be used for expression-based *drug scoring*. One type, say *a priori scores*, evaluates the abilities of a certain drug to restore the normal status of the

---

[1] In that one could find an analogy between the situation and very popular now case, called "domain adaptation" [5].

expression-based pattern, or to terminate the physiological process that is considered pathogenic for a certain disease (e.g. cell proliferation for cancer etc.) [6]. Another possible type of drug scoring exploits precedent-based learning. These *a posteriori drug scores* are the result of a machine learning on a training dataset, which contains expression-based features extracted for the patients, who were treated with a certain drug (for each patient, the outcome of the treatment, whether it is a positive response or lack of it, is also known). The machine learning methods may be applied to distinguish between the responder and non-responder groups in the space of expression-based features. The methods require hundreds or thousands points for the training dataset to provide the adequate coverage of the space [7]: a condition that lies far beyond the current capacity of gene expression profiles for the cancer patients with the case histories that specify both treatment method and the clinical response. For most anti-cancer drugs it is extremely difficult (if ever possible) to find hundreds of gene expression that were obtained using the same investigation platform for the patients that were treated with the same drug with the known clinical outcome of the treatment [8–10]. From the other side, thousands of expression profiling results have been obtained for various cell lines that were used for testing the ability of hundreds of drugs to inhibit the cell proliferation [11].

Thus, we propose here a novel method for the transfer of expression-based data from the more numerous cell lines to less abundant cases of real patients for subsequent application of machine-learning that predict the clinical efficiency of anti-cancer drugs. Practitioners from medicine and pharmaceutics are very interested in those researches [6, 12].

According to the standard approaches to validation [12] of machine leaning methods for analysis of expression-based features, we have used the leave-one-out procedure and AUC metric with a predefined threshold as main algorithms to select appropriate predictors. To make validation tests stronger, we also did parallel analysis with using three different machine-learning methods (SVM [13, 14], binary trees [14] and random forests [15]) to build predictor-classifiers.

2   Data Sources of Cell Lines and Patients to Design, Test and Validate Our Method

We have organized the experimental analysis based on one expression dataset of cell lines and three datasets of patients, each corresponding to specific pair of *a kind of cancer-like disease* together with *a sort of drug used to treat this disease*. The cell lines data set of 227 elements was borrowed from the CancerRxGene study [11]. The three patients' datasets were taken for renal carcinoma (current study), lung adenocarcinoma [8] and chronic myeloid leukemia [9] comprising 28, 38, and 28 case histories, respectively. Those datasets were analyzed by several preprocessing procedures for normalization and extracting informative features. The fundamental procedure to extract informative features was organized based on the OncoFinder method that transforms gene expression data into features o activations of particular signaling pathways [16–19].

According to the OncoFinder approach, the *pathway activation score* (PAS) for a given sample n and a given pathway p is obtained as follows,

$$PAS_p = \sum_n ARR_{np} \cdot \log(CNR_n).$$

Here the case-to-normal ratio, CNR_n, for a gene n is the ratio of the expression level of gene n in the sample under investigation to the average expression level of that gene in the control, or normal, group of samples. In the current work, we have used the geometric mean over expression levels samples in a dataset as a normal expression value for each gene. ARR_{np} is the discrete value of the activator/repressor role equals the following fixed values: -1, when the gene/protein n is a repressor of molecular pathway; 1, if the gene/protein n is an activator of pathway; 0, when the gene/protein n is known to be both an activator and a repressor of the pathway; and 0.5 and -0.5, respectively, tends to be an activator or a repressor of the pathway p, respectively. A positive PAS value indicates activation of a pathway, and a negative value indicates repression. The validity of the PAS concept was confirmed [18] on the calculations according to a low-level kinetic model of EGFR pathway activation [18, 20].

During the PAS calculations, we used three types of normal (control) expression values. First, it was the *auto-normalization (geometric-mean normalization)*, where the normal expression value for each gene was calculated as a geometric mean over all cell line/disease samples for this gene. Also, we have tried the *tissue-based normalization*, when the normal samples were taken from the organs/tissues of healthy persons [21–23]. For CancerRxGene cell lines, as the tissue-based normalization, we have used the non-tumor glial brain tissue [24] (samples GSM362995 to GSM363004) and smooth aortic muscles [25] (samples GSM530379 and GSM530381). The overview of the data sources for the cell lines and patients is given in Table 1.

3 Data Transfer Method: From the Cell Lines to Patients

As was mentioned above, every cell out of 227 was treated by a given drug. The treatment result was presented by the well-known efficiency coefficient IC_{50}, the drug concentration, which allows to inhibit the cell division process by 50% [26]. To quantify drug efficiencies for cell lines for a certain drug, these lines were first sorted by the descending order according to their IC_{50} values. After such sorting, all the lines were divided into five quintiles with equal number of members, with the first quintile contains the highest IC_{50}, and the fifth contains the lowest. Within each quintile i, the drug efficiency Y was assigned as follows, $Y = (i - 1) \cdot 25$, with $Y = 0$ for weakest responders, and $Y = 100$ for the strongest.

Also, every cell line was supported by gene expression profile, which was transformed, as mentioned before, into much shorter profile of activations of signaling pathways (PAS). For each drug type, only those pathways, which contain molecular targets of this drug, were taken into account. The total dataset for each cell line comprises its individual activation profile of targeted pathways and a quantized drug efficiency (Y).

Table 1. Data sources for the cell line and patient datasets

Case samples	Normal samples for tissue-based normalization	Cell type	Drug	Experimental platform	# of samples
GSE68950 [11]	GSE14805 [24] GSE21212 [25]	Cancer-RxGene cell lines	Imatinib Sorafenib	Affymetrix HT Human Genome U133A mRNA array	227
Current study	GSE49972 [21]	Renal carcinoma	Sorafenib	Illumina HumanHT-12v4 Expression BeadChip and CustomArray ECD 4X2K/12K	28 (13 responders, 15 non-respondres)
GSE31428 [8]	GSE43458 [22]	Lung adeno-carcinoma	Sorafenib	Affymetrix Human Gene 1.0 ST Array	37 (23 responders, 14 non-responders)
GSE2535 [9]	GSE2191 [23]	Chronic myeloid leukemia	Imatinib	Affymetrix Human Genome U95 Version 2 Array	28 (16 responders, 12 non-responders)

For the patients, the data set in similar: the set of PAS for targeted pathways, and a clinical outcome of a treatment, which, as we have considered, may be either positive (responder) or negative (non-responder). The overall datasets for the cell lines and patients for all the normalization methods used, are given in Supplementary Table 1.

Using the cell line data, we determine a set of regression models, which characterize a dependency between features of the activation of signaling pathways and the drug efficiency coefficient Y. Individual regression model was calculated based on the data for a particular subset of cell lines, determined by the corresponding patient. Moreover, the individual (patient depended) regression is determined in a particular subspace of features also depend upon of a patient (!). This patient-depended subset of cell lines in the corresponding subspace was defined the following procedure:

(a) on axis for every feature in the space of PAS we fix a point, associated with the chosen patient;

(b) for predefined integer M check if there exist on the axis at least M cell's points above the chosen patient's point, and also at least M cell's points below it. If this condition is satisfied, we keep the feature as relevant to the patient; all set of relevant features forms the subspace, where we determine subset of cell lines associated with the chosen patient;

(c) in the relevant subspace and for the predefined integer K we find the K nearest cell lines to the chosen patient's point [27]; that K cell line point in extracted relevant subspace is the dataset, on which the mentioned above individual regression model is constructed.

As result of that analysis, we get for every patient two values: predicted drug score (*DS*), as well as prior knowledge, that he (or she) gave the positive response on the treatment. Taking some threshold (τ), by which to decide the drug score is high enough to classify this patient as a responder, we are able to calculate AUC value for the whole set of patients. Taking into account that predicted drug scores for patients depend upon chosen constants M and K, we can check the AUC values over the lattice of all feasible pairs (M, K) – to find the largest connected area and the highest AUC within this area. That best pair (M_{opt}, K_{opt}). Also we search the threshold (τ) to find the best (τ_{opt}).

Those three best constants (M_{opt}, K_{opt}) and τ_{opt} determine completely the final predictor:

For a given new patient based on his or her set of gene depended activation features we determine

(1) its unique subset of featured using M_{opt} constant, than,
(2) its regression model in that subspace of features using K_{opt} constant; finaly,
(3) the constant τ_{opt} allows to assign it by a label *responder* or *non-responder*.

4 Validation of the Data Transfer Procedure

The best results for the data transfer procedure described in the previous section were obtained using SVM with linear kernel (Table 2, Figs. 1 and 2, Supplementary Table 3). Nine panels show different disease/drug types and normalization method for PAS calculations. Similarly to columns and rows of Table 2, columns of panels in Figs. 1 and 2 stay for different normalizations, from left to right: auto-normalization (the normal expression value for each gene was calculated as a geometric mean from all cell line/disease samples for this gene over this set of samples), and tissue-based

Table 2. Performance test for data transfer form CancerRxGene data [11] to cancer/leukemia patients for regression-mode SVM with linear kernel. The AUC value was calculated for drug response prediction as a marker for observed response to the treatment.

Disease type, Drug	AUC for optimized (M, K) parameters for auto-normalization (geometric mean)	AUC for optimized (M, K) parameters for tissue-based glial [24] normalization	AUC for optimized (M, K) parameters for tissue-based aortic [25] normalization
Renal carcinoma, sorafenib (current study)	0.81	0.77	0.82
Lung adenocarcinoma, sorafenib [8]	0.72	0.77	0.78
Chromic myeloid leukemia, imatinib [9]	0.76	0.77	0.78

normalization with glial [24] and aortic [25] samples used as normal for CancerRxGene [11] cell lines. Rows of panels show the results for different disease/drug combinations, from top to bottom: renal cancer treated with sorafenib (current study), lung cancer treated with sorafenib [8] and chronic myelogenous leukemia treated with imatinib [9].

To distinguish between clinical responders and non-responders, the threshold (τ) value is determined to *minimizes the sum of false negative and false positive predictions* (Fig. 2): when $DS > \tau$, a patient is classified as a drug responder, whereas the value of $DS < \tau$ indicates a non-responding case.

Those optimal *data transfer* parameters ((M_{opt}, K_{opt}); see Fig. 1), defined as the *top of the biggest connected area in the map of AUC plotted as a function of M and K,*

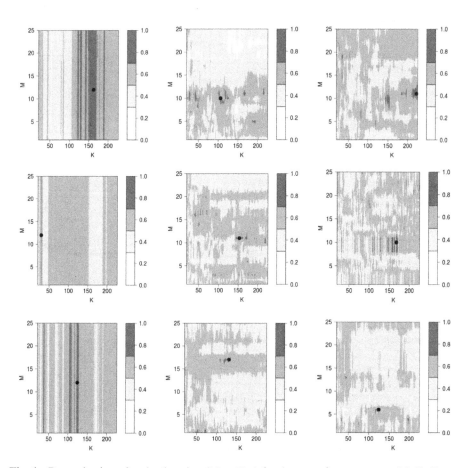

Fig. 1. Determination of optimal setting (M_{opt}, K_{opt}) for data transfer parameters (M, K). Rows and columns of panels correspond to disease types and PAS normalizations, respectively, similarly to rows and columns of Table 2. Within each panel, the color map displays the distribution of AUC values for linear SVM predictor of drug efficiency. The darkest shade of grey shows the areas where AUC > θ for the quality threshold $\theta = 0.7$. The top of the biggest connected area ("island") where AUC > θ (in the lower right corner) is marked with a boldface black point.

when *AUC exceed a threshold* ($\theta = 0.7$), appear to be stable according to the leave-one-out quality assurance procedure for the patient dataset. In fact, these parameters (M_{opt}, K_{opt}) always lie within this biggest connected area ("island"), *which was defined for the whole set of patients*, if we perform similar optimization procedure *for all but one patients* (Supplementary Table 4). Note also that for the auto-normalization of gene expression levels during PAS calculations for cell line sand patients makes all the points of the patient set lie in the central area of the cloud of the points of the cell line dataset. Therefore, the performance of data transfer does not depend on the parameter M (Fig. 1, left column of panels).

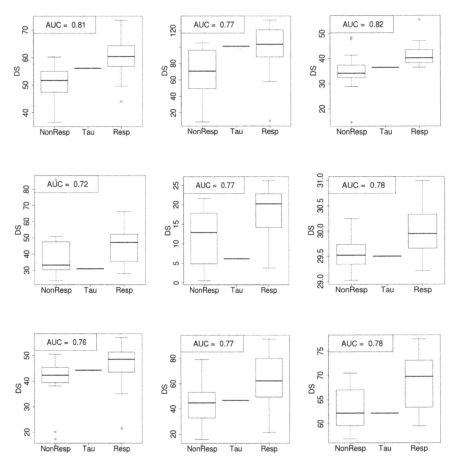

Fig. 2. Data transfer-assisted separation of responders from non-responders to clinical treatment of cancer-like diseases with certain drugs. Rows and columns of panels correspond to disease types and PAS normalizations, respectively, similarly to rows and columns of Table 2. Within each panel, left and right boxes represent continuous value of *DS* (drug score), obtained using the SVM-based regression procedure for non-responding (NonResp) and responding (Resp) patients, respectively. The central line shows the discrimination threshold value (τ) that minimizes the sum of false positive and false negative predictions; all patients with $DS > \tau$ are classified responders and vice versa.

The method has three basic hyper-parameters (M, K and τ), which adjustment took the largest part of the total time budget. After this adjustment, further classification of patients with unknown clinical outcome becomes quite fast. Figure 3 illustrates it on the example of two patients (labeled as P1 and P2) with clear cell renal carcinoma. The optimal data transfer parameters (M_{opt}, K_{opt} and τ_{opt}) for three PAS normalization modes (auto-normalization, glial and aortic tissue-based normalizations) were chosen using the data for patients with known treatment outcome (see Table 2, Figs. 1 and 2, Supplementary Table 3). These parameters were then used for assessment of clinical outcome of possible sorafenib treatment for new patients, P1 and P2. The boxes NR (non-responders) and R (responders), respectively, show the DS value for the patients, who had previously shown negative and outcome of the sorafenib treatment. Although the PAS features for those patients strongly depend on the normalization mode (Supplementary Table 5), the evaluation of sorafenib efficiency for all normalization modes consensually indicates negative clinical outcome for the patient P2 and at least very doubtful positive, if not strongly negative, outcome for the patient P1 (Fig. 3 and Supplementary Table 6).

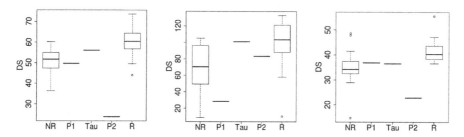

Fig. 3. Prediction of sorafenib clinical efficiency for two clear cell renal carcinoma patients (labeled as P1 and P2) with *unknown* clinical outcome. Panels of the figure, from left to right, respectively, display different modes of PAS normalizations: auto-normalization, as well as glial and aortic tissue-based normalizations.

5 Discussion

The clinical cases with both known treatment outcome and gene expression profile for most drugs and disease types are limited. Taking into account high dimensionality of expression features, it may sound attractive to involve as training datasets for machine learning the profiling results for cell line cultures treated with different drugs, since these cases are generally more numerous than the patient data with known clinical outcome. In the current work, we have used as the training set the CancerRxGene data [11] for cell lines treated with different kinase inhibitors.

In our analysis, the SVM method demonstrates stable good results. Another possible way to increase results is the use of pathway activation strength (PAS) [16, 18] for the ensembles of cell signaling pathways instead of expression levels of distinct genes.

Keeping the main goal to extract (from the cell lines data) essential information for subsequent transfer onto the patients, and to prevent extrapolation during such data transfer, we have proposed a method that excludes from expression-based feature space any dimension, which does not provide the significant (say, M) number of points in the cell line dataset both below and above each point of the patient dataset. Following that individual dimensionality reduction, we than select, similarly to the k nearest neighbor (kNN) method [27], K points in the cell line dataset that are proximal a certain point in patient dataset, to build the SVM model. The main difference was that in our case K can't be small (lower then 10); contrary, it should be enough to build a regression model. In other words, for each patient-point, our data transfer algorithm filters the cell line data using a *floating window* that surrounds every point in the patient dataset, in order to preclude extrapolation and to neglect the influence of too distant points of the cell line dataset.

This data transfer was checked on three cancer-type diseases (chronic myeloid leukemia, as well as renal carcinoma and lung adenocarcinoma). For each disease the optimal parameters (M_{opt}, K_{opt}) allow separation of clinically responding patients from non-responding ones with AUC > 0.70. These appear to be stable according to the leave-one-out quality assurance procedure for the patient dataset.

Let us repeat again how was done all hyper-parameters choices, needed to achieve good result (from constants to find the best to procedures from different kinds). It was combinatorically brute force by leave-one-out procedure. That was the biggest bottleneck of the experimental part of work, which takes about one day of computations (using R of version 3.2.3 for 64-bit Ubuntu 16.04 LTS operational system at a personal computer with four 2.50 GHz Intel® Core™ i7-6500U CPUs and 15.5 GiB of RAM). Very recently, we found a modification of the method, which should dramatically reduce its time consuming, and that work will be the next step in our near future development.

Acknowledgements. This work was supported by the Russian Science Foundation grant 18-15-00061.

Disclosure of Interests
The authors declare no conflicts of interests.

Appendices: Materials and Methods

Transcriptome Profiling for Renal Cancer Samples
The details of experimental procedure at Illumina HumanHT-12v4 and CustomArray ECD 4X2K/12K platform were reported previously [19]. Raw expression data were deposited in the GEO database (http://www.ncbi.nlm.nih.gov/geo/), accession numbers GSE52519 and GSE65635.

Harmonization of Illumina and Custom Array Expression Profiles for Renal Cancer

To cross-harmonize the results for the Illumina and CustomArray gene expression profiling, all expression profiles were transformed with the XPN method [28] using the R package CONOR [29].

SVM, Binary Tree and Random Forest Machine Learning Procedures

All the SVM calculations were performed using the R package 'e1071' [30], that employs the C++ library 'libsvm' [31]. Calculations according to binary tree [14] and random forest [15] methods were done with the R packages 'rpart' and 'randomForest', respectively.

References

1. Vapnik, V., Izmailov, R.: Learning using privileged information: similarity control and knowledge transfer. J. Mach. Learn. Res. **16**, 2023–2049 (2015)
2. Lopez-Paz, D., Bottou, L., Schölkopf, B., Vapnik, V.: Unifying distillation and privileged information. In: ICLR 2016, San Juan, Puerto Rico (2016)
3. Xu, X., Zhou, J.T., Tsang, I., Qin, Z., Goh, R.S.M., Liu, Y.: Simple and efficient learning using privileged information (2016)
4. Celik, Z.B., Izmailov, R., McDaniel, P.: Proof and implementation of algorithmic realization of learning using privileged information (LUPI). In: Paradigm: SVM+. Institute of Networking and Security Research (INSR) (2015)
5. Csurka, G.: Domain Adaptation in Computer Vision Applications. Springer, Cham (2017). https://doi.org/10.1007/978-3-319-58347-1
6. Artemov, A., et al.: A method for predicting target drug efficiency in cancer based on the analysis of signaling pathway activation. Oncotarget **6**, 29347–29356 (2015)
7. Minsky, M.L., Papert, S.A.: Perceptrons - Expanded Edition: An Introduction to Computational Geometry. MIT Press, Boston (1987)
8. Blumenschein, G.R., et al.: Comprehensive biomarker analysis and final efficacy results of sorafenib in the BATTLE trial. Clin. Cancer Res **19**, 6967–6975 (2013). Off. J. Am. Assoc. Cancer Res.
9. Crossman, L.C., et al.: In chronic myeloid leukemia white cells from cytogenetic responders and non-responders to imatinib have very similar gene expression signatures. Haematologica **90**, 459–464 (2005)
10. Mulligan, G., et al.: Gene expression profiling and correlation with outcome in clinical trials of the proteasome inhibitor bortezomib. Blood **109**, 3177–3188 (2007)
11. Yang, W., et al.: Genomics of Drug Sensitivity in Cancer (GDSC): a resource for therapeutic biomarker discovery in cancer cells. Nucleic Acids Res. **41**, D955–D961 (2013)
12. Robin, X., Turck, N., Hainard, A., Lisacek, F., Sanchez, J.-C., Müller, M.: Bioinformatics for protein biomarker panel classification: what is needed to bring biomarker panels into *in vitro* diagnostics? Expert Rev. Proteomics **6**, 675–689 (2009)
13. Osuna, E., Freund, R., Girosi, F.: An improved training algorithm for support vector machines, pp. 276–85. IEEE (1997). http://ieeexplore.ieee.org/document/622408/. Accessed 23 May 2017
14. Bartlett, P., Shawe-Taylor, J.: Generalization performance of support vector machines and other pattern classifiers. In: Advances in Kernel Methods. Support Vector Learn, pp. 43–54 (1999)

15. Toloşi, L., Lengauer, T.: Classification with correlated features: unreliability of feature ranking and solutions. Bioinformatics **27**, 1986–1994 (2011)
16. Buzdin, A.A., et al.: Oncofinder, a new method for the analysis of intracellular signaling pathway activation using transcriptomic data. Front Genet. **5**, 55 (2014)
17. Buzdin, A.A., Prassolov, V., Zhavoronkov, A.A., Borisov, N.M.: Bioinformatics meets biomedicine: oncofinder, a quantitative approach for interrogating molecular pathways using gene expression data. Methods Mol. Biol. **1613**, 53–83 (2017). Clifton NJ.
18. Aliper, A.M., et al.: Mathematical justification of expression-based pathway activation scoring (PAS). Methods Mol. Biol. **1613**, 31–51 (2017). Clifton NJ
19. Borisov, N., et al.: Data aggregation at the level of molecular pathways improves stability of experimental transcriptomic and proteomic data. Cell Cycle **16**(19), 1810–1823 (2017). Georget Tex
20. Kuzmina, N.B., Borisov, N.M.: Handling complex rule-based models of mitogenic cell signaling (On the example of ERK activation upon EGF stimulation). Int. Proc. Chem. Biol. Env. Eng. **5**, 76–82 (2011)
21. Karlsson, J., et al.: Clear cell sarcoma of the kidney demonstrates an embryonic signature indicative of a primitive nephrogenic origin. Genes Chromosomes Cancer **53**, 381–391 (2014)
22. Kabbout, M., et al.: ETS2 mediated tumor suppressive function and MET oncogene inhibition in human non-small cell lung cancer. Clin. Cancer Res **19**, 3383–3395 (2013). Off. J. Am. Assoc. Cancer Res.
23. Yagi, T., et al.: Identification of a gene expression signature associated with pediatric AML prognosis. Blood **102**, 1849–1856 (2003)
24. Hodgson, J.G., et al.: Comparative analyses of gene copy number and mRNA expression in glioblastoma multiforme tumors and xenografts. Neuro-Oncology **11**, 477–487 (2009)
25. Bhasin, M., Yuan, L., Keskin, D.B., Otu, H.H., Libermann, T.A., Oettgen, P.: Bioinformatic identification and characterization of human endothelial cell-restricted genes. BMC Genom. **11**, 342 (2010)
26. Cheng, Y., Prusoff, W.H.: Relationship between the inhibition constant (K1) and the concentration of inhibitor which causes 50 per cent inhibition (I50) of an enzymatic reaction. Biochem. Pharmacol. **22**, 3099–3108 (1973)
27. Altman, N.S.: An introduction to kernel and nearest-neighbor nonparametric regression. Am. Stat. **46**, 175–185 (1992)
28. Shabalin, A.A., Tjelmeland, H., Fan, C., Perou, C.M., Nobel, A.B.: Merging two gene-expression studies via cross-platform normalization. Bioinformatics **24**, 1154–1160 (2008)
29. Rudy, J., Valafar, F.: Empirical comparison of cross-platform normalization methods for gene expression data. BMC Bioinform. **12**, 467 (2011)
30. Wang, Q., Liu, X.: Screening of feature genes in distinguishing different types of breast cancer using support vector machine. OncoTargets Ther. **8**, 2311–2317 (2015)
31. Chang, C.-C., Lin, C.-J.: LIBSVM: a library for support vector machines. ACM Trans. Intell. Syst. Technol. **2**, 1–27 (2011)

On One Approach to Robot Motion Planning

Vladimir Lumelsky[(✉)]

University of Wisconsin-Madison, Madison, WI 53706, USA
vjlumels@wisc.edu

Abstract. As this essay is meant for a book devoted to the memory of Dr. E. Braverman, a few words about my relation to him seem in order. Professor Braverman was my PhD thesis advisor in 1964–1967 at the Institute of Control Problems of the USSR National Academy, in Moscow. After graduation I had stayed at the same Institute, first as junior and then as senior research fellow, and we remained close, up until my emigration to USA in 1975. His sharp intellect, incredible human warmth, and wide cultural and scientific interests were there for us, a circle of friends which I, a younger man, was generously accepted to. This essay presents a big-picture description of a scientific approach to theory and practice of automatic moving machinery such as robots, and directions it has led to - all developed later on, in my years in America. What connects it to my time in the circle around M. Braverman is the skill for structured thinking that I was lucky to have been schooled in while associated with that circle.

Keywords: Robotics · Topology · Sensing · Motion planning algorithms

1 Preface

Note 1: Professor Emmanuil Markovich Braverman was my PhD thesis advisor in 1964–1967 at the Institute of Control Problems of the USSR National Academy, in Moscow. After graduation I had stayed at the same Institute, first as junior researcher, then as senior research fellow, and we remained close, up until my emigration to USA in 1975. His being only six or seven years my older helped - for me he was Misha Braverman, a friend, with our science being but one of many interests that tied us. For me he was a person always to listen to, to discuss and bounce ideas, and to entrust things that I might've hesitated to discuss with others - of which in those uneasy Soviet times there were plenty. His sharp intellect, incredible human warmth, and wide cultural and scientific interests were there for us, a circle of friends which I, a younger man, was generously accepted to.

When my wife and I started thinking of leaving the USSR, a highly critical step wrought with known and unknown dangers - of which one was an immediate loss of our jobs upon letting this idea out - was the procedure of applying to the state authorities for permission to emigrate. While no preparation would guarantee success - not rarely permissions were denied for unfathomable reasons - there was also enough room for mistakes. A thoughtful advice was essential, and so on a bitterly cold evening in December 1974, when no one else, not even our families, had an inkling of our idea, a crucial conversation with loyal friends took place. Talking in the street was too cold,

© Springer Nature Switzerland AG 2018
L. Rozonoer et al. (Eds.): Braverman Readings in Machine Learning, LNAI 11100, pp. 213–228, 2018.
https://doi.org/10.1007/978-3-319-99492-5_10

in other places - dangerous. Four of us - Misha, myself, and two other friends, Andrei Malishevski and Lev Rozonoer - had crammed into Misha's tiny Russian-made car, the engine was turned on to supply badly needed heat, and that's how our emigration idea became a thought-through plan.

Note 2: So why did my later work, the topic of this essay, moved so far away from the topics of my PhD work with M. Braverman? Not by plan. Emigration, especially in circumstances like ours - as political refugees, with no passports and no citizenship, with zero knowledge about the world beyond the Iron Curtain - tends to throw a person around mercilessly, including of course professionally. (Think of a PhD-educated scientist who upon arrival to America is bewildered by strange concepts like a checkbook and credit cards.) No job option, wherever it led, could be neglected. The age of computer programming as an easy backup for a job had not arrived yet. Our first child was born three weeks after our arrival. I needed a job, quickly. I thought my only choice was menial jobs, perhaps fixing refrigerators or TV sets. It didn't come to it, but I had to switch from the academic work to industry, from the fancy intellectual topics I dealt with in Moscow (machine intelligence, pattern recognition...) to down-to-earth technical domains.

Luckily I soon found myself in the nascent field of robotics, first in large industrial research centers and later in academia. Variations of this domain had thereafter occupied me for many years, expanding from industrial system development to algorithms and rather abstract theory, electronic design, and even studies of human psychology and spatial reasoning - all rather far from our topics back in Moscow. I'm quite sure, however, that my unexpectedly smooth path into the wonderful world of Western science was largely helped by the years of learning how to think from the trio of remarkable minds, Misha Braverman one of them - the same three that helped work out our emigration plan in a shoddy Russian-made car on that freezing night in December 1974.

Note 3: The text below presents a summary of a line of research that I started early on in my career in US. Given the purpose of this volume - to commemorate a scholar and show his legacy and impact rather than to inform per se - the wide professional spread of the volume's authors carries an obvious danger of losing the reader. This calls for breadth at the expense of depth. With this in mind, this text is written as an essay, focusing on ideas and omitting technical details. References below are few and far between, more for the sake of volume uniformity than completeness. Readers interested in more detail may see my book [10] and my university site, https://directory.engr.wisc.edu/me/faculty/lumelsky_vladimir.

2   Introduction

Applications with a need for robots operating in an uncertain environment, in close proximity to humans, or with direct human-robot interaction are today in high demand. When planning a machine's motion, the primary concern is *uncertainty* due to *unstructured environment* - the lack of complete input information needed for decision-making. A robot working near a human by definition operates with a high level of uncertainty. The opposite situation, operating in a *structured environment*, is faced by an assembly line robot at the automotive plant: here the robot's surroundings are carefully designed, often at great cost, to eliminate uncertainty completely. It is intuitively clear that operating with uncertainty puts a premium on the robot's sensing abilities and intelligence.

While simpler versions of robot systems operating with some level of uncertainty have been implemented (e.g. for packaging and storage in warehouses [1]), tougher areas remain stubbornly resistive to robotics, such as robots preparing the Mars surface for human arrival, robots for assembly of large telescopes in space, robot mailmen, robot helpers for the elderly etc. Advances in this area, while impressive, are rare. Difficulties are many, both on the side of the robot technology and on the human side: robots have hard time adjusting to uncertainty, while human operators exhibit serious limitations in grasping spatial information and manipulating complex motion.

Successful systems developed up to now are as a rule based on making uncertainty "less uncertain" - which is to say, limiting the task in question to some special case with less uncertainty. Say, since a surgery robot would be of no use without a human surgeon working alongside it, the robot is designed so as to never generate motion so fast as to present difficulty to the pace of human thinking. This effectively means the robot *does not have to be aware* of the humans around it, thus of the dangers it may present to them. That's much easier to design.

While effective in some applications, this approach also heavily limits the potential areas of automation, such as in physical spaces that humans cannot reach - planetary space, nuclear reactors, oil wells etc. Unless a fundamental solution is found, the current slow rate of automation of tasks with uncertainty will continue.

When sometime in early 1970s researchers in the nascent field of robotics realized the centrality of the motion planning problem, the natural idea to handle it was - the robot gathers information about all objects in the scene, builds a map, in it figures out the whole path, and then starts moving. Out of this concept of *motion planning with complete information* came the well known *Piano Movers Problem* approach [2]. Its principle is simple: the piano movers look over the scene, observe what's there to know, define the desired route - and off they go, moving the bulky oddly-shaped piano this way and that until they arrive at the destination. When being implemented computationally, the approach is geometric in nature - it relies on a *connectivity graph* as description of the robot and its surroundings, with graph edges presenting obstacle-free potential route segments, and its vertices being route deflection points usually tied to points on surfaces of obstacles. A typical preprocessing operation here is approximation of real world (perhaps oddly-shaped) obstacles with mathematically manageable entities. The graph is naturally highly dependent on shapes of objects in the scene. Its complexity, and thus the approach's computational load, grows exponentially with the complexity of those objects and their approximation.

The Movers approach has been an important achievement in the theory of algorithms (it was the first example of a physical system with exponential computational complexity). It generated hundreds of algorithm variations and scholarly publications, some even attempting to handle limited uncertainty via various heuristics. It found its uses in various applications, mostly in domains where certainty rules (as e.g. printed circuit design), thus outside of robotics. We don't need to go into details to see why the Piano Movers approach can't be of much help for robot systems in an unstructured environment. First and foremost, in the latter one can't count on complete information about the robot's surroundings. This means, no maps can be built beforehand and no paths can be calculated before the motion starts.

Second, if living in an unstructured environment implies reacting to the surroundings quickly, in real time, the related computation must involve small amounts of data at the time; it must be *sequential*, not a one-time calculation. Building a connectivity graph of the space in question is not realistic. Third, it's clear intuitively that a successful motion planning procedure must not depend on the geometry of space, such as object shapes - that's too computationally expensive. It must depend on general topological properties of space - I can't cross it, I can go around, this sort of thinking. We walk around the lake without much thinking whether it's oval or octagonal or convex or concave. If the doors to a building are locked, we know that most likely no walking around it would help us get inside. That's topology, not geometry.

To recap, we are looking for a motion planning strategy whose all premises are opposite to those of the Movers approach: high level of uncertainty, real-time sequential computation, low step-by-step computational load, reliance on topology rather than geometry of space. But is such a procedure even feasible? And if so, given the uncertainty, is there a hope for *convergence* - a guarantee that the procedure delivers the solution if one exists, and reports its impossibility if it doesn't? The answer turns out to be Yes. The words "one approach" in our title refers to such a strategy, called *sensor-based motion planning* (SBMP), and its ramifications for robotics. Similar to the Piano Movers approach, by now a good number of variations of the SBMP approach have been proposed.

In practice, as a robot (or a human) moves around, information about its surroundings arrive via its sensors - hence the words "sensor-based" in the approach's name. That is, realization of this approach calls for supplying the robot with sensing hardware and related intelligence, The robot will utilize sensing data in real time to produce purposeful collision-free motion. No matter how a specific algorithm works, we want robot intelligence and sensing hardware to be intimately tied in a synergy - better hardware should allow better quality motion, and more powerful intelligence (software) should tease more from a subpar hardware.

One interesting property of SBMP is that in order to guarantee convergence it demands *whole-body sensing* - i.e. robot sensing hardware must allow sensing its surroundings simultaneously at every point of the robot body. As we will see, realizing such sensing calls for novel large-area flexible arrays, sheets of *sensitive skin* covering the whole robot body - an engineering trend somewhat opposite to that of electronics miniaturization.

As an aside, note that whole-body sensing is what all live creatures have developed through the millennia of evolution. Some sensors may be "partial", but the overall sensing apparatus must be "whole-body". For example, human vision, which we value so much, is a partial sensor - when we look ahead we can't see things behind us, whereas our skin is a whole-body sensor. According to SBMP, vision is not sufficient for survival in an unstructured world.

No contradiction here - however important vision is, some humans do live without it. On the other hand, our skin sensing is continuous and simultaneous-sensing for all points of our body. By SBMP "ideology", a skin-like sensing (be it touch or distance sensing, doesn't matter) is necessary and sufficient for one to survive in an unstructured world. Same in nature - skin sensing mechanisms vary widely, but no creature can live without continuous sensing in its skin. As an indication of its importance, skin is the biggest part of the human body; its rapid cell regeneration makes it the most renewable organ. One can speculate that if nature ever tried creatures with no whole-body sensing, evolution had killed them. (In turtles, the closest approximation to such partially-sensitive animals, surface sensing is replaced by a hard shell - with a price being extremely slow motion.)

As another interesting fact, we will see that human engineering intuition and experience turn out to be of limited use in controlling a real-time motion of a machine operating with uncertainty. Studies and tests of human spatial reasoning suggest that while in some motion planning tasks - say, reaching a specific place in the forest - the evolution had equipped us with good skills, more complex spatial reasoning tasks, such as moving a kinematic structure in three-dimensional space, are not our forte. This calls for theoretical analysis and algorithm convergence assurance.

Thus designed combination of whole-body sensing and relevant artificial intelligence, AI, produces systems with interesting, even unexpected, properties and side effects: a robot becomes inherently safe to itself and its surroundings, while operating at relatively high speeds; in turn, the human operator can control it with "natural" speeds, relying on the robot's own ability to foresee potential problems and react to them in real time; the resulting robot motion strategies exceed human spatial reasoning skills; human-robot teams exhibit natural synergy, with each side focusing on what it's best at; there is much leeway for a mix of supervised and unsupervised robot operation. Below we review the algorithmic, hardware (materials, electronics, computing), cognitive science, and control issues involved in realizing such systems.

3   The Problem

Consider a little game: you see on your computer screen the labyrinth shown in Fig. 1a, and you are asked to determine if a collision-free path is feasible from point S (Start) to point T (Target). You do it by moving with your cursor a "point robot" in the labyrinth's plane, thus creating a path. You therefore have two control variables, x and y. As in a typical labyrinth, you can touch its walls but not penetrate them. After a few seconds of thinking you'll likely see a number of options, for example the one shown in Fig. 1b, the dotted line.

Note that here you heavily relied on the advantage of a bird's-eye view of the scene. You see the whole labyrinth, so you can assess a complete path option even before you leave point S. Real labyrinths of course don't give one a luxury of the bird's-eye view: at each moment of your path there you see only the closest walls around you, so you are forced to plan your path on the fly, as you proceed. Global optimization of the path is thus ruled out; the goal is a "reasonable path".

To simulate this situation in our game, at any point of your path the screen will show nothing but the pieces of walls near your current position, plus target T, giving you an idea of the direction toward T and your distance from it. Figure 1c shows one example of the generated path in a test with human subjects.

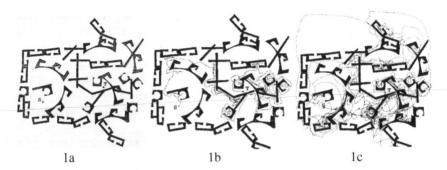

1a   1b   1c

Fig. 1. (a) The labyrinth task: starting at point S (*Start*), reach point T (*Target*). (b) Here the human subject has an advantage of a bird's eye view of the scene, making the task quite easy; typically it takes one a couple seconds to assess the scene, and less than a minute to finish the task. (c) Here the situation is more akin to moving in a real labyrinth - the subject "moves in the dark", feeling the nearby walls and knowing the Target location; the path becomes longer and takes over 5 min to accomplish.

While this performance looks appallingly bad, it is actually close to an average. Some people do better and some do worse than this. The important point here is that while one's performance deteriorates as the task becomes harder - from moving with a bird's-eye view to moving "in the dark" - most of us will nevertheless exhibit a good grasp of the task and ways to solve it. Looking at some path example, we know what and why we may want to move differently.

In fact, a robot equipped with an algorithm for *sensor-based motion planning* [4], with no beforehand information about the scene and operating "in the dark", solely with real-time sensing, will produce paths that are typically much better than the human solutions under the same conditions. Unlike with the human, the algorithm will guarantee a solution (or a conclusion that the destination is not reachable if true). While applicable to real-life tasks, this strategy's bigger value is in more complex cases, considered next, where human spatial reasoning gets weaker.

Consider a different setup of our game, a simple two-link model of a robot arm manipulator shown in Fig. 2a. The arm can move in its two-dimensional (2D) plane; its motion is complicated by four black obstacles shown. Using your cursor, you can move each link separately, or both links simultaneously, by leading the arm endpoint, as you would with a physical arm. As in the labyrinth task of Fig. 1, here we also have two independent control variables, rotation angles $\theta1$ and $\theta2$ of the arm links. The axis of angle $\theta1$ is fixed at point 0.

For simplicity let us assume that short of interfering with obstacles each arm link can rotate indefinitely. Adding a full rotation to a link will obviously bring it to the same position: that is, for a given arm endpoint position $P(\theta1, \theta2)$, we have $P(\theta1, \theta2) = P(2\pi n + \theta1, 2\pi n + \theta2)$, $n = 1, 2, 3...$ You are asked to move the arm, collision-free, from its current position S to a new position T. As before, touching obstacles is allowed, penetrating them is not possible.

Although in both tasks, moving in a labyrinth or moving the arm, you deal with two linear control variables, you'll likely find the task in Fig. 2 significantly harder and less

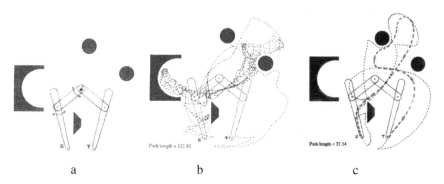

Fig. 2. (a) Task with a two-link planar manipulator: starting at position S (*Start*), move both links to reach position T (*Target*), without collisions with the obstacles shown in black. While being formally of the same complexity as the labyrinth of Fig. 1 - two control variables x,y there, and two here, θ_1, θ_2 - this task happen to present much more difficulty to humans. (b) Example of the human performance; the dotted line is the trajectory of the 2nd link endpoint on its way from S to T. With the bird's eye view of the scene it takes a subject 12–15 min to complete the task; some subjects can't finish it; in the version equivalent to "moving in the dark", as in Fig.1c, no subject has been able to perform the task. (c) Performance of the motion planning algorithm, in the version "moving in the dark": thin dotted line is trajectory when the links have touch sensing; the thick dotted line - trajectory when the links can sense obstacles at some distance from their bodies (e.g. as we sense temperature).

intuitive than that in Fig. 1. A typical human-generated path of the arm endpoint from S to T (in a test with about 40 human subjects) is shown in Fig. 2b, dotted line. While the human solution in the labyrinth, Fig. 1b, usually takes a minute or so, a solution like the one in Fig. 2a took the subjects on average 13–15 min. Some subjects declared it impossible and gave up. This is surprising: we are talking of producing a collision-free path for a trivial two-link (in robotics parlance, two degrees of freedom) planar manipulator. The whole workspace is here before you, there's no uncertainty. If that's so hard with a two-link 2D arm, what would it be with a realistic three-dimensional manipulator with 5–6 degrees of freedom! The task hardness does not seem to correlate with the subjects' age, gender, amount of education, or profession. Further, when shown a path done by someone else, subjects cannot grasp what goes on and cannot repeat it, even approximately.

As with the human performance in labyrinth "in the dark", Fig. 1c, it's reasonable to ask, how then will one perform in the task of Fig. 2a if one has no luxury of seeing the whole scene at once? (For a human operator controlling a real multi-link three-dimensional (3D) manipulator this is reality - obstacles may move around, they occlude each other or arm links...) Simulating this situation for our two-link planar arm means that instead of seeing the whole scene, at a given moment the subject sees only the arm's whole body and those points of the obstacles that happen to be in close proximity to it. In our experiments no human subject was able to solve the task in this version. Worse yet, repetition and practice were of little help, and when shown a seemingly simple successful algorithm, no subject was able to use it.

How an automatic strategy works in this task will be addressed below in Sects. 4 and 5. For now, to compare with the human performance, performance of that strategy in the same task is shown in Fig. 2c. Here the thin dotted line presents the path of the arm's endpoint in the case of tactile sensing; the thick dotted line presents the case with the algorithm modified to take advantage of proximal sensing (e.g. infrared sensing). As expected, better sensing produces better paths.

Recall again the crucial difference between the human and algorithm performance here: while in Fig. 2b the human subject was enjoying a bird's-eye view of the whole scene (that is, operating with complete information), the algorithm-generated path in Fig. 2c was produced "in the dark", with no prior information about the obstacles, and with motion planning done on the fly, based solely on the arm's real-time sensing - in other words, under much harder conditions, which human subjects could not handle at all. Clearly, if the human performance in the arm motion planning is so inferior to that of the robot, something is to be done in human-robot systems.

(Why human subjects find the task of Fig. 2a so much harder than the task of Fig. 1a is an interesting question on mechanisms of human spatial reasoning; it is beyond our topic here.)

4   Who Needs Such Robots and What It Takes to Realize Them?

People have thought of such machines long ago. Jewish mysticism, with its Cabbala teaching and literary imagery such as the Golem story, is a good example. In the Hebrew Bible (Psalms 139:16) and in the Talmud, Golem is a substance without form. Later on in the Middle Ages the idea took form: it was said that through magic a wise man can instill life into an effigy thus creating a Golem with legs and arms and a head and mighty muscles. In today's terms Golem was a human-like automaton, a robot.

One such well-known story features Rabbi Loew of the 16th-century Prague, in Czechia. (The Rabbi's somewhat scary gravestone still greets you in the Jewish cemetery in downtown Prague.) Rabbi Loew had made his Golem from clay, to be his servant and to protect the Jews of Prague from enemies and other calamities (of which there was no shortage). For a while Golem was doing its job well, using its strength and skills on the service of the Prague Jews. With time, however - all machines break eventually - Golem got out of hand and became dangerous to its creator and other Jews. At some point Rabbi Loew decided to return the Golem back to its clay immobility. He achieved this via a secret Cabbalistic formula, and promptly exiled Golem to the attic of his Prague synagogue. (Where it presumably still dwells, two street blocks from the Rabbi's grave). The story became popular through the 1915 novel Der Golem by Gustav Meyrink and the 1920 German film under the same title (still available today).

Golems are in high demand today. Consider this task: Currently new space telescopes are delivered into space fully folded, and unfolded umbrella-like on the spot. But astronomers want bigger space telescopes, so big that they cannot be set up in this fashion, they must be assembled in place from components delivered from Earth by rockets. And astronomers want them in places of space where human builders won't be able to come any time soon, such as the so-called Lagrangian points of the Earth-Sun system, the first of which is at a distance 1.5 million kilometers from Earth. For a human operator to handle a telescope-assembling robot manipulator distantly from Earth is not realistic - the combination of insufficient information, a complex multi-link manipulator and continuously changing three-dimensional scene, and human inherent difficulties in spatial reasoning mentioned above would make this operation so slow as to require dozens of years to finish.

The solution is to supply a space assembly robot manipulator with enough sensing and intelligence to do the obstacle-free motion planning job on its own. The operator may still monitor the work, change plans and schedules, and interfere in individual operations if needed, but the bulk of the work, the actual real-time assembly, with all the intelligence needed for producing requisite movements, must be done by the robot. Or, think of a robot helper for disable persons, big and strong enough to give one a hand, to assist in standing up etc. To make this robot helper really safe and reasonably autonomous, it has to be able to sense the surrounding objects all around its body, the way humans and dogs do, and have enough intelligence to react to this information accordingly. There are myriads of other scenarios dictating such needs.

That's the applications side. On the robot side, realizing such systems requires solving some new theoretical and technological issues. As mentioned above, the popular motion planning strategy called *Piano Mover's Problem,* while producing much new knowledge in algorithm theory and computational complexity [2, 3], does not fit systems of the kind discussed here. The root of the trouble is of course the approach's assumption of complete information about the scene in question. Unwieldy computations aside, in reality complete information is never available except in specially designed environments. Humans and animals don't rely on that assumption, yet somehow they operate reasonably successfully in this uncertain world. One might think the success may come at a price of loose heuristics resulting in approximate and hard to guarantee solutions. Surprisingly, this is not so for quite a few cases important to theory and practice.

5 Topology Provides an Alternative

To clarify the idea, consider the following simple two-dimensional example: You walk in the forest, aiming to reach some well-defined location and trying to stay more or less on a straight line. Suddenly you find yourself at the shore of a lake. Since continuing on a straight line is not possible, you walk around the lake, reach the opposite shore, and then continue as before toward your destination. Notice that the shape of the lake - whether it's circular, rectangular, or of some irregular form - is immaterial to you. By keeping the lake, say, at your left, you can be confident that eventually you will come to the lake's other side, closer to your destination, or at the least get back to the point where you first hit the lake.

While making a full circle around the lake may not appeal to you as a way to get to your destination, the essential topological fact here is that the lake's shore presents a simple closed curve, i.e. a curve with no self-intersections, so passing around it can be done with zero knowledge about its shape and dimensions. All you need is your sensing, here your eyes, making sure you stick to the lake shore as a kind of an

Ariadne's thread. In other words, computationally walking around the lake is a very cheap operation, unlike in the Piano Mover's algorithmic approach where the data on the lake's shape and dimensions must be acquired and processed, likely with much data grinding, before you embark on the journey.

That's the central idea of the approach of *sensor-based motion planning* - combining real-time feedback, such as from sensing, with concepts from topology, which ignores particular shapes, unlike in computational geometry where one deals with shapes. As a price for this advantage, algorithmic solutions obtained here are not as universal as in the Piano Mover's approach. They are general enough, however, for many cases in robotics, and are computationally concise to provide interesting theory and technologically elegant solutions for autonomous robot control.

6   How Does It Work?

Taking first the task of motion planning in a labyrinth, Fig. 1, let us assume that our tiny point robot always knows its own coordinates *(x, y)* and those of its target *T*. With this information in hand, it can for example always figure out its distance from point *T* and the direction toward it. On the sensing side, let us assume the robot has a simple touch sensing that allows it to recognize the fact of touching an obstacle. This useful skill will e.g. allow the robot to walk along an obstacle boundary if needed. The important information-theoretical detail here is that the robot knows nothing about the objects it may encounter on its way. It does not even know if its destination is reachable - who knows, point *T* may be completely encircled by a loop-like obstacle.

Having a motion planning strategy for this model would clearly provide tremendous advantages: there would be no need in a database or a map of the terrain, objects may be of any exotic shapes, they may even move around in ways unknown to the robot. In other words, this would allow the robot's world to be as real as the world that humans and animals live in every day. As a bonus, the strategy would be free of the prohibitive computational load typical of the Piano Mover's model. Can such a motion planning algorithm with proven convergence be designed? - meaning by "proven convergence" that if target *T* is reachable, the algorithm guarantees reaching it in finite time, and if *T* is not reachable, the algorithm will come to this conclusion, also in finite time.

As we will see below, once even a simple algorithm based on the above "simplistic" model is designed and its convergence is proven, it can be expanded to more complex systems and environments, such as robot arm manipulators, motion in three-dimensional space, dealing with objects of arbitrary shapes, and to more complex sensing media, such as vision. As expected, adding richer sensing should result in better paths (with a standard proviso of dealing with uncertainty: as global information is not available, global optimization is by definition out of the question).

To build such a motion planning strategy, notice that a straight line between points *S* and *T* in Fig. 1 cuts any labyrinth wall in a special way: at a point where the line enters a wall there are exactly three lines coming together - one from our line (*S, T*) and the other two belonging to the obstacle boundary. (The line (*S, T*) being tangential to a wall presents a special case that is easy to detect and handle.) Our general strategy will be to make the robot move straight toward T when feasible, and deal with encountered obstacles by passing around them in some organized manner. When hitting a wall, one can turn either left or right to continue walking along it. Given the uncertainty, either one is a valid choice, the strategy says; choose one and stick to it for the rest of the journey.

It turns out that two very distinct strategies emerge here (more if counting variations), both providing a guaranteed convergence [4]. They differ in how they treat the obstacles on the robot's way, and this affects their worst-case performance profoundly. Without going into details, the first of these strategies, known as *Bug1*, has a wonderful worst-case performance - which it delivers, unfortunately, by exploring fully every obstacle on its way. Except in special applications, for example to build maps, no sane user will choose this strategy to reach a destination in the forest. The other strategy, *Bug2*, has a much worse worst-case performance due to its being "more adventurous", but it tends to produce more natural paths in realistic scenes. It has a feel of a reasonable risk-taker, making it a better choice in most but specially designed killer labyrinths.

An example of *Bug2* performance in the labyrinth of Fig. 1 is shown in Fig. 3a. In fairness of comparison, note that while in terms of input information the "robot" was here in the situation of a human struggling with the labyrinth "in the dark" (as in Fig. 1c), *Bug2*'s path is better than that in Fig. 1b, where the human subject had an advantage of a bird's-eye view. Giving *Bug2* sensing slightly better than tactile, say within some radius r_v around it, produces a still better path, Fig. 3b.

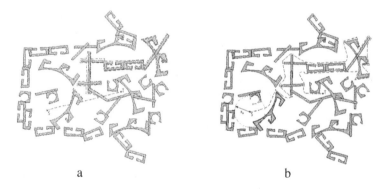

a   b

Fig. 3. (a) Performance of the sensor-based algorithm in the labyrinth of Fig. 1a. Here the point robot has touch sensing, knows its own and its target position, and has no information about the labyrinth. (b) Same as in (a), except here the robot has sensing in radius r_v, such as with an ultrasound sensor.

7 The Arm Manipulator Case

Let us turn now to motion planning for a simple arm manipulator of Fig. 2. It is known from topology that if angles θ_1 and θ_2 present two independent variables, this pair is equivalent to a point on the surface of a simple torus. This is shown in Fig. 4a, where angle θ_1 goes along the torus and angle θ_2 across the torus. Points S and T represent the arm positions S and T in Fig. 2a. To emphasize the equivalence of Figs. 2a and 4a, we say that the torus of Fig. 4a is the *configuration space* (or *C-space*) of the *workspace* (or *W-space*) of Fig. 2a.

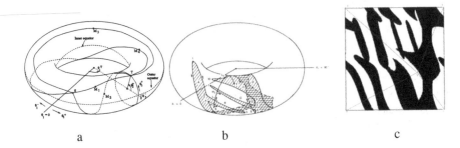

a   b   c

Fig. 4. Shaping up a sensor-based motion planning algorithm for a two-link robot manipulator of Fig. 2. (a) Two angular variables (θ_1, θ_2) are presented as a torus, forming the robot's configuration space (*C-space*). (b) The *C-space* of the example of Fig. 2a. (c) Same torus cut flat for demonstration purposes; in reality the robot will not need to know the obstacles.

As said, to make the strategy work, our point robot in *C-space* needs sensing that helps it detect obstacles on its way. For a real mobile robot, such as a Google car, this means providing it with a sensing ability at every point of its body - so that no object in its surroundings will ever catch it unprepared. (That's not so for the cars on the drawing boards today - they are designed for highways and streets, not for a crowded backyard or Mars surface.) For the arm manipulator of Fig. 2a, every point of its two links must have sensing of its own.

It is not hard to show that arm positions in Fig. 2a where one or more points of the arm body touch one or more points on the workspace obstacles form in *C-space* one or two closed curves. Together those positions define the *C-space* equivalent of the workspace obstacle boundaries. Positions that the arm cannot occupy due to the interference with obstacles will correspond to points inside those *C-space obstacles*. For the scene of Fig. 2a these are shown in Fig. 4b. This projection is highly nonlinear. Note for example that the four distinct convex obstacles in Fig. 2a became a single non-convex rather complex obstacle image in *C-space*.

As clear from Fig. 4b, since the torus floats in 3D space, some parts of that *C*-obstacle are hidden from the viewer, being occluded by its other parts or by the torus body. To make things more visible, let us cut the torus along two mutually perpendicular lines along θ_1 and θ_2 directions intersecting at point T. The result is known as a *flattened torus*, a square of size 2π, Fig. 4c, where point S is somewhere inside the square and all four corners correspond to point T. (In topology parlance, those four corners are *identified*.) In other words, the scene of Fig. 2a becomes just another labyrinth not unlike that of Fig. 1a, and the motion planning task for the arm becomes that of moving a "point robot" in that labyrinth from point S to point T. One difference is of course that, unlike in Fig. 1a, here we have four distinct (*C-space*) straight-line paths between *S* and *T* (see Fig. 4c). These four correspond to the four combinations of link rotation, clockwise and counter-clockwise, for each link.

More technical details have to be taken care of - for example, the case when the arm position *T* is not reachable along one or more of the four straight lines *(S, T)*. Overall, though, the motion planning strategy outlined above for the labyrinth of Fig. 1a is readily expandable to our two-link arm of Fig. 2a [5], and to other types of kinematics [6]. The required whole-body sensing becomes equivalent to covering the arm with a continuous sensitive skin. Further improvements in sensing, such as supplying the arm with a vision cameras or other distance sensors, result in advantages of local path optimization similar to that in Fig. 3b.

8 Moving the Approach to Real-Life Applications

As clear from the above, unlike in the Piano Movers approach, in sensor-based motion planning the complexity of obstacle shapes is not an issue. More or bigger obstacles encountered by the robot will add to the computations only linearly, typically a negligible load. But what about the growth in complexity of the robot itself? That's a different matter. The difficulty here is not in the amounts of computations but in the algorithm convergence. In fact, even a change in the topological properties of the robot of the same complexity, typically due to its kinematic design, can affect the algorithm convergence. It is not so in the Piano Mover's approach where the assumption of complete information always guarantees a solution.

Similar to the scheme outlined in Sect. 6, an arm manipulator of any complexity can be represented by an appropriate *C-space*, where the arm becomes a point robot and obstacles form a "labyrinth" for that robot to navigate in. That *C-space* depends not only on the obstacles in the scene but also on the robot kinematic characteristics. Say, the allowed motion of a sliding arm link interacting with an obstacle will differ from that of a revolute link. In our terms here, while an arm with two revolute links in the scene of Fig. 2a forms a (flatten) *C-space* of Fig. 4c, an arm with two links sliding relative to each other (it's called the *Cartesian arm*) in the same scene with four obstacles will form quite a different (though still flat) *C-space*. (A three-link three-dimensional Cartesian arm is shown in Fig. 5, arm (c)).

While the same sensor-based motion planning algorithm can be used in both cases, the question of its convergence in case of a Cartesian kinematics has to be studied separately. The same is true for other kinematic structures. Once this issue is assessed for a given kinematics, the approach is good for this arm operating in any scene. This additional chore - will my new kinematic system operates as expected? - is no surprise, that's how it is in nature. Say, bipedal animals deal with obstacles very differently from the quadrupeds; what it easy for one may be hard for the other.

It turns out that in terms of algorithm convergence various kinematics can be "folded" in groups where all members of the same group become sub-cases of one "most complex" kinematics of that group. In other words, a motion planning algorithm that works for the "most complex" case of the group will work for all other kinematics in that group. For example, the above approach readily extends to 3D arm manipulators with sliding and rotating joints shown in Fig. 5 [7].

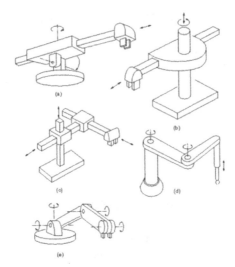

Fig. 5. With simple modifications, three-dimensional arm manipulators of various kinematics shown here are amenable to the same sensor-based motion planning algorithm.

9 Sensing System Hardware

What requirements does the theory and algorithms outlined above postulate on the sensitive skin hardware [8]? A human skin is a good model here. A good robot sensitive skin should present an easily bendable and stretchable surface densely populated by sensors. The latter can be tactile sensors or sensors capable of distance sensing, such as infrared or ultra-sound, or combinations of different sensors. How densely should they be spaced on the skin? Other things being equal, the higher the sensor density the crisper and more dexterous motion can the robot achieve. The millennia of evolution made the density of tactile sensing on the human skin vary along its surface - it is denser on your palm, where maximum localization and precision is required, it is less dense on your elbows where precision of sensing and motion is less of an issue. The same is true for the robot. In reality today's sensor density is limited by the technological constraints.

To work well, the sheet of material that forms the sensitive skin should cling closely to the robot body. This dictates a bendable thin material capable of accommodating many individual sensors, which are typically small hard bodies of fixed shape, as well as auxiliary hardware and printed wiring between them. This is the domain of material science and large-sheet electronics, where objectives are, one might say, opposite to the miniaturization trends in electronic chip production - the larger the sheet and the more sensors it can accommodate the better. Designers talk of printing those sheets via processes similar to newspaper printers. Some such materials exist already.

A number of versions of such systems with bendable sensitive skin, most in small dozen-sensors size imitating the expected large-sheet format, have been designed in laboratories around the world - in US, England, Japan, Italy. An example of a skin module with infrared sensing developed in this author's laboratory at the University of Wisconsin-Madison is shown in Fig. 6a. When installed on a human-size industrial robot manipulator, the system comprised about 1000 infrared sensors, each sensor being a combination of an LED light source and a detector. An earlier model of this system developed in the author's lab at Yale University, with about 500 more sparsely spaced sensors, appears in Fig. 6b.

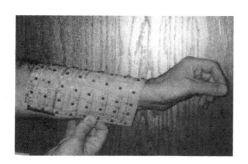 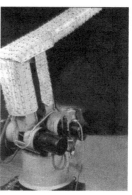

a   b

Fig. 6. Implementations of sensitive skin for realizing sensor-based motion planning algorithms for robot arm manipulators. (a) A skin module, with all local sensing and computational machinery, with 8 by 8 infrared sensors per module (University of Wisconsin-Madison). (b) A robot manipulator clad into sensitive skin (Yale University).

Since the skin may be covering surfaces of complex shape or manipulator links moving relative to each other, skin's bending about a single axis may not be enough. Ideally the skin material should be able to stretch and shrink the way our skin does at the elbow as we move our arm. Again, interaction of such stretchable skin with hard-body sensor electronics on it puts a premium on the system design. This is an active area of research and development.

A sensing system consisting of the expected tens of thousands or millions sensors and their control "nerves" stretched over a large bendable and stretchable surface is bound to suffer from eventual local failures. With time individual sensors will mal-function or die. To prevent an expensive replacement of the whole skin due to failure of one or a few sensors, various algorithmic schemes of graceful degradation step in [9], whereby neighboring sensors take over a fallen neighbor to maintain acceptable if degraded performance.

Assume, for example, that our robot's skin features infrared proximity sensors, and the distance between neighboring sensors on it is 0.5 cm. Each infrared sensor sends in space in front of it a narrow cone of light; when reflected from an object (obstacle) it informs the arm's "brain" of the obstacle at this spot of its body.

As the cones of sensitivity of neighboring sensors overlap in space around the skin, forming continuous sensing coverage, the cones of sensors in the neighborhood 1–2 cm likely overlap. When some sensor dies, the robot intelligence will use that overlap to compensate for the failed sensor. Or, if the overlap is not sufficient - or if the robot skin is built, say, on tactile sensing - the measured value of the failed sensor can be interpolated from the values of its neighbors. This implies system degradation: with more sensors dying or malfunctioning, the skin's sensing accuracy and thus dexterity of robot motion go down. But the degradation is not abrupt, it is "graceful". The higher the sensor density, the more of them are in the system, the more "graceful" the degradation, the system can be in use for longer, until the losses due to elemental failures become intolerable.

Some videos of experiments with the system of Figs. 6a and b appear at https://directory.engr.wisc.edu/me/faculty/lumelsky_vladimir.

For the viewer watching these videos, keep in mind the implemented "ideology" of dealing with uncertainty: nowhere in these experiments was the robot motion pre-programmed. The robot moves to its destination fully autonomously, avoiding collisions with previously unknown (moving or still, alive or dead) objects on its way, relying solely on its own sensing and its own intelligence.

References

1. Knight, W.: Inside Amazon's Warehouse, Human-Robot Symbiosis. MIT Technology Review (2015)
2. Reif, J.H.: Complexity of the *movers problem* and generalizations. In: Proceedings of Symposium on Foundations of Computer Science (1979)
3. Reif, J.H.: A survey on advances in the theory of computational robotics. In: Proceedings of 4th Workshop on Adaptive Systems Control Theory as Adaptive and Learning Systems. Plenum Press, New York (1986)
4. Lumelsky, V., Stepanov, A.: Algorithmic and complexity issues of robot motion in an uncertain environment. J. Complex. **3**, 146–182 (1987)
5. Lumelsky, V.: Dynamic path planning for a planar articulated robot arm moving amidst unknown obstacles. Automatica **23**(5), 551–570 (1987)
6. Lumelsky, V.: Effect of kinematics on dynamic path planning for planar robot arms moving amidst unknown obstacles. IEEE J. Robot. Autom. **3**(3), 207–223 (1987)
7. Lumelsky, V., Sun, K.: A unified methodology for motion planning with uncertainty for 2D and 3D two-link robot arm manipulators. Int. J. Robot. Res. **9**(5), 89–104 (1990)
8. Lumelsky, V., Shur, M., Wagner, S.: Sensitive skin. IEEE Sens. J. **1**(1), 41–51 (2000)
9. Um, D., Lumelsky, V.: Fault Tolerance via analytic redundancy for a modularized sensitive skin. Int. J. Robot. Autom. **15**(4) (2000)
10. Lumelsky, V.: Sensing, Intelligence, Motion. Wiley, Hoboken (2006)

Geometrical Insights for Implicit Generative Modeling

Leon Bottou[1(✉)], Martin Arjovsky[2], David Lopez-Paz[3], and Maxime Oquab[4]

[1] Facebook AI Research, New York, USA
leon@bottou.org
[2] New York University, New York, USA
martinarjovsky@gmail.com
[3] Facebook AI Research, Paris, France
dlp@fb.com
[4] Inria, Paris, France
maxime.oquab@inria.fr

Abstract. Learning algorithms for implicit generative models can optimize a variety of criteria that measure how the data distribution differs from the implicit model distribution, including the Wasserstein distance, the Energy distance, and the Maximum Mean Discrepancy criterion. A careful look at the geometries induced by these distances on the space of probability measures reveals interesting differences. In particular, we can establish surprising approximate global convergence guarantees for the 1-Wasserstein distance, even when the parametric generator has a nonconvex parametrization.

1   Introduction

Instead of representing the model distribution with a parametric density function, implicit generative models directly describe how to draw samples of the model distribution by first drawing a sample z from a fixed random generator and mapping into the data space with a parametrized generator function $G_\theta(z)$. The reparametrization trick [13,43], Variational Auto-Encoders (VAEs) [27], and Generative Adversarial Networks (GANs) [20] are recent instances of this approach.

Many of these authors motivate implicit modeling with the computational advantage that results from the ability of using the efficient back-propagation algorithm to update the generator parameters. In contrast, our work targets another, more fundamental, advantage of implicit modeling.

Although unsupervised learning is often formalized as estimating the data distribution [24, Sect. 14.1], the practical goal of the learning process rarely consists in recovering actual probabilities. Instead, the probability models are often structured in a manner that is interpretable as a physical or causal model of the data. This is often achieved by defining an interpretable density $p(y)$ for well chosen latent variables y and letting the appearance model $p(x|y)$ take the slack.

© Springer Nature Switzerland AG 2018
L. Rozonoer et al. (Eds.): Braverman Readings in Machine Learning, LNAI 11100, pp. 229–268, 2018.
https://doi.org/10.1007/978-3-319-99492-5_11

This approach is well illustrated by the *inverse graphics* approach to computer vision [30,31,44]. Implicit modeling makes this much simpler:

- The structure of the generator function $G_\theta(z)$ could be directly interpreted as a set of equations describing a physical or causal model of the data [28].
- There is no need to deal with latent variables, since all the variables of interest are explicitly computed by the generator function.
- Implicit modeling can easily represent simple phenomena involving a small set of observed or inferred variables. The corresponding model distribution cannot be represented with a density function because it is supported by a low-dimensional manifold. But nothing prevents an implicit model from generating such samples.

Unfortunately, we cannot fully realize these benefits using the popular Maximum Likelihood Estimation (MLE) approach, which asymptotically amounts to minimizing the Kullback-Leibler (KL) divergence $D_{KL}(Q, P_\theta)$ between the data distribution Q and the model distribution P_θ,

$$D_{KL}(Q, P_\theta) = \int \log\left(\frac{q(x)}{p_\theta(x)}\right) q(x) d\mu(x) \tag{1}$$

where p_θ and q are the density functions of P_θ and Q with respect to a common measure μ. This criterion is particularly convenient because it enjoys favorable statistical properties [14] and because its optimization can be written as an expectation with respect to the data distribution,

$$\underset{\theta}{\text{argmin}}\, D_{KL}(Q, P_\theta) \;=\; \underset{\theta}{\text{argmin}}\, \mathbb{E}_{x \sim Q}[-\log(p_\theta(x))] \;\approx\; \underset{\theta}{\text{argmax}} \prod_{i=1}^{n} p_\theta(x_i).$$

which is readily amenable to computationally attractive stochastic optimization procedures [10]. First, this expression is ill-defined when the model distribution cannot be represented by a density. Second, if the likelihood $p_\theta(x_i)$ of a single example x_i is zero, the dataset likelihood is also zero, and there is nothing to maximize. The typical remedy is to add a noise term to the model distribution. Virtually all generative models described in the classical machine learning literature include such a noise component whose purpose is not to model anything useful, but merely to make MLE work.

Instead of using ad-hoc noise terms to coerce MLE into optimizing a different similarity criterion between the data distribution and the model distribution, we could as well explicitly optimize a different criterion. Therefore it is crucial to understand how the selection of a particular criterion will influence the learning process and its final result.

Section 2 reviews known results establishing how many interesting distribution comparison criteria can be expressed in adversarial form, and are amenable to tractable optimization algorithms. Section 3 reviews the statistical properties of two interesting families of distribution distances, namely the family of the Wasserstein distances and the family containing the Energy Distances and the Maximum Mean Discrepancies. Although the Wasserstein distances have far worse statistical properties, experimental evidence shows that it can deliver better performances in meaningful applicative setups. Section 4 reviews essential concepts about geodesic geometry in metric spaces. Section 5 shows how different probability distances induce different geodesic geometries in the space of probability measures. Section 6 leverages these geodesic structures to define various flavors of convexity for parametric families of generative models, which can be used to prove that a simple gradient descent algorithm will either reach or approach the global minimum regardless of the traditional nonconvexity of the parametrization of the model family. In particular, when one uses implicit generative models, minimizing the Wasserstein distance with a gradient descent algorithm offers much better guarantees than minimizing the Energy distance.

2   The Adversarial Formulation

The adversarial training framework popularized by the Generative Adversarial Networks (GANs) [20] can be used to minimize a great variety of probability comparison criteria. Although some of these criteria can also be optimized using simpler algorithms, adversarial training provides a common template that we can use to compare the criteria themselves.

This section presents the adversarial training framework and reviews the main categories of probability comparison criteria it supports, namely Integral Probability Metrics (IPM) (Sect. 2.4), f-divergences (Sect. 2.5), Wasserstein distances (WD) (Sect. 2.6), and Energy Distances (ED) or Maximum Mean Discrepancy distances (MMD) (Sect. 2.7).

2.1   Setup

Although it is intuitively useful to consider that the sample space $\mathcal{X}$ is some convex subset of $\mathbb{R}^d$, it is also useful to spell out more precisely which properties are essential to the development. In the following, we assume that $\mathcal{X}$ is a *Polish metric space*, that is, a complete and separable space whose topology is defined by a distance function

$$d: \begin{cases} \mathcal{X} \times \mathcal{X} & \to \quad \mathbb{R}_+ \cup \{+\infty\} \\ (x,y) & \mapsto \quad d(x,y) \end{cases}$$

satisfying the properties of a metric distance:

$$\forall x,y,z \in \mathcal{X} \begin{cases} (o) & d(x,x) = 0 & \text{(zero)} \\ (i) & x \neq y \Rightarrow d(x,y) > 0 & \text{(separation)} \\ (ii) & d(x,y) = d(y,x) & \text{(symmetry)} \\ (iii) & d(x,y) \leq d(x,z) + d(z,y) & \text{(triangular inequality)} \end{cases} \tag{2}$$

Let $\mathfrak{U}$ be the Borel σ-algebra generated by all the open sets of $\mathcal{X}$. We use the notation $\mathcal{P}_\mathcal{X}$ for the set of probability measures μ defined on $(\mathcal{X}, \mathfrak{U})$, and the notation $\mathcal{P}_{\mathcal{X}}^p \subset \mathcal{P}_{\mathcal{X}}$ for those satisfying $\mathbb{E}_{x,y\sim\mu}[d(x,y)^p] < \infty$. This condition is equivalent to $\mathbb{E}_{x\sim\mu}[d(x,x_0)^p] < \infty$ for an arbitrary origin x_0 when d is finite, symmetric, and satisfies the triangular inequality.

We are interested in criteria to compare elements of $\mathcal{P}_{\mathcal{X}}$,

$$D: \begin{cases} \mathcal{P}_{\mathcal{X}} \times \mathcal{P}_{\mathcal{X}} & \to & \mathbb{R}_+ \cup \{+\infty\} \\ (Q,P) & \mapsto & D(Q,P) \end{cases}.$$

Although it is desirable that D also satisfies the properties of a distance (2), this is not always possible. In this contribution, we strive to only reserve the word *distance* for criteria that satisfy the properties (2) of a metric distance. We use the word *pseudodistance*[1] when a nonnegative criterion fails to satisfy the separation property (2.i). We use the word *divergence* for criteria that are not symmetric (2.ii) or fail to satisfy the triangular inequality (2.iii).

We generally assume in this contribution that the distance d defined on $\mathcal{X}$ is finite. However we allow probability comparison criteria to be infinite. When the distributions Q,P do not belong to the domain for which a particular criterion D is defined, we take that $D(Q,P) = 0$ if $Q = P$ and $D(Q,P) = +\infty$ otherwise.

2.2 Implicit Modeling

We are particularly interested in model distributions P_θ that are supported by a low-dimensional manifold in a large ambient sample space (recall Sect. 1). Since such distributions do not typically have a density function, we cannot represent the model family $\mathcal{F}$ using a parametric density function. Following the example of Variational Auto-Encoders (VAE) [27] and Generative Adversarial Networks (GAN) [20], we represent the model distributions by defining how to produce samples.

Let z be a random variable with known distribution μ_z defined on a suitable probability space $\mathcal{Z}$ and let G_θ be a measurable function, called the *generator*, parametrized by $\theta \in \mathbb{R}^d$,

$$G_\theta: \quad z \in \mathcal{Z} \quad \mapsto \quad G_\theta(z) \in \mathcal{X}.$$

The random variable $G_\theta(Z) \in \mathcal{X}$ follows the *push-forward* distribution[2]

$$G_\theta(z)\#\mu_Z(z): \quad A \in \mathfrak{U} \quad \mapsto \quad \mu_z(G_\theta^{-1}(A)).$$

By varying the parameter θ of the generator G_θ, we can change this push-forward distribution and hopefully make it close to the data distribution Q according to the criterion of interest.

---

[1] Although failing to satisfy the separation property (2.i) can have serious practical consequences, recall that a pseudodistance always becomes a full fledged distance on the quotient space $\mathcal{X}/\mathcal{R}$ where $\mathcal{R}$ denotes the equivalence relation $x\mathcal{R}y \Leftrightarrow d(x,y) = 0$. All the theory applies as long as one never distinguishes two points separated by a zero distance.

[2] We use the notation $f\#\mu$ or $f(x)\#\mu(x)$ to denote the probability distribution obtained by applying function f or expression $f(x)$ to samples of the distribution μ.

This *implicit modeling approach* is useful in two ways. First, unlike densities, it can represent distributions confined to a low-dimensional manifold. Second, the ability to easily generate samples is frequently more useful than knowing the numerical value of the density function (for example in image superresolution or semantic segmentation when considering the conditional distribution of the output image given the input image). In general, it is computationally difficult to generate samples given an arbitrary high-dimensional density [38].

Learning algorithms for implicit models must therefore be formulated in terms of two sampling oracles. The first oracle returns training examples, that is, samples from the data distribution Q. The second oracle returns generated examples, that is, samples from the model distribution $P_\theta = G_\theta \# \mu_Z$. This is particularly easy when the comparison criterion $D(Q, P_\theta)$ can be expressed in terms of expectations with respect to the distributions Q or P_θ.

2.3   Adversarial Training

We are more specifically interested in distribution comparison criteria that can be expressed in the form

$$D(Q, P) = \sup_{(f_Q, f_P) \in \mathcal{Q}} \mathbb{E}_Q[f_Q(x)] - \mathbb{E}_P[f_P(x)]. \tag{3}$$

The set $\mathcal{Q}$ defines which pairs (f_Q, f_P) of real-valued *critic* functions defined on $\mathcal{X}$ are considered in this maximization. As discussed in the following subsections, different choices of $\mathcal{Q}$ lead to a broad variety of criteria. This formulation is a mild generalization of the Integral Probability Metrics (IPMs) [37] for which both functions f_Q and f_P are constrained to be equal (Sect. 2.4).

Finding the optimal generator parameter θ^* then amounts to minimizing a cost function $C(\theta)$ which itself is a supremum,

$$\min_\theta \left\{ C(\theta) \stackrel{\Delta}{=} \max_{(f_Q, f_P) \in \mathcal{Q}} \mathbb{E}_{x \sim Q}[f_Q(x)] - \mathbb{E}_{z \sim \mu_z}[f_P(G_\theta(z))] \right\}. \tag{4}$$

Although it is sometimes possible to reformulate this cost function in a manner that does not involve a supremum (Sect. 2.7), many algorithms can be derived from the following variant of the envelope theorem [35].

Theorem 2.1. *Let C be the cost function defined in (4) and let θ_0 be a specific value of the generator parameter. Under the following assumptions,*

a. *there is $(f_Q^*, f_P^*) \in \mathcal{Q}$ such that $C(\theta_0) = \mathbb{E}_Q[f_Q^*(x)] - \mathbb{E}_{\mu_z}[f_P^*(G_{\theta_0}(z))]$,*
b. *the function C is differentiable in θ_0,*
c. *the functions $h_z = \theta \mapsto f_P^*(G_\theta(z))$ are μ_z-almost surely differentiable in θ_0,*
d. *and there exists an open neighborhood $\mathcal{V}$ of θ_0 and a μ_z-integrable function $D(z)$ such that $\forall \theta \in \mathcal{V}, |h_z(\theta) - h_z(\theta_0)| \leq D(z) \|\theta - \theta_0\|$,*

we have the equality $\mathrm{grad}_\theta C(\theta_0) = -\mathbb{E}_{z \sim \mu_z}[\mathrm{grad}_\theta h_z(\theta_0)]$.

This result means that we can compute the gradient of $C(\theta_0)$ without taking into account the way f_P^* changes with θ_0. The most important assumption here is the differentiability of the cost C. Without this assumption, we can only assert that $-\mathbb{E}_{z\sim\mu_z}[\mathrm{grad}_\theta h_z(\theta_0)]$ belongs to the "local" subgradient

$$\partial^{\mathrm{loc}} C(\theta_0) \triangleq \{\, g \in \mathbb{R}^d \;:\; \forall \theta \in \mathbb{R}^d \;\; C(\theta) \geq C(\theta_0) + \langle g, \theta - \theta_0 \rangle + o(\|\theta - \theta_0\|) \,\} \;.$$

Proof. Let $\lambda \in \mathbb{R}_+$ and $u \in \mathbb{R}^d$ be an arbitrary unit vector. From (3),

$$C(\theta_0 + \lambda u) \geq \mathbb{E}_{z\sim Q}\left[f_Q^*(x)\right] - \mathbb{E}_{z\sim\mu_z}[f_P^*(G_{\theta_0+\lambda u}(z))]$$
$$C(\theta_0 + \lambda u) - C(\theta_0) \geq -\mathbb{E}_{z\sim\mu_z}[h_z(\theta_0 + \lambda u) - h_z(\theta_0)]\,.$$

Dividing this last inequality by λ, taking its limit when $\lambda \to 0$, recalling that the dominated convergence theorem and assumption (d) allow us to take the limit inside the expectation operator, and rearranging the result gives

$$Au \geq 0 \quad \text{with} \quad A : u \in \mathbb{R}^d \mapsto \langle u, \mathrm{grad}_\theta C(\theta_0) + \mathbb{E}_{z\sim\mu_z}[\mathrm{grad}_\theta h_z(\theta_0)]\rangle\,.$$

Writing the same for unit vector $-u$ yields inequality $-Au \geq 0$. Therefore $Au = 0$. ∎

Thanks to this result, we can compute an unbiased[3] stochastic estimate $\hat{g}(\theta_t)$ of the gradient $\mathrm{grad}_\theta C(\theta_t)$ by first solving the maximization problem in (4), and then using the back-propagation algorithm to compute the average gradient on a minibatch $z_1 \ldots z_k$ sampled from μ,

$$\hat{g}(\theta_t) = -\frac{1}{k}\sum_{i=1}^{k} \mathrm{grad}_\theta f_P^*(G_\theta(z_i)).$$

Such an unbiased estimate can then be used to perform a stochastic gradient descent update iteration on the generator parameter

$$\theta_{t+1} = \theta_t - \eta_t\,\hat{g}(\theta_t)\,.$$

Although this algorithmic idea can be made to work relatively reliably [3,22], serious conceptual and practical issues remain:

Remark 2.2. In order to obtain an unbiased gradient estimate $\hat{g}(\theta_t)$, we need to solve the maximization problem in (4) for the true distributions rather than for a particular subset of examples. On the one hand, we can use the standard machine learning toolbox to avoid overfitting the maximization problem. On the other hand, this toolbox essentially works by restricting the family Q in ways that can change the meaning of the comparison criteria itself [5,34].

---

[3] Stochastic gradient descent often relies on unbiased gradient estimates (for a more general condition, see [10, Assumption 4.3]). This is not a given: estimating the Wasserstein distance (14) and its gradients on small minibatches gives severely biased estimates [7]. This is in fact very obvious for minibatches of size one. Theorem 2.1 therefore provides an imperfect but useful alternative.

Remark 2.3. In practice, solving the maximization problem (4) during each iteration of the stochastic gradient algorithm is computationally too costly. Instead, practical algorithms interleave two kinds of stochastic iterations: gradient ascent steps on (f_Q, f_P), and gradient descent steps on θ, with a much smaller effective stepsize. Such algorithms belong to the general class of stochastic algorithms with two time scales [9, 29]. Their convergence properties form a delicate topic, clearly beyond the purpose of this contribution.

2.4   Integral Probability Metrics

Integral probability metrics (IPMs) [37] have the form

$$D(Q, P) = \left|\ \sup_{f \in \mathcal{Q}} \mathbb{E}_Q[f(X)] - \mathbb{E}_P[f(X)]\ \right|.$$

Note that the surrounding absolute value can be eliminated by requiring that $\mathcal{Q}$ also contains the opposite of every one of its functions.

$$D(Q, P) = \sup_{f \in \mathcal{Q}} \mathbb{E}_Q[f(X)] - \mathbb{E}_P[f(X)]$$
$$\text{where } \mathcal{Q} \text{ satisfies} \quad \forall f \in \mathcal{Q},\ -f \in \mathcal{Q}. \tag{5}$$

Therefore an IPM is a special case of (3) where the critic functions f_Q and f_P are constrained to be identical, and where $\mathcal{Q}$ is again constrained to contain the opposite of every critic function. Whereas expression (3) does not guarantee that $D(Q, P)$ is finite and is a distance, an IPM is always a pseudodistance.

Proposition 2.4. *Any integral probability metric D, (5) is a pseudodistance.*

Proof. To establish the triangular inequality (2.*iii*), we can write, for all $Q, P, R \in \mathcal{P}_\mathcal{X}$,

$$
\begin{aligned}
D(Q, P) + D(P, R) &= \sup_{f_1, f_2 \in \mathcal{Q}} \mathbb{E}_Q[f_1(X)] - \mathbb{E}_P[f_1(X)] + \mathbb{E}_P[f_2(X)] - \mathbb{E}_R[f_2(X)] \\
&\geq \sup_{f_1 = f_2 \in \mathcal{Q}} \mathbb{E}_Q[f_1(X)] - \mathbb{E}_P[f_1(X)] + \mathbb{E}_P[f_2(X)] - \mathbb{E}_R[f_2(X)] \\
&= \sup_{f \in \mathcal{Q}} \mathbb{E}_Q[f(X)] - \mathbb{E}_R[f(X)] \quad = \quad D(Q, R).
\end{aligned}
$$

The other properties of a pseudodistance are trivial consequences of (5).   ∎

The most fundamental IPM is the Total Variation (TV) distance.

$$D_{TV}(Q, P) \overset{\Delta}{=} \sup_{A \in \mathfrak{U}} |P(A) - Q(A)| = \sup_{f \in C(\mathcal{X}, [0,1])} \mathbb{E}_Q[f(x)] - \mathbb{E}_P[f(x)], \tag{6}$$

where $C(\mathcal{X}, [0, 1])$ is the space of continuous functions from $\mathcal{X}$ to $[0, 1]$.

2.5 f-Divergences

Many classical criteria belong to the family of f-divergences

$$D_f(Q, P) \triangleq \int f\left(\frac{q(x)}{p(x)}\right) p(x)\, d\mu(x) \tag{7}$$

where p and q are respectively the densities of P and Q relative to measure μ and where f is a continuous convex function defined on R_+^* such that $f(1) = 0$.

Expression (7) trivially satisfies (2.o). It is always nonnegative because we can pick a subderivative $u \in \partial f(1)$ and use the inequality $f(t) \geq u(t-1)$. This also shows that the separation property (2.i) is satisfied when this inequality is strict for all $t \neq 1$.

Proposition 2.5 ([39, 40, informal]). *Usually,*[4]

$$D_f(Q, P) \quad = \quad \sup_{\substack{g \text{ bounded, measurable} \\ g(\mathcal{X}) \subset \mathrm{dom}(f^*)}} \quad \mathbb{E}_Q[g(x)] - \mathbb{E}_P[f^*(g(x))].$$

where f^* denotes the convex conjugate of f.

Table 1 provides examples of f-divergences and provides both the function f and the corresponding conjugate function f^* that appears in the variational formulation. In particular, as argued in [40], this analysis clarifies the probability comparison criteria associated with the early GAN variants [20].

Table 1. Various f-divergences and the corresponding f and f^*.

	$f(t)$	$\mathrm{dom}(f^*)$	$f^*(u)$
Total variation (6)	$\frac{1}{2}\lvert t-1\rvert$	$[-\frac{1}{2}, \frac{1}{2}]$	u
Kullback-Leibler (1)	$t\log(t)$	$\mathbb{R}$	$\exp(u-1)$
Reverse Kullback-Leibler	$-\log(t)$	$\mathbb{R}_-$	$-1 - \log(-u)$
GAN's Jensen Shannon [20]	$t\log(t) - (t+1)\log(t+1)$	$\mathbb{R}_-$	$-\log(1-\exp(u))$

Despite the elegance of this framework, these comparison criteria are not very attractive when the distributions are supported by low-dimensional manifolds that may not overlap. The following simple example shows how this can be a problem [3].

---

[4] The statement holds when there is an $M > 0$ such that $\mu\{x : \lvert f(q(x)/p(x))\rvert > M\} = 0$ Restricting μ to exclude such subsets and taking the limit $M \to \infty$ may not work because $\lim\sup \neq \sup\lim$ in general. Yet, in practice, the result can be verified by elementary calculus for the usual choices of f, such as those shown in Table 1.

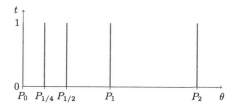

Fig. 1. Let distribution P_θ be supported by the segment $\{(\theta,t)\ t \in [0,1]\}$ in $\mathbb{R}^2$. According to both the TV distance (6) and the f-divergences (7), the sequence of distributions $(P_{1/i})$ does not converge to P_0. However this sequence converges to P_0 according to either the Wasserstein distances (8) or the Energy distance (15).

Example 2.6 Let U be the uniform distribution on the real segment $[0,1]$ and consider the distributions $P_\theta = (\theta, x)\#U(x)$ defined on $\mathbb{R}^2$. Because P_0 and P_θ have disjoint support for $\theta \neq 0$, neither the total variation distance $D_{TV}(P_0, P_\theta)$ nor the f-divergence $D_f(P_0, P_\theta)$ depend on the exact value of θ. Therefore, according to the topologies induced by these criteria on $\mathcal{P}_\mathcal{X}$, the sequence of distributions $(P_{1/i})$ does not converge to P_0 (Fig. 1).

The fundamental problem here is that neither the total variation distance (6) nor the f-divergences (7) depend on the distance $d(x,y)$ defined on the sample space $\mathcal{X}$. The minimization of such a criterion appears more effective for adjusting the probability values than for matching the distribution supports.

2.6   Wasserstein Distance

For any $p \geq 1$, the p-Wasserstein distance (WD) is the p-th root of

$$\forall Q, P \in \mathcal{P}_\mathcal{X}^p \qquad W_p(Q,P)^p \overset{\Delta}{=} \inf_{\pi \in \Pi(Q,P)} \mathbb{E}_{(x,y)\sim\pi}[\,d(x,y)^p\,], \qquad (8)$$

where $\Pi(Q,P)$ represents the set of all measures π defined on $\mathcal{X} \times \mathcal{X}$ with marginals $x\#\pi(x,y)$ and $y\#\pi(x,y)$ respectively equal to Q and P. Intuitively, $d(x,y)^p$ represents the cost of transporting a grain of probability from point x to point y, and the joint distributions $\pi \in \Pi(Q,P)$ represent transport plans.

Since $d(x,y) \leq d(x,x_0) + d(x_0,y) \leq 2\max\{d(x,x_0), d(x_0,y)\}$,

$$\forall Q, P \in \mathcal{P}_\mathcal{X}^p \qquad W_p(Q,P)^p \leq \mathbb{E}_{\substack{x\sim Q \\ y\sim P}}[d(x,y)^p] < \infty. \qquad (9)$$

Example 2.7 Let P_θ be defined as in Example 2.6. Since it is easy to see that the optimal transport plan from P_0 to P_θ is $\pi^* = ((0,t),(\theta,t))\#U(t)$, the Wassertein distance $W_p(P_0, P_\theta) = |\theta|$ converges to zero when θ tends to zero. Therefore, according to the topology induced by the Wasserstein distance on $\mathcal{P}_\mathcal{X}$, the sequence of distributions $(P_{1/i})$ converges to P_0 (Fig. 1).

Thanks to the Kantorovich duality theory, the Wasserstein distance is easily expressed in the variational form (3). We summarize below the essential results useful for this work and we direct the reader to [56, Chapters 4 and 5] for a full exposition.

Theorem 2.8 ([56, **Theorem 4.1**]). *Let $\mathcal{X}, \mathcal{Y}$ be two Polish metric spaces and $c : \mathcal{X} \times \mathcal{Y} \to \mathbb{R}_+ \cup \{+\infty\}$ be a nonnegative continuous cost function. Let $\Pi(Q, P)$ be the set of probablity measures on $\mathcal{X} \times \mathcal{Y}$ with marginals $Q \in \mathcal{P}_\mathcal{X}$ and $P \in \mathcal{P}_\mathcal{Y}$. There is a $\pi^* \in \Pi(Q, P)$ that minimizes $\mathbb{E}_{(x,y) \sim \pi}[c(x, y)]$ over all $\pi \in \Pi(Q, P)$.*

Definition 2.9. Let $\mathcal{X}, \mathcal{Y}$ be two Polish metric spaces and $c : \mathcal{X} \times \mathcal{Y} \to \mathbb{R}_+ \cup \{+\infty\}$ be a nonnegative continuous cost function. The pair of functions $f : \mathcal{X} \to \mathbb{R}$ and $g : \mathcal{Y} \to \mathbb{R}$ is c-conjugate when

$$\forall x \in \mathcal{X} \; f(x) = \inf_{y \in \mathcal{Y}} g(y) + c(x, y) \quad \text{and} \quad \forall y \in \mathcal{Y} \; g(y) = \sup_{x \in \mathcal{X}} f(x) - c(x, y) . \quad (10)$$

Theorem 2.10. (Kantorovich duality [56, Theorem 5.10]**).** *Let $\mathcal{X}$ and $\mathcal{Y}$ be two Polish metric spaces and $c : \mathcal{X} \times \mathcal{Y} \to \mathbb{R}_+ \cup \{+\infty\}$ be a nonnegative continuous cost function. For all $Q \in \mathcal{P}_\mathcal{X}$ and $P \in \mathcal{P}_\mathcal{Y}$, let $\Pi(Q, P)$ be the set of probability distributions defined on $\mathcal{X} \times \mathcal{Y}$ with marginal distributions Q and P. Let $\mathcal{Q}_c$ be the set of all pairs (f_Q, f_P) of respectively Q and P-integrable functions satisfying the property $\forall x \in \mathcal{X} \; y \in \mathcal{Y}, \; f_Q(x) - f_P(y) \le c(x, y)$.*

(i) We have the duality

$$\min_{\pi \in \Pi(Q,P)} \mathbb{E}_{(x,y) \sim \pi}[c(x, y)] = \quad (11)$$

$$\sup_{(f_Q, f_P) \in \mathcal{Q}_c} \mathbb{E}_{x \sim Q}[f_Q(x)] - \mathbb{E}_{y \sim P}[f_P(y)] . \quad (12)$$

(ii) Further assuming that $\mathbb{E}_{x \sim Q \; y \sim P}[c(x, y)] < \infty$,
  (a) Both (11) and (12) have solutions with finite cost.
  (b) The solution (f_Q^*, f_P^*) of (12) is a c-conjugate pair.

Corollary 2.11 ([56, **Particular case 5.16**]). *Under the same conditions as Theorem 2.10.ii, when $\mathcal{X} = \mathcal{Y}$ and when the cost function c is a distance, that is, satisfies (2), the dual optimization problem (12) can be rewritten as*

$$\max_{f \in \mathrm{Lip1}} \mathbb{E}_Q[f(x)] - \mathbb{E}_P[f(x)] ,$$

where Lip1 *is the set of real-valued 1-Lipschitz continuous functions on $\mathcal{X}$.*

Thanks to Theorem 2.10, we can write the p-th power of the p-Wasserstein distance in variational form

$$\forall Q, P \in \mathcal{P}_\mathcal{X}^p \quad W_p(Q, P)^p = \sup_{(f_Q, f_P) \in \mathcal{Q}_c} \mathbb{E}_Q[f_Q(x)] - \mathbb{E}_P[f_P(x)] , \quad (13)$$

where Q_c is defined as in Theorem 2.10 for the cost $c(x, y) = d(x, y)^p$. Thanks to Corollary 2.11, we can also obtain a simplified expression in IPM form for the 1-Wasserstein distance.

$$\forall Q, P \in \mathcal{P}_{\mathcal{X}}^1 \qquad W_1(Q, P) = \sup_{f \in \text{Lip1}} \mathbb{E}_Q[f(x)] - \mathbb{E}_P[f(x)]. \qquad (14)$$

Let us conclude this presentation of the Wassertein distance by mentioning that the definition (8) immediately implies several distance properties: zero when both distributions are equal (2.o), strictly positive when they are different (2.i), and symmetric (2.ii). Property 2.4 gives the triangular inequality (2.iii) for the case $p = 1$. In the general case, the triangular inequality can also be established using the Minkowsky inequality [56, Chapter 6].

2.7   Energy Distance and Maximum Mean Discrepancy

The Energy Distance (ED) [54] between the probability distributions Q and P defined on the Euclidean space $\mathbb{R}^d$ is the square root of [5]

$$\mathcal{E}(Q, P)^2 \triangleq 2\mathbb{E}_{\substack{x \sim Q \\ y \sim P}}[\|x - y\|] - \mathbb{E}_{\substack{x \sim Q \\ x' \sim Q}}[\|x - x'\|] - \mathbb{E}_{\substack{y \sim P \\ y' \sim P}}[\|y - y'\|], \qquad (15)$$

where, as usual, $\| \cdot \|$ denotes the Euclidean distance.

Let $\hat{q}$ and $\hat{p}$ represent the characteristic functions of the distribution Q and P respectively. Thanks to a neat Fourier transform argument [53,54],

$$\mathcal{E}(Q, P)^2 = \frac{1}{c_d} \int_{\mathbb{R}^d} \frac{|\hat{q}(t) - \hat{p}(t)|^2}{\|t\|^{d+1}} dt \quad \text{with } c_d = \frac{\pi^{\frac{d+1}{2}}}{\Gamma(\frac{d+1}{2})}. \qquad (16)$$

Since there is a one-to-one mapping between distributions and characteristic functions, this relation establishes an isomorphism between the space of probability distributions equipped with the ED distance and the space of the characteristic functions equipped with the weighted L_2 norm given in the right-hand side of (16). As a consequence, $\mathcal{E}(Q, P)$ satisfies the properties (2) of a distance.

Since the squared ED is expressed with a simple combination of expectations, it is easy to design a stochastic minimization algorithm that relies only on two oracles producing samples from each distribution [7,11]. This makes the energy distance a computationally attractive criterion for training the implicit models discussed in Sect. 2.2.

Generalized ED. It is therefore natural to ask whether we can meaningfully generalize (15) by replacing the Euclidean distance $\|x - y\|$ with a symmetric function $d(x, y)$.

$$\mathcal{E}_d(Q, P)^2 = 2\mathbb{E}_{\substack{x \sim Q \\ y \sim P}}[d(x, y)] - \mathbb{E}_{\substack{x \sim Q \\ x' \sim Q}}[d(x, x')] - \mathbb{E}_{\substack{y \sim P \\ y' \sim P}}[d(y, y')]. \qquad (17)$$

---

[5] We take the square root because this is the quantity that behaves like a distance.

The right-hand side of this expression is well defined when $Q, P \in \mathcal{P}_{\mathcal{X}}^1$. It is obviously symmetric (2.ii) and trivially zero (2.o) when both distributions are equal. The first part of the following theorem gives the necessary and sufficient conditions on $d(x, y)$ to ensure that the right-hand side of (17) is nonnegative and therefore can be the square of $\mathcal{E}_d(Q, P) \in \mathbb{R}_+$. We shall see later that the triangular inequality (2.iii) comes for free with this condition (Corollary 2.19). The second part of the theorem gives the necessary and sufficient condition for satisfying the separation property (2.i).

Theorem 2.12 ([57]). *The right-hand side of definition (17) is:*

(i) *nonnegative for all P, Q in $\mathcal{P}_{\mathcal{X}}^1$ if and only if the symmetric function d is a negative definite kernel, that is,*

$$\forall n \in \mathbb{N} \quad \forall x_1 \ldots x_n \in \mathcal{X} \quad \forall c_1 \ldots c_n \in \mathbb{R}$$
$$\sum_{i=1}^{n} c_i = 0 \implies \sum_{i=1}^{n} \sum_{j=1}^{n} d(x_i, x_j) c_i c_j \leq 0 . \tag{18}$$

(ii) *strictly positive for all $P \neq Q$ in $\mathcal{P}_{\mathcal{X}}^1$ if and only if the function d is a strongly negative definite kernel, that is, a negative definite kernel such that, for any probability measure $\mu \in \mathcal{P}_{\mathcal{X}}^1$ and any μ-integrable real-valued function h such that $\mathbb{E}_\mu[h(x)] = 0$,*

$$\mathbb{E}_{\substack{x \sim \mu \\ y \sim \mu}}[d(x, y)h(x)h(y)] = 0 \implies h(x) = 0 \quad \mu - almost\ everywhere.$$

Remark 2.13. The definition of a strongly negative kernel is best explained by considering how its meaning would change if we were only considering probability measures μ with finite support $\{x_1 \ldots x_n\}$. This amounts to requiring that (18) is an equality only if all the c_is are zero. However, this weaker property is not sufficient to ensure that the separation property (2.i) holds.

Remark 2.14. The relation (16) therefore means that the Euclidean distance on $\mathbb{R}^d$ is a strongly negative definite kernel. In fact, it can be shown that $d(x, y) = \|x - y\|^\beta$ is a strongly negative definite kernel for $0 < \beta < 2$ [53]. When $\beta = 2$, it is easy to see that $\mathcal{E}_d(Q, P)$ is simply the distance between the distribution means and therefore cannot satisfy the separation property (2.i).

Proof of Theorem 2.12. Let $E(Q, P)$ be the right-hand side of (17) and let $S(\mu, h)$ be the quantity $\mathbb{E}_{x,y \sim \mu}[d(x, y)h(x)h(y)]$ that appears in clause (ii). Observe:

(a) Let $Q, P \in \mathcal{P}_{\mathcal{X}}^1$ have respective density functions $q(x)$ and $p(x)$ with respect to measure $\mu = (Q + P)/2$. Function $h = q - p$ then satisfies $\mathbb{E}_\mu[h] = 0$, and

$$E(Q, P) = \mathbb{E}_{\substack{x \sim \mu \\ y \sim \mu}}\left[(q(x)p(y) + q(y)p(x) - q(x)q(y) - p(x)p(y)) d(x, y) \right] = -S(\mu, h).$$

(b) With $\mu \in \mathcal{P}_{\mathcal{X}}^1$, any h such that $\mu\{h = 0\} < 1$ (ie., non-μ-almost-surely-zero) and $\mathbb{E}_\mu[h] = 0$ can be written as a difference of two nonnegative functions $h = \tilde{q} - \tilde{p}$ such that $\mathbb{E}_\mu[\tilde{q}] = \mathbb{E}_\mu[\tilde{p}] = \rho^{-1} > 0$. Then, $Q = \rho\tilde{q}\,\mu$ and $P = \rho\tilde{p}\,\mu$ belong to $\mathcal{P}_{\mathcal{X}}^1$, and

$$E(Q, P) = -\rho\, S(\mu, h) \ .$$

We can then prove the theorem:

(i) From these observations, if $E(Q, P) \geq 0$ for all P, Q, then $S(\mu, h) \leq 0$ for all μ and h such that $\mathbb{E}_\mu[h(x)] = 0$, implying (18). Conversely, assume there are $Q, P \in \mathcal{P}_{\mathcal{X}}^1$ such that $E(Q, P) < 0$. Using the weak law of large numbers [26] (see also Theorem 3.3 later in this document,) we can find finite support distributions Q_n, P_n such that $E(Q_n, P_n) < 0$. Proceeding as in observation (a) then contradicts (18) because $\mu = (Q_n + P_n)/2$ has also finite support.

(ii) By contraposition, suppose there is μ and h such that $\mu\{h = 0\} < 1$, $\mathbb{E}_\mu[h(x)] = 0$, and $S(\mu, h) = 0$. Observation (b) gives $P \neq Q$ such that $E(Q, P) = 0$. Conversely, suppose $E(Q, P) = 0$. Observation (a) gives μ and $h = q - p$ such that $S(\mu, h) = 0$. Since h must be zero, $Q = P$. ∎

Requiring that d be a negative definite kernel is a quite strong assumption. For instance, a classical result by Schoenberg [46] establishes that a squared distance is a negative definite kernel if and only if the whole metric space induced by this distance is isometric to a subset of a Hilbert space and therefore has a Euclidean geometry:

Theorem 2.15 (Schoenberg, [46]). *The metric space $(\mathcal{X}, d)$ is isometric to a subset of a Hilbert space if and only if d^2 is a negative definite kernel.*

Requiring d to be negative definite (not necessarily a squared distance anymore) has a similar impact on the geometry of the space $\mathcal{P}_{\mathcal{X}}^1$ equipped with the Energy Distance (Theorem 2.17). Let x_0 be an arbitrary origin point and define the symmetric *triangular gap* kernel K_d as

$$K_d(x, y) \overset{\Delta}{=} \tfrac{1}{2}\left(d(x, x_0) + d(y, x_0) - d(x, y)\right) \ . \tag{19}$$

Proposition 2.16. *The function d is a negative definite kernel if and only if K_d is a positive definite kernel, that is,*

$$\forall n \in \mathbb{N} \ \ \forall x_1 \ldots x_n \in \mathcal{X} \ \ \forall c_1 \ldots c_n \in \mathbb{R} \ \ \sum_{i=1}^{n}\sum_{j=1}^{n} c_i c_j K_d(x_i, x_j) \geq 0 \ .$$

Proof. The proposition directly results from the identity

$$2\sum_{i=1}^{n}\sum_{j=1}^{n} c_i c_j K_d(x_i, x_j) = -\sum_{i=0}^{n}\sum_{j=0}^{n} c_i c_j d(x_i, x_j) \ ,$$

where x_0 is the chosen origin point and $c_0 = -\sum_{i=1}^{n} c_i$. ∎

Positive definite kernels in the machine learning literature have been extensively studied in the context of the so-called *kernel trick* [47]. In particular, it is well known that the theory of the Reproducing Kernel Hilbert Spaces (RKHS) [1,4] establishes that there is a unique Hilbert space $\mathcal{H}$, called the RKHS, that contains all the functions

$$\Phi_x : y \in \mathcal{X} \mapsto K_d(x, y)$$

and satisfies the *reproducing property*

$$\forall x \in \mathcal{X} \ \ \forall f \in \mathcal{H} \ \ \ \langle f, \Phi_x \rangle = f(x) \ . \tag{20}$$

We can then relate $\mathcal{E}_d(Q, P)$ to the RKHS norm.

Theorem 2.17 ([48] [41, Chapter 21]). *Let d be a negative definite kernel and let $\mathcal{H}$ be the RKHS associated with the corresponding positive definite triangular gap kernel* (19). *We have then*

$$\forall Q, P \in \mathcal{P}_{\mathcal{X}}^1 \ \ \ \mathcal{E}_d(Q, P) \ = \ \| \, \mathbb{E}_{x \sim Q}[\Phi_x] - \mathbb{E}_{y \sim P}[\Phi_y] \, \|_{\mathcal{H}} \, .$$

Proof. We can write directly

$$\begin{aligned}
\mathcal{E}_d(Q, P)^2 &= \mathbb{E}_{\substack{x,x' \sim Q \\ y,y' \sim P}}[d(x, y) + d(x', y') - d(x, x') - d(y, y')] \\
&= \mathbb{E}_{\substack{x,x' \sim Q \\ y,y' \sim P}}[K_d(x, x') + K_d(y, y') - K_d(x, y) - K_d(x', y')] \\
&= \langle \mathbb{E}_Q[\Phi_x], \mathbb{E}_Q[\Phi_x] \rangle + \langle \mathbb{E}_P[\Phi_y], \mathbb{E}_P[\Phi_y] \rangle - 2 \langle \mathbb{E}_Q[\Phi_x], \mathbb{E}_P[\Phi_y] \rangle \\
&= \| \mathbb{E}_{x \sim Q}[\Phi_x] - \mathbb{E}_{y \sim P}[\Phi_y] \|_{\mathcal{H}}^2 \, ,
\end{aligned}$$

where the first equality results from (19) and where the second equality results from the identities $\langle \Phi_x, \Phi_y \rangle = K_d(x, y)$ and $\mathbb{E}_{x,y}[\langle \Phi_x, \Phi_y \rangle] = \langle \mathbb{E}_x[\Phi_x], \mathbb{E}_y[\Phi_y] \rangle$. ∎

Remark 2.18. In the context of this theorem, the relation (16) is simply an analytic expression of the RKHS norm associated with the triangular gap kernel of the Euclidean distance.

Corollary 2.19. *If d is a negative definite kernel, then $\mathcal{E}_d$ is a pseudodistance, that is, it satisfies all the properties* (2) *of a distance except maybe the separation property (2.i).*

Corollary 2.20. *The following three conditions are then equivalent:*

(i) $\mathcal{E}_d$ satisfies all the properties (2) of a distance.
(ii) d is a strongly negative definite kernel.
(iii) the map $P \in \mathcal{P}_{\mathcal{X}}^1 \mapsto \mathbb{E}_P[\Phi_x] \in \mathcal{H}$ is injective (characteristic kernel [21].)

Maximum Mean Discrepancy. Following [21], we can then write $\mathcal{E}_d$ as an IPM:

$$
\begin{aligned}
\mathcal{E}_d(Q,P) &= \| \mathbb{E}_Q[\Phi_x] - \mathbb{E}_P[\Phi_x] \|_{\mathcal{H}} \\
&= \sup_{\|f\|_{\mathcal{H}} \leq 1} \langle f, \mathbb{E}_P[\Phi_x] - \mathbb{E}_Q[\Phi_x] \rangle \\
&= \sup_{\|f\|_{\mathcal{H}} \leq 1} \mathbb{E}_P[\langle f, \Phi_x \rangle] - \mathbb{E}_Q[\langle f, \Phi_x \rangle] \\
&= \sup_{\|f\|_{\mathcal{H}} \leq 1} \mathbb{E}_P[f(x)] - \mathbb{E}_Q[f(x)] .
\end{aligned} \tag{21}
$$

This last expression (21) is also called the Maximum Mean Discrepancy (MMD) associated with the positive definite kernel K_d [21]. Conversely, for any positive definite kernel K, the reader will easily prove that the symmetric function

$$
d_K(x,y) = \| \Phi_x - \Phi_y \|_{\mathcal{H}}^2 = K(x,x) + K(y,y) - 2K(x,y) ,
$$

is a negative definite kernel, that $d_{K_d} = d$, and that

$$
\| \mathbb{E}_Q[\Phi_x] - \mathbb{E}_P[\Phi_x] \|_{\mathcal{H}}^2 = \mathcal{E}_{d_K}(Q,P)^2. \tag{22}
$$

Therefore the ED and MMD formulations are essentially equivalent [48]. Note however that the negative definite kernel d_K defined above may not satisfy the triangular inequality (its square root does.)

Remark 2.21. Because this equivalence was not immediately recognized, many important concepts have been rediscovered with subtle technical variations. For instance, the notion of characteristic kernel [21] depends subtly on the chosen domain for the map $P \mapsto \mathbb{E}_P[\Phi_x]$ that we want injective. Corollary 2.20 gives a simple necessary and sufficient condition when this domain is $\mathcal{P}_{\mathcal{X}}^1$ (with respect to the distance d). Choosing a different domain leads to complications [52].

3   Energy Distance vs. 1-Wasserstein Distance

The dual formulation of the 1-Wasserstein (14) and the MMD formulation of the Energy Distance (21) only differ by the use of a different family of critic functions: for all $Q, P \in \mathcal{P}_{\mathcal{X}}^1$,

$$
W_1(Q,P) = \sup_{f \in \mathrm{Lip1}} \mathbb{E}_Q[f(x)] - \mathbb{E}_P[f(x)] ,
$$

$$
\mathcal{E}_d(Q,P) = \sup_{\|f\|_{\mathcal{H}} \leq 1} \mathbb{E}_P[f(x)] - \mathbb{E}_Q[f(x)].
$$

At first sight, requiring that the functions f are 1-Lipschitz or are contained in the RKHS unit ball seem to be two slightly different ways to enforce a smoothness constraint. Nevertheless, a closer comparison reveals very important differences.

3.1   Three Quantitative Properties

Although both the WD [56, Theorem 6.9] and the ED/MMD [50, Theorem 3.2] metrize the weak convergence topology, they may be quantitatively very different and therefore hard to compare in practical situations. The following upper bound provides a clarification.

Proposition 3.1. *Let $\mathcal{X}$ be equipped with a distance d that is also a negative definite kernel. Let the 1-Wasserstein distance W_1 and the Energy Distance $\mathcal{E}_d$ be defined as in (8) and (17).*

$$\mathcal{E}_d(Q, P)^2 \leq 2W_1(Q, P) .$$

This inequality is tight. It is indeed easy to see that it becomes an equality when both P and Q are Dirac distributions.

The proof relies on an elementary geometrical lemma:

Lemma 3.2. *Let A, B, C, D be four points in $\mathcal{X}$ forming a quadrilateral. The perimeter length $d(A, B)+d(B, C)+d(C, D)+d(D, A)$ is longer than the diagonal lenghts $d(A, C) + d(B, D)$.*

Proof of the Lemma. Summing the following triangular inequalities yields the result.

$$d(A, C) \leq d(A, B) + d(B, C) \qquad d(A, C) \leq d(C, D) + d(D, A)$$
$$d(B, D) \leq d(B, C) + d(C, D) \qquad d(B, D) \leq d(D, A) + d(A, B) \qquad \blacksquare$$

Proof of Proposition 3.1. Let (x, y) and (x', y') be two independent samples of the optimal transport plan π with marginals Q and P. Since they are independent,

$$2\,\mathbb{E}_{\substack{x \sim Q \\ y \sim P}}[d(x, y)] = \mathbb{E}_{\substack{(x,y) \sim \pi \\ (x',y') \sim \pi}}[d(x, y') + d(x', y)] .$$

Applying the lemma and rearranging

$$2\mathbb{E}_{\substack{x \sim Q \\ y \sim P}}[d(x, y)] \leq \mathbb{E}_{\substack{(x,y) \sim \pi \\ (x',y') \sim \pi}}[d(x, y) + d(y, y') + d(y', x') + d(x', x)]$$
$$= W_1(Q, P) + \mathbb{E}_{\substack{y \sim P \\ y' \sim P}}[d(y, y')] + W_1(Q, P) + \mathbb{E}_{\substack{x \sim Q \\ x' \sim Q}}[d(x, x')] .$$

Moving the remaining expectations to the left-hand side gives the result. $\blacksquare$

In contrast, the following results not only show that $\mathcal{E}_d$ can be very significantly smaller than the 1-Wasserstein distance, but also show that this happens in the particularly important situation where one approximates a distribution with a finite sample.

Theorem 3.3. *Let* $Q, P \in \mathcal{P}_{\mathcal{X}}^1$ *be two probability distributions on* $\mathcal{X}$*. Let* $x_1 \ldots x_n$ *be* n *independent* Q*-distributed random variables, and let* $Q_n = \frac{1}{n} \sum_{i=1}^n \delta_{x_i}$ *be the corresponding empirical probability distribution. Let* $\mathcal{E}_d$ *be defined as in* (17) *with a kernel satisfying* $d(x, x) = 0$ *for all* x *in* $\mathcal{X}$*. Then,*

$$\mathbb{E}_{x_1 \ldots x_n \sim Q} \left[\mathcal{E}_d(Q_n, P)^2 \right] = \mathcal{E}_d(Q, P)^2 + \frac{1}{n} \mathbb{E}_{x, x' \sim Q}[d(x, x')],$$

and

$$\mathbb{E}_{x_1 \ldots x_n \sim Q} \left[\mathcal{E}_d(Q_n, Q)^2 \right] = \frac{1}{n} \mathbb{E}_{x, x' \sim Q}[d(x, x')] = \mathcal{O}(n^{-1}).$$

Therefore the effect of replacing Q by its empirical approximation disappears quickly, like $\mathcal{O}(1/n)$, when n grows. This result is not very surprising when one notices that $\mathcal{E}_d(Q_n, P)$ is a V-statistic [36, 49]. However it gives a precise equality with a particularly direct proof.

Proof. Using the following equalities in the definition (17) gives the first result.

$$\mathbb{E}_{x_1 \ldots x_n \sim Q} \left[\mathbb{E}_{\substack{x \sim Q_n \\ y \sim P}}[d(x, y)] \right] = \mathbb{E}_{x_1 \ldots x_n \sim Q} \left[\frac{1}{n} \sum_{i=1}^n \mathbb{E}_{y \sim P}[d(x_i, y)] \right]$$

$$= \frac{1}{n} \sum_i \mathbb{E}_{\substack{x \sim Q \\ y \sim P}}[d(x, y)] = \mathbb{E}_{\substack{x \sim Q \\ y \sim P}}[d(x, y)] .$$

$$\mathbb{E}_{x_1 \ldots x_n \sim Q} \left[\mathbb{E}_{\substack{x \sim Q_n \\ x' \sim Q_n}}[d(x, x')] \right] = \mathbb{E}_{x_1 \ldots x_n \sim Q} \left[\frac{1}{n^2} \sum_{i \neq j} d(x_i, x_j) \right]$$

$$= \frac{1}{n^2} \sum_{i \neq j} \mathbb{E}_{\substack{x \sim Q \\ y \sim Q}}[d(x, y)] = \left(1 - \frac{1}{n} \right) \mathbb{E}_{\substack{x \sim Q \\ y \sim Q}}[d(x, y)].$$

Taking $Q = P$ then gives the second result. ∎

Comparable results for the 1-Wasserstein distance describe a convergence speed that quickly becomes considerably slower with the dimension $d > 2$ of the sample space $\mathcal{X}$ [16, 18, 51].

Theorem 3.4 ([18]). *Let* $\mathcal{X}$ *be* $\mathbb{R}^d$*,* $d > 2$*, equipped with the usual Euclidean distance. Let* $Q \in \mathcal{P}_{\mathbb{R}^d}^2$ *and let* Q_n *be defined as in Theorem 3.3. Then,*

$$\mathbb{E}_{x_1 \ldots x_n \sim Q}[W_1(Q_n, Q)] = \mathcal{O}(n^{-1/d}).$$

The following example, inspired by [5], illustrates this slow rate and its consequences.

Example 3.5. Let Q be a uniform distribution supported by the unit sphere in $\mathbb{R}^d$ equipped with the Euclidean distance. Let $x_1 \ldots x_n$ be n points sampled independently from this distribution and let Q_n be the corresponding empirical distribution. Let x be an additional point sampled from Q. It is well known[6]

---

[6] The curious reader can pick an expression of $F_d(t) = P\{\|x - x_i\| < t\}$ in [23], then derive an asymptotic bound for $P\{\min_i \|x - x_i\| < t\} = 1 - (1 - F_d(t))^n$.

that $\min_i \|x - x_i\|$ remains arbitrarily close to $\sqrt{2}$, say, greater than 1.2, with arbitrarily high probability when $d \gg \log(n)$. Therefore,

$$W_1(Q_n, Q) \geq 1.2 \quad \text{when} n \ll \exp(d).$$

In contrast, observe

$$W_1(Q, \delta_0) = W_1(Q_n, \delta_0) = 1 \ .$$

In other words, as long as $n \ll \exp(d)$, a Dirac distribution in zero is closer to the empirical distribution than the actual distribution [5].

Theorem 3.3 and Example 3.5 therefore show that $\mathcal{E}_d(Q_n, Q)$ can be much smaller than $W_1(Q_n, Q)$. They also reveal that the statistical properties of the 1-Wasserstein distance are very discouraging. Since the argument of Example 3.5 naturally extends to the p-Wasserstein distance for all $p \geq 1$, the problem seems shared by all Wasserstein distances.

Remark 3.6. In the more realistic case where the 1-Lipschitz critic is constrained to belong to a parametric family with sufficient regularity, the bound of theorem 3.4 can be improved to $\mathcal{O}(\sqrt{\log(n)/n})$ with a potentially large constant [5]. On the other hand, constraining the critic too severely might prevent it from distinguishing distributions that differ in meaningful ways.

3.2 WD and ED/MMD in Practice

Why should we consider the Wasserstein Distance when the Energy Distance and Maximum Mean Discrepancy offer better statistical properties (Sect. 3.1) and more direct learning algorithms [11,17,33]?

The most impressive achievement associated with the implicit modeling approach certainly is the generation of photo-realistic random images that resemble the images provided as training data [15,25,42]. In apparent contradiction with the statistical results of the previous section, and with a couple notable exceptions discussed later in this section, the visual quality of the images generated using models trained by directly minimizing the MMD [17] usually lags behind those obtained with the WD [3,22,25] and with the original Generative Adversarial Network formulation[7] [42].

Before discussing the two exceptions, it is worth recalling that the visual quality of the generated images is a peculiar way to benchmark generative models. This is an incomplete criterion because it does not ensure that the model generates images that cover all the space covered by the training data. This is an interesting criterion because common statistical metrics, such as estimates of the negative log-likelihood, are generally unable to indicate which models generate the better-looking images [55]. This is a finicky criterion because, despite efforts to quantify visual quality with well-defined scores [45], the evaluation of the image quality fundamentally remains a beauty contest. Figure 2 nevertheless shows a clear difference.

---

[7] Note that it is then important to use the $\log(D)$ trick succinctly discussed in the original GAN paper [20].

A sample of 64 training examples

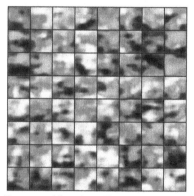

Generated by the ED trained model   Generated by the WD trained model

Fig. 2. Comparing images generated by a same implicit model trained with different criteria. The top square shows a sample of 64 training examples represening bedroom pictures. The bottom left square shows the images generated by a model trained with ED using the algorithm of [11]. The bottom right square shows images generated by a model trained using the WGAN-GP approach [22].

A few authors report good image generation results by using the ED/MMD criterion in a manner that substantially changes its properties:

– The AE+GMMN approach [33] improves the pure MMD approach by training an implicit model that does not directly generate images but targets the compact representation computed by a pretrained auto-encoder network. This changes a high-dimensional image generation problem into a comparatively low-dimensional code generation problem with a good notion of distance. There is independent evidence that low-dimensional implicit models work relatively well with ED/MMD [11].

– The CramérGAN approach [7] minimizes the Energy Distance[8] computed on the representations produced by an adversarially trained 1-Lipschitz continuous *transformation layer* $T_\phi(x)$. The resulting optimization problem

$$\min_\theta \left\{ \max_{T_\phi \in \text{Lip1}} \mathcal{E}(T_\phi \# Q, T_\phi \# P_\theta) \right\},$$

can then be re-expressed using the IPM form of the energy distance

$$\min_\theta \left\{ D(Q, P) = \max_{T_\phi \in \text{Lip1}} \max_{\|f\|_\mathcal{H} \leq 1} \mathbb{E}_{x \sim Q}[f(T_\phi(x))] - \mathbb{E}_{x \sim P_\theta}[f(T_\phi(x))] \right\}.$$

The cost $D(Q, P)$ above is a new IPM that relies on critic functions of the form $f \circ T_\phi$, where f belongs to the RKHS unit ball, and T_ϕ is 1-Lipschitz continuous. Such hybrid critic functions still have smoothness properties comparable to that of the Lipschitz-continuous critics of the 1-Wasserstein distance. However, since these critic functions do not usually form a RKHS ball, the resulting IPM criterion no longer belongs to the ED/MMD family.

– The same hybrid approach gives comparable results in GMMN-C [32] where the authors replace autoencoder of GMMN+AE with an adversarially trained transformer layer.

On the positive side, such hybrid approaches may lead to more efficient training algorithms than those described in Sect. 2.3. The precise parametric structure of the transformation layer also provides the means to match what WGAN models achieve by selecting a precise parametric structure for the critic. Yet, in order to understand these subtle effects, it remains useful to clarify the similarities and differences between pure ED/MMD training and pure WD training.

4   Length Spaces

This section gives a concise review of the elementary metric geometry concepts useful for the rest of our analysis. Readers can safely skip this section if they are already familiar with metric geometry textbooks such as [12].

Rectifiable Curves. A continuous mapping $\gamma : t \in [a, b] \subset \mathbb{R} \mapsto \gamma_t \in \mathcal{X}$ defines a curve connecting γ_a and γ_b. A curve is said to be *rectifiable* when its *length*

$$L(\gamma, a, b) \triangleq \sup_{n > 1} \sup_{a = t_0 < t_1 < \cdots < t_n = b} \sum_{i=1}^n d(\gamma_{t_{i-1}}, \gamma_t) \tag{23}$$

is finite. Intuitively, thanks to the triangular inequality, dividing the curve into n segments $[\gamma_{t-1}, \gamma_t]$ and summing their sizes yields a quantity that is greater than $d(\gamma_a, \gamma_b)$ but smaller than the curvilinear length of the curve. By construction, $L(\gamma, a, b) \geq d(\gamma_a, \gamma_b)$ and $L(\gamma, a, c) = L(\gamma, a, b) + L(\gamma, b, c)$ for all $a \leq b \leq c$.

---

[8] See [54] for the relation between Energy Distance and Cramér distance.

Constant Speed Curves. Together with the continuity of γ, this additivity property implies that the function $t \in [a, b] \mapsto L(\gamma, a, t)$ is nondecreasing and continuous [12, Prop. 2.3.4]. Thanks to the intermediate value theorem, when a curve is rectifiable, for all $s \in [0, 1]$, there is $t_s \in [a, b]$ such that $L(\gamma, a, t_s) = s\, L(\gamma, a, b)$. Therefore, we can construct a new curve $\bar{\gamma} : s \in [0, 1] \mapsto \bar{\gamma}_s = \gamma_{t_s}$ that visits the same points in the same order as curve γ and satisfies the property $\forall s \in [0, 1]$, $L(\bar{\gamma}, 0, s) = sL(\bar{\gamma}, 0, 1)$. Such a curve is called a *constant speed curve*.

Length Spaces. It is easy to check that the *distance induced by d*,

$$\hat{d} : (x, y) \in \mathcal{X}^2 \quad \mapsto \quad \inf_{\substack{\gamma:[a,b]\to\mathcal{X} \\ \text{s.t.}\,\gamma_a=x\,\gamma_b=y}} \quad L(\gamma, a, b) \ \in \mathbb{R}_+^* \cup \{\infty\}, \qquad (24)$$

indeed satisfies all the properties (2) of a distance. It is also easy to check that the distance induced by $\hat{d}$ coincides with $\hat{d}$ [12, Prop. 2.3.12]. For this reason, a distance that satisfies $\hat{d} = d$ is called an *intrinsic distance*. A Polish metric space equipped with an intrinsic distance is called an *intrinsic Polish space*. A metric space $\mathcal{X}$ equipped with an intrinsic distance d is called a *length space*.

Minimal Geodesics. A curve $\gamma : [a, b] \to \mathcal{X}$ that achieves the infimum in (24) is called a *shortest path* or a *minimal geodesic* connecting γ_a and γ_b.

When the distance d is intrinsic, the length of a minimal geodesic γ satisfies the relation $L(\gamma, a, b) = \hat{d}(\gamma_a, \gamma_b) = d(\gamma_a, \gamma_b)$. When such a curve exists between any two points x, y such that $d(x, y) < \infty$, the distance d is called *strictly intrinsic*. A Polish space equipped with a strictly intrinsic distance is called a *strictly intrinsic Polish space*.

Conversely, a rectifiable curve $\gamma : [a, b] \to \mathcal{X}$ of length $d(\gamma_a, \gamma_b)$ is a minimal geodesic because no curve joining γ_a and γ_b can be shorter. If there is such a curve between any two points x, y such that $d(x, y) < \infty$, then d is a strictly intrinsic distance.

Characterizing Minimal Geodesics. Let $\gamma : [a, b] \to \mathcal{X}$ be a minimal geodesic in a length space $(\mathcal{X}, d)$. Using the triangular inequality and (23),

$$\forall a \le t \le b \quad d(\gamma_a, \gamma_b) \le d(\gamma_a, \gamma_t) + d(\gamma_t, \gamma_b) \le L(\gamma, a, b) = d(\gamma_a, \gamma_b) . \qquad (25)$$

This makes clear that every minimal geodesic in a length space is made of points γ_t for which the triangular inequality is an equality. However, as shown in Fig. 3, this is not sufficient to ensure that a curve is a minimal geodesic. One has to consider two intermediate points:

Theorem 4.1. *Let* $\gamma : [a, b] \to \mathcal{X}$ *be a curve joining two points* γ_a, γ_b *such that* $d(\gamma_a, \gamma_b) < \infty$. *This curve is a minimal geodesic of length* $d(\gamma_a, \gamma_b)$ *if and only if* $\forall\, a \le t \le t' \le b, \quad d(\gamma_a, \gamma_t) + d(\gamma_t, \gamma_{t'}) + d(\gamma_{t'}, \gamma_b) = d(\gamma_a, \gamma_b) .$

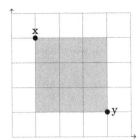 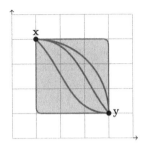 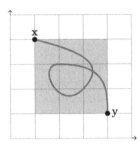

Fig. 3. Consider $\mathbb{R}^2$ equipped with the L_1 distance. Left: all points z in the gray area are such that $d(x, z) + d(z, y) = d(x, y)$. Center: all minimal geodesics connecting x and y live in the gray area. Right: but not all curves that live in the gray area are minimal geodesics.

Corollary 4.2. *Let* $\gamma : [0, 1] \to \mathcal{X}$ *be a curve joining two points* $\gamma_0, \gamma_1 \in \mathcal{X}$ *such that* $d(\gamma_0, \gamma_1) < \infty$. *The following three assertions are equivalent:*

(a) The curve γ is a constant speed minimal geodesic of length $d(\gamma_0, \gamma_1)$.
(b) $\forall \, t, t' \in [0, 1], \quad d(\gamma_t, \gamma_{t'}) = |t - t'| \, d(\gamma_0, \gamma_1)$.
(c) $\forall \, t, t' \in [0, 1], \quad d(\gamma_t, \gamma_{t'}) \leq |t - t'| \, d(\gamma_0, \gamma_1)$.

Proof. The necessity ($\Rightarrow$) is easily proven by rewriting (25) with two points t and t' instead of just one. The sufficiency ($\Leftarrow$) is proven by induction. Let

$$h_n = \sup_{a = t_0 \leq t_1 \leq \cdots \leq t_n \leq b} \sum_{i=1}^{n} d(\gamma_{t_{i-1}}, \gamma_{t_i}) \, + \, d(\gamma_{t_n}, \gamma_b) \, .$$

The hypothesis implies that $h_2 = d(\gamma_a, \gamma_b)$. We now assuming that the induction hypothesis $h_n = d(\gamma_a, \gamma_b)$ is true for some $n \geq 2$. For all partition $a = t_0 \leq t_1 \ldots t_n \leq b$, using twice the triangular inequality and the induction hypothesis,

$$d(\gamma_a, \gamma_b) \leq d(\gamma_a, \gamma_{t_n}) + d(\gamma_{t_n}, \gamma_b) \leq \sum_{i=1}^{n} d(\gamma_{t_{i-1}}, \gamma_{t_i}) + d(\gamma_{t_n}, \gamma_b) \leq h_n = d(\gamma_a, \gamma_b) \, .$$

Therefore $\sum_{i=1}^{n} d(\gamma_{t_{i-1}}, \gamma_{t_i}) = d(\gamma_a, \gamma_{t_n})$. Then, for any $t_{n+1} \in [t_n, b]$,

$$\sum_{i=1}^{n+1} d(\gamma_{t_{i-1}}, \gamma_{t_i}) + d(\gamma_{t_{n+1}}, \gamma_b) = d(\gamma_a, \gamma_{t_n}) + d(\gamma_{t_n}, \gamma_{t_{n+1}}) + d(\gamma_{t_{n+1}}, \gamma_b) = d(\gamma_a, \gamma_b) \, .$$

Since this is true for all partitions, $h_{n+1} = d(\gamma_a, \gamma_b)$. We just have proved by induction that $h_n = d(\gamma_a, \gamma_b)$ for all n. Therefore $L(\gamma, a, b) = \sup_n h_n = d(\gamma_a, \gamma_b)$. ∎

5   Minimal Geodesics in Probability Space

We now assume that $\mathcal{X}$ is a strictly intrinsic Polish space and we also assume that its distance d is never infinite. Therefore any pair of points in $\mathcal{X}$ is connected by at least one minimal geodesic. When the space $\mathcal{P}_{\mathcal{X}}$ of probability distributions is equipped with one of the probability distances discussed in Sect. 2, it often becomes a length space itself and inherits some of the geometrical properties of $\mathcal{X}$. Since this process depends critically on how the probability distance compares different distributions, understanding the geodesic structure of $\mathcal{P}_{\mathcal{X}}$ reveals fundamental differences between probability distances.

This approach is in fact quite different from the celebrated work of Amari on *Information Geometry* [2]. We seek here to understand the geometry of the space of all probability measures equipped with different distances. Information Geometry characterizes the Riemannian geometry of a parametric family of probability measures under the Kullback-Leibler distance. This difference is obviously related to the contrast between *relying on good distances* versus *relying on good model families* discussed in Sect. 1. Since we are particularly interested in relatively simple models that have a physical or causal interpretation but cannot truly represent the actual data distribution, we cannot restrict our geometrical insights to what happens within the model family.

5.1 Mixture Geodesics

For any two distributions $P_0, P_1 \in \mathcal{P}_{\mathcal{X}}$, the mixture distributions

$$\forall t \in [0,1] \quad P_t = (1-t)P_0 + tP_1 \tag{26}$$

form a curve in the space of distributions $\mathcal{P}_{\mathcal{X}}$.

Theorem 5.1. *Let $\mathcal{P}_{\mathcal{X}}$ be equipped with a distance D that belongs to the IPM family (5). Any mixture curve (26) joining two distributions $P_0, P_1 \in \mathcal{P}_{\mathcal{X}}$ such that $D(P_0, P_1) < \infty$ is a constant speed minimal geodesic, making D a strictly intrinsic distance.*

Proof. The proof relies on Corollary 4.2: for all $t, t' \in [0,1]$,

$$
\begin{aligned}
D(P_t, P_{t'}) &= \sup_{f \in \mathcal{Q}} \left\{ \mathbb{E}_{(1-t)P_0+tP_1}[f(x)] - \mathbb{E}_{(1-t')P_0+t'P_1}[f(x)] \right\} \\
&= \sup_{f \in \mathcal{Q}} \left\{ -(t-t')\mathbb{E}_{P_0}[f(x)] + (t-t')\mathbb{E}_{P_1}[(|f(x))] \right\} \\
&= |t-t'| \sup_{f \in \mathcal{Q}} \left\{ \mathbb{E}_{P_0}[f(x)] - \mathbb{E}_{P_1}[f(x)] \right\}.
\end{aligned}
$$

where the last equality relies on the fact that if $f \in \mathcal{Q}$, then $-f \in \mathcal{Q}$. By Corollary 4.2, the mixture curve is a constant speed minimal geodesic. Since this is true for any $P_0, P_1 \in \mathcal{P}_{\mathcal{X}}$ such that $D(P_0, P_1) < \infty$, the distance D is strictly intrinsic. ∎

Remark 5.2. Although Theorem 5.1 makes $(\mathcal{P}_{\mathcal{X}}, D)$ a length space, it does not alone make it a strictly intrinsic Polish space. One also needs to establish the completeness and separability[9] properties of a Polish space. Fortunately, these properties are true for both $(\mathcal{P}_{\mathcal{X}}^1, W_1)$ and $(\mathcal{P}_{\mathcal{X}}^1, \mathcal{E}_d)$ when the ground space is Polish.[10]

Since both the 1-Wasserstein distance W_1 and the Energy Distance or MMD $\mathcal{E}_d$ belong to the IPM family, $\mathcal{P}_{\mathcal{X}}$ equipped with either distance is a strictly intrinsic Polish space. Any two probability measures are connected by at least one minimal geodesic, the mixture geodesic. We shall see later that the 1-Wasserstein distance admits many more minimal geodesics. However, in the case of ED/MMD distances, mixture geodesics are the only minimal geodesics.

Theorem 5.3. *Let K be a characteristic kernel and let $\mathcal{P}_{\mathcal{X}}$ be equipped with the MMD distance $\mathcal{E}_{d_K}$. Then any two probability measures $P_0, P_1 \in \mathcal{P}_{\mathcal{X}}$ such that $\mathcal{E}_{d_K}(P_0, P_1) < \infty$ are joined by exactly one constant speed minimal geodesic, the mixture geodesic* (26).

Note that $\mathcal{E}_{d_K}$ is also the ED for the strongly negative definite kernel d_K.

Proof. Theorem 5.1 already shows that any two measures $P_0, P_1 \in \mathcal{P}_{\mathcal{X}}$ are connected by the mixture geodesic P_t. We only need to show that it is unique. For any $t \in [0,1]$, the measure P_t belongs to the set

$$\left\{ P \in \mathcal{P}_{\mathcal{X}} : \mathcal{E}_{d_K}(P_0, P) = tD \text{ and } \mathcal{E}_{d_K}(P, P_1) = (1-t)D \right\} \subset \mathcal{P}_{\mathcal{X}} \quad (27)$$

where $D = \mathcal{E}_{d_K}(P_0, P_1)$. Thanks to Theorem 2.17, $\mathbb{E}_{P_t}[\Phi_x]$ must belong to the set

$$\left\{ \Psi \in \mathcal{H} : \|\mathbb{E}_{P_0}[\Phi_x] - \Psi\|_{\mathcal{H}} = tD \text{ and } \|\Psi - \mathbb{E}_{P_1}[\Phi_x]\|_{\mathcal{H}} = (1-t)D \right\} \subset \mathcal{H}. \quad (28)$$

with $D = \|\mathbb{E}_{P_0}[\Phi_x] - \mathbb{E}_{P_1}[\Phi_x]\|_{\mathcal{H}}$. Since there is only one point Ψ that satisfies these conditions in $\mathcal{H}$, and since Corollary 2.20 says that the map $P \mapsto \mathbb{E}_P[\Phi_x]$ is injective, there can only be one P satisfying (27) and this must be P_t. Therefore the mixture geodesic is the only one.   ∎

5.2   Displacement Geodesics

Displacement Geodesics in the Euclidean Case. Let us first assume that $\mathcal{X}$ is a Euclidean space and $\mathcal{P}_{\mathcal{X}}$ is equipped with the p-Wasserstein distance W_p. Let $P_0, P_1 \in \mathcal{P}_{\mathcal{X}}^p$ be two distributions with optimal transport plan π. The *displacement curve* joining P_0 to P_1 is formed by the distributions

$$\forall t \in [0,1] \quad P_t = \left((1-t)x + ty\right) \# \pi(x, y). \quad (29)$$

---

[9] For instance the set of probability measures on $\mathbb{R}$ equipped with the total variation distance (6) is not separable because any dense subset needs one element in each of the disjoint balls $B_x = \{ P \in \mathcal{P}_{\mathbb{R}} : D_{TV}(P, \delta_x) < 1/2 \}$.

[10] For the Wasserstein distance, see [56, Theorem 6.18]. For the Energy distance, both properties can be derived from Theorem 2.17 after recaling that $\Phi_{\mathcal{X}} \subset \mathcal{H}$ is both complete and separable because it is isometric to $\mathcal{X}$ which is Polish.

Intuitively, whenever the optimal transport plan specifies that a grain of proba-bility mass must be transported from x to y in $\mathcal{X}$, we follow the shortest path connecting x and y, that is, in a Euclidean space, a straight line, but we drop the grain after performing a fraction t of the journey.

Proposition 5.4. *Let $\mathcal{X}$ be a Euclidean space and let $\mathcal{P}_{\mathcal{X}}$ be equipped with the p-Wasserstein distance (8) for some $p \geq 1$. Any displacement curve (29) joining two distributions P_0, P_1 such that $W_p(P_0, P_1) < \infty$ is a constant speed minimal geodesic, making W_p a strictly intrinsic distance.*

Proof. Let π_{01} be the optimal transport plan between P_0 and P_1. For all $t, t' \in [0, 1]$, define a tentative transport plan $\pi_{tt'}$ between P_t and $P_{t'}$ as

$$\pi_{tt'} = \big((1-t)x + ty, \, (1-t')x + t'y \big) \# \pi_{01}(x, y) \ \in \ \Pi(P_t, P_{t'}).$$

Then

$$
\begin{aligned}
W_p(P_t, P_{t'})^p &\leq \mathbb{E}_{(x,y) \sim \pi_{tt'}}[\, \|x - y\|^p \,] \\
&= \mathbb{E}_{(x,y) \sim \pi}[\, \|(1-t)x + ty - (1-t')x - t'y\|^p \,] \\
&= |t - t'|^p \, \mathbb{E}_{(x,y) \sim \pi}[\, \|x - y\| \,] = |t - t'|^p \, W_p(P_0, P_1)^p.
\end{aligned}
$$

By Corollary 4.2, the displacement curve is a constant speed minimal geodesic. Since this is true for any $P_0, P_1 \in \mathcal{P}_{\mathcal{X}}$ such that $W_p(P_0, P_1) < \infty$, the distance W_p is strictly intrinsic.   ∎

When $p > 1$, it is a well-known fact that the displacement geodesics are the only geodesics of $\mathcal{P}_{\mathcal{X}}$ equipped with the W_p distance.

Proposition 5.5. *The displacement geodesics (29) are the only constant speed minimal geodesics of $\mathcal{P}_{\mathcal{X}}$ equipped with the p-Wasserstein distance W_p with $p > 1$.*

This is a good opportunity to introduce a very useful lemma.

Lemma 5.6 (Gluing). *Let $\mathcal{X}_i$, $i = 1, 2, 3$ be Polish metric spaces. Let probabil-ity measures $\mu_{12} \in \mathcal{P}_{\mathcal{X}_1 \times \mathcal{X}_2}$ and $\mu_{23} \in \mathcal{P}_{\mathcal{X}_2 \times \mathcal{X}_3}$ have the same marginal distribu-tion μ_2 on $\mathcal{X}_2$. Then there exists $\mu \in \mathcal{P}_{\mathcal{X}_1 \times \mathcal{X}_2 \times \mathcal{X}_3}$ such that $(x, y) \# \mu(x, y, z) = \mu_{12}$ and $(y, z) \# \mu(x, y, z) = \mu_{23}$.*

Proof Notes for Lemma 5.6. At first sight, this is simply $P(x, y, z) = P(x|y)P(z|y)P(y)$ with $\mu_{12} = P(x, y)$, $\mu_{23} = P(y, z)$. Significant technical diffi-culties arise when $P(y) = 0$. This is where one needs the topological properties of a Polish space [8].

Proof of Proposition 5.5. Let $t \in [0, 1] \mapsto P_t$ be a constant speed minimal geodesic. Any point P_t must satisfy the equality

$$W_p(P_0, P_t) + W_p(P_t, P_1) = W_p(P_0, P_1)$$

Let π_0 and π_1 be the optimal transport plans associated with $W_p(P_0, P_t)$ and $W_p(P_t, P_1)$ and construct $\pi_3 \in \mathcal{P}_{\mathcal{X}^3}$ by gluing them. Then we must have

$$\left(\mathbb{E}_{(x,y,z)\sim\pi_3}[\,\|x-y\|^p\,]\right)^{1/p} + \left(\mathbb{E}_{(x,y,z)\sim\pi_3}[\,\|y-z\|^p\,]\right)^{1/p}$$
$$= W_p(P_0, P_1) \leq \left(\mathbb{E}_{(x,y,z)\sim\pi_3}[\,\|x-z\|^p\,]\right)^{\frac{1}{p}}.$$

Thanks to the properties of the Minkowski's inequality, this can only happen for $p > 1$ if there exists $\lambda \in [0,1]$ such that, π_3-almost surely, $\|x-y\| = \lambda\|x-z\|$ and $\|y-z\| = (1-\lambda)\|x-z\|$. This constant can only be t because $W_P(P_0, P_t) = tW_p(P_0, P_1)$ on a constant speed minimal geodesic. Therefore $y = tx + (1-t)y$, π_3-almost surely. Therefore $P_t = y\#\pi(x,y,z)$ describes a displacement curve as defined in (29). ∎

Note however that the displacement geodesics are not the only minimal geodesics of the 1-Wasserstein distance W_1. Since W_1 is an IPM (14), we know that the mixture geodesics are also minimal geodesics (Theorem 5.1). There are in fact many more geodesics. Intuitively, whenever the optimal transport plan from P_0 to P_1 transports a grain of probability from x to y, we can drop the grain after a fraction t of the journey (displacement geodesics), we can randomly decide whether to transport the grain as planned (mixture geodesics), we can also smear the grain of probability along the shortest path connecting x to y, and we can do all of the above using different t in different parts of the space.

Displacement Geodesics in the General Case. The rest of this section reformulates these results to the more general situation where $\mathcal{X}$ is a strictly intrinsic Polish space. Rather than following the random curve approach described in [56, Chapter 7], we chose a more elementary approach because we also want to characterize the many geodesics of W_1. Our definition is equivalent for $p>1$ and subtly weaker for $p = 1$.

The main difficulties are that we may no longer have a single shortest path connecting two points $x, y \in \mathcal{X}$, and that we may not be able to use the push-forward formulation (29) because the function that returns the point located at position t along a constant speed minimal geodesic joining x to y may not satisfy the necessary measurability requirements.

Definition 5.7 (Displacement geodesic). Let $\mathcal{X}$ be a strictly intrinsic Polish metric space and let $\mathcal{P}_{\mathcal{X}}^p$ be equipped with the p-Wasserstein distance W_p. The curve $t \in [0,1] \mapsto P_t \in \mathcal{P}_{\mathcal{X}}^p$ is called a displacement geodesic if, for all $0 \leq t \leq t' \leq 1$, there is a distribution $\pi_4 \in \mathcal{P}_{\mathcal{X}^4}$ such that

(i) The four marginals of π_4 are respectively equal to $P_0, P_t, P_{t'}, P_1$.
(ii) The pairwise marginal $(x,z)\#\pi_4(x,u,v,z)$ is an optimal transport plan

$$W_p(P_0, P_1)^p = \mathbb{E}_{(x,u,v,z)\sim\pi_4}[d(x,z)^p].$$

(iii) The following relations hold $\pi_4(x,u,v,z)$-almost surely:

$$d(x,u) = t\,d(x,z), \quad d(u,v) = (t'-t)\,d(x,z), \quad d(v,z) = (1-t')\,d(x,z).$$

Proposition 5.8. *Definition 5.7 indeed implies that P_t is a constant speed minimal geodesic of length $W_p(P_0, P_1)$. Furthermore, for all $0 \leq t \leq t' \leq 1$, all the pairwise marginals of π_4 are optimal transport plans between their marginals.*

Proof. For all $0 \leq t \leq t' \leq 1$, we have

$$W_p(P_t, P_{t'})^p \leq \mathbb{E}_{(x,u,v,z)\sim\pi_4}[d(u,v)^p]$$
$$= (t - t')^p \, \mathbb{E}_{(x,u,v,z)\sim\pi_4}[d(x,z)^p] = (t - t')^p \, W_p(P_0, P_1)^p.$$

By Corollary 4.2, the curve P_t is a constant speed minimal geodesic. We can then write

$$t' \, W_p(P_0, P_1) = W_p(P_0, P_{t'}) \leq \left(\mathbb{E}_{(x,u,v,z)\sim\pi_4}[d(x,v)^p]\right)^{1/p}$$
$$\leq \left(\mathbb{E}_{(x,u,v,z)\sim\pi_4}[(d(x,u) + d(u,v))^p]\right)^{1/p}$$
$$\leq \left(\mathbb{E}_{(x,u,v,z)\sim\pi_4}[d(x,u)^p]\right)^{1/p} + \left(\mathbb{E}_{(x,u,v,z)\sim\pi_4}[d(u,v)^p]\right)^{1/p}$$
$$\leq t \left(\mathbb{E}_{(x,u,v,z)\sim\pi_4}[d(x,z)^p]\right)^{1/p} + (t' - t) \left(\mathbb{E}_{(x,u,v,z)\sim\pi_4}[d(x,z)^p]\right)^{1/p}$$
$$= t' \, W_p(P_0, P_1),$$

where the third inequality is Minkowski's inequality. Since both ends of this chain of inequalities are equal, these inequalities must be equalities, implying that $(x,v)\#\pi_4$ is an optimal transport plan between P_0 and $P_{t'}$. We can do likewise for all pairwise marginals of π_4. ∎

The proposition above does not establish that a displacement geodesic always exists. As far as we know, this cannot be established without making an additional assumption such as the local compacity of the intrinsic Polish space $\mathcal{X}$. Since it is often easy to directly define a displacement geodesic as shown in (29), we omit the lengthy general proof.

Theorem 5.9. *Let $\mathcal{X}$ be a strictly intrinsic Polish metric space and let P_0, P_1 be two distributions of $\mathcal{P}_{\mathcal{X}}^p$ equipped with the p-Wasserstein with $p > 1$. The only constant speed minimal geodesics of length $W_p(P_0, P_1)$ joining P_0 and P_1 are the displacement geodesics.*

Proof. Let P_t be a constant speed minimal geodesic of length $W_p(P_0, P_1)$. By Theorem 4.1, for all $0 \leq t \leq t' \leq 1$,

$$W_p(P_0, P_t) + W_p(P_t, P_{t'}) + W_p(P_{t'}, P_1) = W_p(P_0, P_1) \ .$$

Let π_4 be constructed by gluing optimal transport plans associated with the three distances appearing on the left hand side of the above equality. We can then write

$$W_p(P_0, P_1) \leq \left(\mathbb{E}_{(x,u,v,z)\sim\pi_4}[d(x,z)^p]\right)^{1/p}$$
$$\leq \left(\mathbb{E}_{(x,u,v,z)\sim\pi_4}[(d(x,u) + d(u,v) + d(v,z))^p]\right)^{1/p}$$
$$\leq \left(\mathbb{E}_{(x,u,v,z)\sim\pi_4}[d(x,u)^p]\right)^{1/p} + \left(\mathbb{E}_{(x,u,v,z)\sim\pi_4}[d(u,v)^p]\right)^{1/p}$$
$$+ \left(\mathbb{E}_{(x,u,v,z)\sim\pi_4}[d(v,z)^p]\right)^{1/p}$$
$$= W_p(P_0, P_t) + W_p(P_t, P_{t'}) + W_p(P_{t'}, P_1) = W_p(P_0, P_1).$$

Since this chain of inequalities has the same value in both ends, all these inequalities must be equalities. The first one means that π_4 is an optimal transport plan for $W_p(P_0, P_1)$. The second one means that $(d(x,u) + d(u,v) + d(v,z) = d(x,z)$, π_4-almost surely. When $p > 1$, the third one, Minkowski's inequality can only be an inequality if there are scalars $\lambda_1 + \lambda_2 + \lambda_3 = 1$ such that, π_4-almost surely, $d(x,u) = \lambda_1 d(x,z)$, $d(x,u) = \lambda_2 d(x,z)$, and $d(v,z) = \lambda_3 d(x,z)$. Since P_t must satisfy Corollary 4.2, these scalars can only be $\lambda_1 = t$, $\lambda_2 = t' - t$, and $\lambda_3 = 1 - t'$. ∎

Minimal Geodesics for the 1-Wasserstein Distance. We can characterize the many minimal geodesics of the 1-Wasserstein distance using a comparable strategy.

Theorem 5.10. *Let $\mathcal{X}$ be a strictly intrinsic Polish space and let $\mathcal{P}_{\mathcal{X}}^1$ be equipped with the distance W_1. A curve $t \in [a,b] \mapsto P_t \in \mathcal{P}_{\mathcal{X}}^1$ joining P_a and P_b is a minimal geodesic of length $W_1(P_a, P_b)$ if and only if, for all $a \leq t \leq t' \leq b$, there is a distribution $\pi_4 \in \mathcal{P}_{\mathcal{X}^4}$ such that*

(i) *The four marginals of π_4 are respectively equal to P_a, P_t, $P_{t'}$, P_b.*
(ii) *The pairwise marginal $(x,z) \# \pi_4(x,u,v,z)$ is an optimal transport plan*

$$W_p(P_a, P_b) = \mathbb{E}_{(x,u,v,z) \sim \pi_4}[d(x,z)].$$

(iii) *The following relation holds $\pi_4(x,u,v,z)$-almost surely:*

$$d(x,u) + d(u,v) + d(v,z) = d(x,z).$$

It is interesting to compare this condition to Theorem 4.1. Instead of telling us that two successive triangular inequalities in the probability space $(\mathcal{P}_{\mathcal{X}}^1, W_1)$ must be an equality, this result tells us that the same holds almost-surely in the sample space $(\mathcal{X}, d)$. In particular, this means that x, u, v, and z must be aligned along a geodesic of $\mathcal{X}$. In the case of a mixture geodesic, u and v coincide with x or z. In the case of a displacement geodesic, u and v must be located at precise positions along a constant speed geodesic joining x to z. But there are many other ways to fulfil these conditions.

Proof. When P_t is a minimal geodesic, Theorem 4.1 states

$$\forall a \leq t \leq t' \leq b \quad W_1(P_a, P_t) + W_1(P_t, P_{t'}) + W_1(P_{t'}, P_b) = W_1(P_a, P_b).$$

Let π_4 be constructed by gluing optimal transport plans associated with the three distances appearing on the left hand side of the above equality. We can then write

$$\mathbb{E}_{(x,u,v,z) \sim \pi_4}[d(x,z)] \leq \mathbb{E}_{(x,u,v,z) \sim \pi_4}[d(x,u) + d(u,v) + d(v,z)]$$
$$= W_1(P_a, P_b) \leq \mathbb{E}_{(x,u,v,z) \sim \pi_4}[d(x,z)].$$

Since this chain of equalities has the same value on both ends, all these inequalities must be equalities. The first one means that $d(x, u) + d(u, v) + d(v, z) = d(x, z)$, π_4-almost surely. The second one means that $(x, z) \# \pi_4$ is an optimal transport plan.

Conversely, assume P_t satisfies the conditions listed in the proposition. We can then write, for all $a \leq t \leq t' \leq b$,

$$
\begin{aligned}
W_1(P_a, P_b) &= \mathbb{E}_{(x,u,v,z) \sim \pi_4}[d(x, z)] \\
&= \mathbb{E}_{(x,u,v,z) \sim \pi_4}[d(x, u) + d(u, v) + d(v, z)] \\
&= W_1(P_a, P_t) + W_1(P_t, P_{t'}) + W_1(P_{t'}, P_b),
\end{aligned}
$$

and we conclude using Theorem 4.1.   ∎

6   Unsupervised Learning and Geodesic Structures

We have seen in the previous section that the geometry of the space $\mathcal{P}_\mathcal{X}$ of probability distributions changes considerably with our choice of a probability distance. Critical aspects of these possible geometries can be understood from the characterization of the shortest paths between any two distributions:

- With the Energy Distance $\mathcal{E}_d$ or the Maximum Mean Discrepancy $\mathcal{E}_{d_K}$, the sole shortest path is the mixture geodesic (Theorem 5.3.)
- With the p-Wasserstein distance W_p, for $p > 1$, the sole shortest paths are displacement geodesics (Theorem 5.9.)
- With the 1-Wasserstein distance W_1, there are many shortest paths, including the mixture geodesic, all the displacement geodesics, and all kinds of hybrid curves (Theorem 5.10.)

The purpose of this section is to investigate the consequences of these geometrical differences on unsupervised learning problems. In the following discussion, $Q \in \mathcal{P}_\mathcal{X}$ represents the data distribution which is only known through the training examples, and $\mathcal{F} \subset \mathcal{P}_\mathcal{X}$ represent the family of parametric models $P_\theta \in \mathcal{P}_\mathcal{X}$ considered by our learning algorithm.

Minimal geodesics in length spaces can sometimes be compared to line segments in Euclidean spaces because both represent shortest paths between two points. This association provides the means to extend the familiar Euclidean notion of convexity to length spaces. This section investigates the geometry of implicit modeling learning problems through the lens of this generalized notion of convexity.

6.1   Convexity à-la-carte

We now assume that $\mathcal{P}_\mathcal{X}$ is a strictly intrinsic Polish space equipped with a distance D. Let $\mathcal{C}$ be a family of smooth constant speed curves in $\mathcal{P}_\mathcal{X}$. Although these curves need not be minimal geodesics, the focus of this section is limited to three families of curves defined in Sect. 5:

- the family $\mathcal{C}_g(D)$ of all minimal geodesics in $(\mathcal{P}_\mathcal{X}, D)$,
- the family $\mathcal{C}_d(W_p)$ of the displacement geodesics in $(\mathcal{P}_\mathcal{X}^p, W_p)$,
- the family $\mathcal{C}_m$ of the mixture curves in $\mathcal{P}_\mathcal{X}$.

Definition 6.1. Let $\mathcal{P}_\mathcal{X}$ be a strictly intrinsic Polish space. A closed subset $\mathcal{F} \subset \mathcal{P}_\mathcal{X}$ is called convex with respect to the family of curves $\mathcal{C}$ when $\mathcal{C}$ contains a curve $t \in [0,1] \mapsto P_t \in \mathcal{X}$ connecting P_0 and P_1 whose graph is contained in $\mathcal{F}$, that is, $P_t \in \mathcal{F}$ for all $t \in [0,1]$.

Definition 6.2. Let $\mathcal{P}_\mathcal{X}$ be a strictly intrinsic Polish space. A real-valued function f defined on $\mathcal{P}_\mathcal{X}$ is called convex with respect to the family of constant speed curves $\mathcal{C}$ when, for every curve $t \in [0,1] \mapsto P_t \in \mathcal{P}_\mathcal{X}$ in $\mathcal{C}$, the function $t \in [0,1] \mapsto f(P_t) \in \mathbb{R}$ is convex.

For brevity we also say that $\mathcal{F}$ or f is *geodesically convex* when $\mathcal{C} = \mathcal{C}_g(D)$, *mixture convex* when $\mathcal{C} = \mathcal{C}_m$, and *displacement convex* when $\mathcal{C} = \mathcal{C}_d(W_p)$.

Theorem 6.3 (Convex optimization à-la-carte). *Let $\mathcal{P}_\mathcal{X}$ be a strictly intrinsic Polish space equipped with a distance D. Let the closed subset $\mathcal{F} \subset \mathcal{P}_\mathcal{X}$ and the cost function $f : \mathcal{P}_\mathcal{X} \mapsto \mathbb{R}$ be both convex with respect to a same family $\mathcal{C}$ of constant speed curves. Then, for all $M \geq \min_\mathcal{F}(f)$,*

(i) *the level set $L(f, \mathcal{F}, M) = \{P \in \mathcal{F} : f(P) \leq M\}$ is connected,*
(ii) *for all $P_0 \in \mathcal{F}$ such that $f(P_0) > M$ and all $\epsilon > 0$, there exists $P \in \mathcal{F}$ such that $D(P, P_0) = \mathcal{O}(\epsilon)$ and $f(P) \leq f(P_0) - \epsilon(f(P_0) - M)$.*

This result essentially means that it is possible to optimize the cost function f over $\mathcal{F}$ with a descent algorithm. Result (i) means that all minima are global minima, and result (ii) means that any neighborhood of a suboptimal distribution P_0 contains a distribution P with a sufficiently smaller cost to ensure that the descent will continue.

Proof. (i): Let $P_0, P_1 \in L(f, \mathcal{F}, M)$. Since they both belong to $\mathcal{F}$, $\mathcal{C}$ contains a curve $t \in [0,1] \mapsto P_t \in \mathcal{F}$ joining P_0 and P_1. For all $t \in [0,1]$, we know that $P_t \in \mathcal{F}$ and, since $t \mapsto f(P_t)$ is a convex function, we can write $f(P_t) \leq (1-t)f(P_0) + tf(P_1) \leq M$. Therefore $P_t \in L(f, \mathcal{F}, M)$. Since this holds for all P_0, P_1, $L(f, \mathcal{F}, M)$ is connected.

(ii): Let $P_1 \in L(f, \mathcal{F}, M)$. Since $\mathcal{F}$ is convex with respect to $\mathcal{C}$, $\mathcal{C}$ contains a constant speed curve $t \in [0,1] \mapsto P_t \in \mathcal{F}$ joining P_0 and P_1. Since this is a constant speed curve, $d(P_0, P_\epsilon) \leq \epsilon D(P_0, P_1)$, and since $t \mapsto f(P_t)$ is convex, $f(P_\epsilon) \leq (1-\epsilon)f(P_0) + \epsilon f(P_1)$, implies $f(P_\epsilon \leq f(P_0) - \epsilon(f(P_0) - M)$.

One particularly striking aspect of this result is that it does not depend on the parametrization of the family $\mathcal{F}$. Whether the cost function $C(\theta) = f(G_\theta \# \mu_z)$ is convex or not is irrelevant: as long as the family $\mathcal{F}$ and the cost function f are convex with respect to a well-chosen set of curves, the level sets of the cost function $C(\theta)$ will be connected, and there will be a nonincreasing path connecting any starting point θ_0 to a global optimum θ^*.

It is therefore important to understand how the definition of C makes it easy or hard to ensure that both the model family $\mathcal{F}$ and the training criterion f are convex with respect to C.

6.2   The Convexity of Implicit Model Families

We are particularly interested in the case of implicit models (Sect. 2.2) in which the distributions P_θ are expressed by pushing the samples $z \in \mathcal{Z}$ of a known source distribution $\mu_z \in \mathcal{P}_\mathcal{Z}$ through a parametrized generator function $G_\theta(z) \in \mathcal{X}$. This push-forward operation defines a deterministic coupling between the distributions μ_z and P_θ because the function G_θ maps every source sample $z \in \mathcal{Z}$ to a single point $G_\theta(z)$ in $\mathcal{X}$. In contrast, a stochastic coupling distribution $\pi_\theta \in \Pi(\mu_z, P_\theta) \subset \mathcal{P}_{\mathcal{Z} \times \mathcal{X}}$ would be allowed to distribute a source sample $z \in \mathcal{Z}$ to several locations in $\mathcal{X}$, according to the conditional distribution $\pi_\theta(x|z)$.

The deterministic nature of this coupling makes it very hard to achieve mixture convexity using smooth generator functions G_θ.

Example 6.4. Let the distributions $P_0, P_1 \in \mathcal{F}$ associated with parameters θ_0 and θ_1 have disjoint supports separated by a distance greater than $D>0$. Is there a continuous path $t \in [0,1] \mapsto \theta_t$ in parameter space such that $G_{\theta_t} \# \mu_z$ is the mixture $P_t = (1-t)P_0 + P_1$?

If we assume there is such a path, we can write

$$\mu_z\{G_{\theta_0}(z) \in \mathrm{supp}(P_0)\} = 1$$

and, for any $\epsilon>0$,

$$\mu_z\{G_{\theta_\epsilon}(z) \in \mathrm{supp}(P_1)\} = \epsilon > 0 .$$

Therefore, for all $\epsilon>0$, there exists $z \in \mathcal{Z}$ such that $d(G_{\theta_0}(z), G_{\theta_\epsilon}(z)) \geq D$. Clearly such a generator function is not compatible with the smoothness requirements of an efficient learning algorithm.

In contrast, keeping the source sample z constant, a small change of the parameter θ causes a small displacement of the generated sample $G_\theta(z)$ in the space $\mathcal{X}$. Therefore we can expect that such an implicit model family has a particular affinity for displacement geodesics.

It is difficult to fully assess the consequences of the quasi-impossibility to achieve mixture convexity with implicit models. For instance, although the Energy Distance $\mathcal{E}_d(Q, P_\theta)$ is a mixture convex function (see Proposition 6.6 in the next section), we cannot expect that a family $\mathcal{F}$ of implicit models will be mixture convex.

Example 6.5. Let μ_z be the uniform distribution on $\{-1, +1\}$. Let the parameter θ be constrained to the square $[-1,1]^2 \subset \mathbb{R}^2$ and let the generator function be

$$G_\theta : z \in \{-1, 1\} \mapsto G_\theta(z) = z\theta .$$

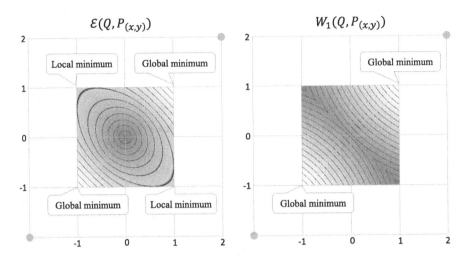

Fig. 4. Level sets for the problems described in Example 6.5.

The corresponding model family is

$$\mathcal{F} = \left\{ P_\theta = \tfrac{1}{2}(\delta_\theta + \delta_{-\theta}) : \theta \in [-1,1] \times [-1,1] \right\}.$$

It is easy to see that this model family is displacement convex but not mixture convex. Figure 4 shows the level sets for both criteria $\mathcal{E}(Q, P_\theta)$ and $W_1(Q, P_\theta)$ for the target distribution $Q = P_{(2,2)} \notin \mathcal{F}$. Both criteria have the same global minima in $(1,1)$ and $(-1,-1)$. However the energy distance has spurious local minima in $(-1,1)$ and $(1,-1)$ with a relatively high value of the cost function.

Constructing such an example in $\mathbb{R}^2$ is nontrivial. Whether such situations arise commonly in higher dimension is not known. However we empirically observe that the optimization of a MMD criterion on high-dimensional image data often stops with unsatisfactory results (Sect. 3.2).

6.3   The Convexity of Distances

Let $Q \in \mathcal{P}_\mathcal{X}$ be the target distribution for our learning algorithm. This could be the true data distribution or the empirical training set distribution. The learning algorithm minimizes the cost function

$$\min_{P_\theta \in \mathcal{F}} C(\theta) \triangleq D(Q, P_\theta). \tag{30}$$

The cost function itself is therefore a distance. Since such a distance function is always convex in a Euclidean space, we can ask whether a distance in a strictly intrinsic Polish space is geodesically convex. This is not always the case. Figure 5 gives a simple counter-example in $\mathbb{R}^2$ equipped with the L_1 distance.

Yet we can give a positive answer for the mixture convexity of IPM distances.

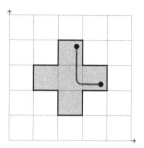 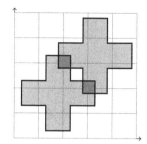 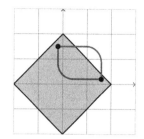

Fig. 5. Geodesic convexity often differs from Euclidean convexity in important ways. There are many different minimal geodesics connecting any two points in $\mathbb{R}^2$ equipped with the L_1 distance (see also Fig. 3). The cross-shaped subset of $\mathbb{R}_2$ shown in the left plot is geodesically convex. The center plot shows that the intersection of two geodesically convex sets is not necessarily convex or even connected. The right plot shows that two points located inside the unit ball can be connected by a minimal geodesic that does not stay in the unit ball. This means that the L_1 distance itself is not convex because its restriction to that minimal geodesic is not convex.

Proposition 6.6. *Let $\mathcal{P}_{\mathcal{X}}$ be equipped with a distance D that belongs to the IPM family (5). Then D is mixture convex.*

Proof. Let $t \in [0,1] \mapsto P_t = (1-t)P_0 + tP_1$ be a mixture curve. Theorem 5.1 tells us that such mixtures are minimal geodesics. For any target distribution Q we can write

$$D(Q, P_t) = \sup_{f \in \mathcal{Q}} \{ \mathbb{E}_Q[f(x)] - \mathbb{E}_{P_t}[f(x)] \}$$
$$= \sup_{f \in \mathcal{Q}} \{ (1-t)\left(\mathbb{E}_Q[f(x)] - \mathbb{E}_{P_0}[f(x)]\right) + t\left(\mathbb{E}_Q[f(x)] - \mathbb{E}_{P_1}[f(x)]\right) \}$$
$$\leq (1-t)\, D(Q, P_0) + t\, D(Q, P_1).$$

The same holds for any segment $t \in [t_1, t_2] \subset [0,1]$ because such segments are also mixture curves up to an affine reparametrization. Therefore $t \mapsto D(Q, P_t)$ is convex. ∎

Therefore, when D is an IPM distance, and when $\mathcal{F}$ is a mixture convex family of generative models, Theorem 6.3 tells us that a simple descent algorithm can find the global minimum of (30). As discussed in Example 6.4, it is very hard to achieve mixture convexity with a family of implicit models. But this could be achieved with nonparametric techniques.

However the same does not hold for displacement convexity. For instance, the Wasserstein distance is not displacement convex, even when the sample space distance d is geodesically convex, and even when the sample space is Euclidean.

Example 6.7. Let $\mathcal{X}$ be $\mathbb{R}^2$ equipped with the Euclidean distance. Let Q be the uniform distribution on the unit circle, and let $P_{\ell,\theta}$ be the uniform distribution on a line segment of length 2ℓ centered on the origin (Fig. 6, left). The distance

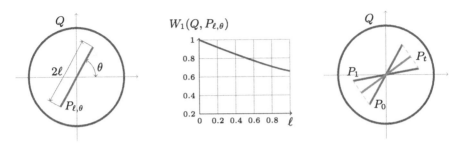

Fig. 6. Example 6.7 considers a target distribution Q that is uniform on the $\mathbb{R}^2$ unit circle and a displacement geodesic between two line segments centered on the origin and with identical length (right plot.)

$W_1(Q, P_{\ell,\theta})$ is independent on θ and decreases when $\ell \in [0,1]$ increases (Fig. 6, center). Consider a displacement geodesic $t \in [0,1] \mapsto P_t$ where $P_0 = P_{\ell,\theta_0}$ and $P_1 = P_{\ell,\theta_1}$ for $0 < \theta_0 < \theta_1 < \pi/2$. Since the space $\mathbb{R}_2$ is Euclidean, displacements occur along straight lines. Therefore the distribution P_t for $0 < t < 1$ is uniform on a slightly shorter line segment (Fig. 6, right), implying

$$W_1(Q, P_t) > W_1(Q, P_0) = W_1(Q, P_1) \ .$$

Therefore the distance function $P \mapsto W_1(Q, P)$ is not displacement convex.

Although this negative result prevents us from invoking Theorem 6.3 for the minimization of the Wasserstein distance, observe that the convexity violation in Example 6.7 is rather small. Convexity violation examples are in fact rather diffi-cult to construct. The following section shows that we can still obtain interesting guarantees by bounding the size of the convexity violation.

6.4   Almost-Convexity

We consider in this section that the distance d is geodesically convex in $\mathcal{X}$: for any point $x \in \mathcal{X}$ and any constant speed geodesic $t \in [0,1] \mapsto \gamma_t \in \mathcal{X}$,

$$d(x, \gamma_t) \le (1-t)\, d(x, \gamma_0) + t\, d(x, \gamma_1) \ .$$

This requirement is of course verified when $\mathcal{X}$ is an Euclidean space. This is also trivially true when $\mathcal{X}$ is a Riemannian or Alexandrov space with nonpositive curvature [12].

The following result bounds the convexity violation:

Proposition 6.8. *Let $\mathcal{X}$ be a strictly intrinsic Polish space equipped with a geodesically convex distance d and let $\mathcal{P}^1_{\mathcal{X}}$ be equipped with the 1-Wasserstein distance W_1. For all $Q \in \mathcal{P}_{\mathcal{X}}$ and all displacement geodesics $t \in [0,1] \mapsto P_t$,*

$$\forall t \in [0,1] \quad W_1(Q, P_t) \le (1-t)\, W_1(Q, P_0) + t\, W_1(Q, P_1) + 2t(1-t)K(Q, P_0, P_1)$$

with $K(Q, P_0, P_1) \le 2 \min_{u_0 \in \mathcal{X}} \mathbb{E}_{u \sim Q}[d(u, u_0)]$.

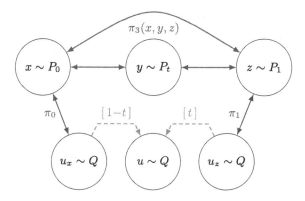

Fig. 7. The construction of $\pi \in \mathcal{P}_{\mathcal{X}^6}$ in the proof of Proposition 6.8.

Proof. The proof starts with the construction of a distribution $\pi \in \mathcal{P}_{\mathcal{X}^6}$ illustrated in Fig. 7. Thanks to Proposition 5.8 we can construct a distribution $\pi_3(x, y, z) \in \mathcal{P}_{\mathcal{X}^3}$ whose marginals are respectively P_0, P_t, P_1, whose pairwise marginals are optimal transport plans, and such that, π_3-almost surely,

$$d(x, y) = t \, d(x, z) \quad d(y, z) = (1 - t) \, d(x, z).$$

We then construct distribution $\pi_5(x, y, z, u_x, u_z)$ by gluing π_3 with the optimal transport plans π_0 between P_0 and Q and π_1 between P_1 and Q. Finally $\pi(x, y, z, u_x, u_z, u)$ is constructed by letting u be equal to u_x with probability $1 - t$ and equal to u_z with probability t. The last three marginals of π are all equal to Q.

Thanks to the convexity of d in $\mathcal{X}$, the following inequalities hold π-almost surely:

$$d(u_x, y) \le (1 - t) \, d(u_x, x) + t \, d(u_x, z)$$
$$\le (1 - t) \, d(u_x, x) + t \, d(u_z, z) + t \, d(u_x, u_z)$$
$$d(u_z, y) \le (1 - t) \, d(u_z, x) + t \, d(u_z, z)$$
$$\le (1 - t) \, d(u_x, x) + t \, d(u_z, z) + (1 - t) \, d(u_x, u_z).$$

Therefore

$$W_1(Q, P_t) \le \mathbb{E}_\pi[d(u, y)]$$
$$= \mathbb{E}_\pi[(1 - t) d(u_x, y) + t d(u_z, y)]$$
$$\le \mathbb{E}_\pi[(1 - t) \, d(u_x, x) + t \, d(u_z, z) + 2t(1 - t) \, d(u_x, u_z)]$$
$$= (1 - t) \, W_1(Q, P_0) + t \, W_1(Q, P_1) + 2t(1 - t) \, \mathbb{E}_\pi[d(u_x, u_z)].$$

For any $u_0 \in \mathcal{X}$, the constant K in the last term can then be coarsely bounded with

$$K(Q, P_0, P_1) = \mathbb{E}_\pi[d(u_x, u_z)]$$
$$\le \mathbb{E}_\pi[d(u_x, u_0)] + \mathbb{E}_\pi[d(u_0, u_z)] = 2\mathbb{E}_{u \sim Q}[d(u, u_0)].$$

Taking the minimum over u_0 gives the final result. ∎

When the optimal transport plan from P_0 to P_1 specifies that a grain of probability must be transported from x to z, its optimal coupling counterpart in Q moves from u_x to u_z. Therefore the quantity $K(Q, P_0, P_1)$ quantifies how much the transport plan from P_t to Q changes when P_t moves along the geodesic. This idea could be used to define a Lipschitz-like property such as

$$\forall P_0, P_1 \in \mathcal{F}_L \subset \mathcal{F} \qquad K(Q, P_0, P_1) \leq L W_1(P_0, P_1) \ .$$

Clearly such a property does not hold when the transport plan changes very suddenly. This only happens in the vicinity of distributions P_t that can be coupled with Q using multiple transport plans.

Unfortunately we have not found an elegant way to leverage this idea into a global description of the cost landscape. Proposition 6.8 merely bounds $K(Q, P_0, P_1)$ by the expected diameter of the distribution Q. We can nevertheless use this bound to describe some level sets of $W_1(Q, P_\theta)$

Theorem 6.9. *Let $\mathcal{X}$ be a strictly intrinsic Polish space equipped with a geodesically convex distance d and let $\mathcal{P}_\mathcal{X}^1$ be equipped with the 1-Wasserstein distance W_1. Let $\mathcal{F} \subset \mathcal{P}_\mathcal{X}^1$ be displacement convex and let $Q \in \mathcal{P}_\mathcal{X}^1$ have expected diameter*

$$D = 2 \min_{u_0 \in \mathcal{X}} \mathbb{E}_{u \sim Q}[d(u, u_0)] \ .$$

Then the level set $L(Q, \mathcal{F}, M) = \{P_\theta \in \mathcal{F} : W_1(Q, P_\theta) \leq M\}$ is connected if

$$M > \inf_{P_\theta \in \mathcal{F}} W_1(Q, P_\theta) + 2D \ .$$

Proof. Choose $P_1 \in \mathcal{F}$ such that $W_1(Q, P_1) < M - 2D$. For any $P_0, P_0' \in L(Q, \mathcal{F}, M)$, let $t \in [0, 1] \mapsto P_t \in \mathcal{F}$ be a displacement geodesic joining P_0 and P_1 without leaving $\mathcal{F}$. Thanks to Proposition 6.8,

$$W_1(Q, P_t) \leq (1 - t) M + t (M - 2D) + 2t(1 - t)D = M - 2t^2 D \leq M \ .$$

Therefore this displacement geodesic is contained in $L(Q, \mathcal{F}, M)$ and joins P_0 to P_1. We can similarly construct a second displacement geodesic that joins P_0' to P_1 without leaving $L(Q, \mathcal{F}, M)$. Therefore there is a continuous path connecting P_0 to P_0' without leaving $L(Q, \mathcal{F}, M)$. ∎

This result means that optimizing the Wasserstein distance with a descent algorithm will not stop before finding a generative model $P \in \mathcal{F}$ whose distance $W_1(Q, P)$ to the target distribution is within $2D$ of the global minimum. Beyond that point, the algorithm could meet local minima and stop progressing. Because we use a rather coarse bound on the constant $K(Q, P_0, P_1)$, we believe that it is possible to give much better suboptimality guarantee in particular cases.

Note that this result does not depend on the parametrization of G_θ and therefore applies to the level sets of potentially very nonconvex neural network parametrizations. Previous results on the connexity of such level sets [6, 19] are very tied to a specific parametric form. The fact that we can give such a result in an abstract setup is rather surprising. We hope that further developments will clarify how much our approach can help these efforts.

Finally, comparing this result with Example 3.5 also reveals a fascinating possibility: a simple descent algorithm might in fact be unable to find that the Dirac distribution at the center of the sphere is a global minimum. Therefore the effective statistical performance of the learning process may be subtantially better than what Theorem 3.4 suggests. Further research is necessary to check whether such a phenomenon occurs in practice.

7   Conclusion

This work illustrates how the geometrical study of probability distances provides useful —but still incomplete— insights on the practical performance of implicit modeling approaches using different distances. In addition, using a technique that differs substantially from previous works, we also obtain surprising global optimization results that remain valid when the parametrization is nonconvex.

Acknowledgments. We would like to thank Joan Bruna, Marco Cuturi, Arthur Gretton, Yann Ollivier, and Arthur Szlam for stimulating discussions and also for pointing out numerous related works.

References

1. Aizerman, M.A., Braverman, É.M., Rozonoér, L.I.: Theoretical foundations of the potential function method in pattern recognition learning. Autom. Remote Control **25**, 821–837 (1964)
2. Amari, S.I., Nagaoka, H.: Methods of Information Geometry, vol. 191. American Mathematical Society (2007)
3. Arjovsky, M., Chintala, S., Bottou, L.: Wasserstein generative adversarial networks. In: Proceedings of the 34nd International Conference on Machine Learning, ICML 2017, Sydney, Australia, 7–9 August 2017
4. Aronszajn, N.: Theory of reproducing kernels. Trans. Am. Mathe. Soc. **68**, 337–404 (1950)
5. Arora, S., Ge, R., Liang, Y., Ma, T., Zhang, Y.: Generalization and equilibrium in generative adversarial nets (gans). arXiv preprint arXiv:1703.00573 (2017)
6. Auffinger, A., Ben Arous, G.: Complexity of random smooth functions of many variables. Ann. Probab. **41**(6), 4214–4247 (2013)
7. Bellemare, M.G., et al.: The cramer distance as a solution to biased Wasserstein gradients. arXiv preprint arXiv:1705.10743 (2017)
8. Berti, P., Pratelli, L., Rigo, P., et al.: Gluing lemmas and skorohod representations. Electron. Commun. Probab. **20** (2015)
9. Borkar, V.S.: Stochastic approximation with two time scales. Syst. Control Lett. **29**(5), 291–294 (1997)
10. Bottou, L., Curtis, F.E., Nocedal, J.: Optimization methods for large-scale machine learning. CoRR abs/1606.04838 (2016)
11. Bouchacourt, D., Mudigonda, P.K., Nowozin, S.: DISCO nets: DISsimilarity cOef-ficients networks. In: Advances in Neural Information Processing Systems, vol. 29, pp. 352–360 (2016)

12. Burago, D., Burago, Y., Ivanov, S.: A Course in Metric Geometry. Volume 33 of AMS Graduate Studies in Mathematics, American Mathematical Society (2001)
13. Challis, E., Barber, D.: Affine independent variational inference. In: Pereira, F., Burges, C.J.C., Bottou, L., Weinberger, K.Q. (eds.) Advances in Neural Information Processing Systems, vol. 25, pp. 2186–2194. Curran Associates, Inc. (2012)
14. Cramér, H.: Mathematical Methods of Statistics. Princeton University Press, Princeton (1946)
15. Denton, E., Chintala, S., Szlam, A., Fergus, R.: Deep generative image models using a laplacian pyramid of adversarial networks. In: Cortes, C., Lawrence, N.D., Lee, D.D., Sugiyama, M., Garnett, R. (eds.) Advances in Neural Information Processing Systems, vol. 28, pp. 1486–1494. Curran Associates, Inc. (2015)
16. Dereich, S., Scheutzow, M., Schottstedt, R.: Constructive quantization: approximation by empirical measures. Annales de l'I.H.P. Probabilités et statistiques **49**(4), 1183–1203 (2013)
17. Dziugaite, G.K., Roy, D.M., Ghahramani, Z.: Training generative neural networks via maximum mean discrepancy optimization. In: Proceedings of the Thirty-First Conference on Uncertainty in Artificial Intelligence, UAI, pp. 258–267 (2015)
18. Fournier, N., Guillin, A.: On the rate of convergence in Wasserstein distance of the empirical measure. Probab. Theor. Relat. Fields **162**(3), 707–738 (2015)
19. Freeman, C.D., Bruna, J.: Topology and geometry of half-rectified network optimization. arXiv preprint arXiv:1611.01540 (2016)
20. Goodfellow, I.J., et al.: Generative adversarial nets. In: Advances in Neural Information Processing Systems, vol. 27, pp. 2672–2680. Curran Associates, Inc. (2014)
21. Gretton, A., Borgwardt, K.M., Rasch, M.J., Schölkopf, B., Smola, A.: A kernel two-sample test. J. Mach. Learn. Res. **13**, 723–773 (2012)
22. Gulrajani, I., Ahmed, F., Arjovsky, M., Dumoulin, V., Courville, A.: Improved training of Wasserstein GANs. arXiv preprint arXiv:1704.00028 (2017)
23. Hammersley, J.M.: The distribution of distance in a hypersphere. Ann. Mathe. Stat. **21**(3), 447–452 (1950)
24. Hastie, T., Tibshirani, R., Friedman, J.: The Elements of Statistical Learning. Springer Series in Statistics, 2nd edn. Springer, New York (2009)
25. Karras, T., Aila, T., Laine, S., Lehtinen, J.: Progressive growing of GANs for improved quality, stability, and variation. arXiv preprint arXiv:1710.10196 (2017)
26. Khinchin, A.Y.: Sur la loi des grandes nombres. Comptes Rendus de l'Académie des Sciences (1929)
27. Kingma, D.P., Welling, M.: Auto-encoding variational bayes. CoRR abs/1312.6114 (2013)
28. Kocaoglu, M., Snyder, C., Dimakis, A.G., Vishwanath, S.: CausalGAN: learning causal implicit generative models with adversarial training. arXiv preprint arXiv:1709.02023 (2017)
29. Konda, V.R., Tsitsiklis, J.N.: Convergence rate of linear two-time-scale stochastic approximation. Ann. Appl. Probab., 796–819 (2004)
30. Kulkarni, T.D., Kohli, P., Tenenbaum, J.B., Mansinghka, V.: Picture: A probabilistic programming language for scene perception. In: Proceedings of the IEEE Conference on Computer Vision And Pattern Recognition, CVPR 2015, pp. 4390–4399 (2015)
31. Lee, M.W., Nevatia, R.: Dynamic human pose estimation using Markov Chain Monte Carlo approach. In: 7th IEEE Workshop on Applications of Computer Vision/IEEE Workshop on Motion and Video Computing (WACV/MOTION 2005), pp. 168–175 (2005)

32. Li, C.L., Chang, W.C., Cheng, Y., Yang, Y., Póczos, B.: MMD GAN: towards deeper understanding of moment matching network. arXiv preprint arXiv:1705.08584 (2017)
33. Li, Y., Swersky, K., Zemel, R.: Generative moment matching networks. In: Proceedings of the 32nd International Conference on International Conference on Machine Learning, ICML 2015, vol. 37, pp. 1718–1727 (2015)
34. Liu, S., Bousquet, O., Chaudhuri, K.: Approximation and convergence properties of generative adversarial learning. arXiv preprint arXiv:1705.08991 (2017). to appear in NIPS 2017
35. Milgrom, P., Segal, I.: Envelope theorems for arbitrary choice sets. Econometrica **70**(2), 583–601 (2002)
36. von Mises, R.: On the asymptotic distribution of differentiable statistical functions. Ann. Mathe. Stat. **18**(3), 309–348 (1947)
37. Müller, A.: Integral probability metrics and their generating classes of functions. Adv. Appl. Probab. **29**(2), 429–443 (1997)
38. Neal, R.M.: Annealed importance sampling. Stat. Comput. **11**(2), 125–139 (2001)
39. Nguyen, X., Wainwright, M.J., Jordan, M.I.: Estimating divergence functionals and the likelihood ratio by convex risk minimization. IEEE Trans. Inf. Theor. **56**(11), 5847–5861 (2010)
40. Nowozin, S., Cseke, B., Tomioka, R.: f-GAN: training generative neural samplers using variational divergence minimization. In: Advances in Neural Information Processing Systems, vol. 29, pp. 271–279 (2016)
41. Rachev, S.T., Klebanov, L., Stoyanov, S.V., Fabozzi, F.: The Methods of Distances in the Theory of Probability and Statistics. Springer, New York (2013)
42. Radford, A., Metz, L., Chintala, S.: Unsupervised representation learning with deep convolutional generative adversarial networks. arXiv preprint arXiv:1511.06434 (2015)
43. Rezende, D.J., Mohamed, S., Wierstra, D.: Stochastic backpropagation and approximate inference in deep generative models. In: Proceedings of the 31st International Conference on Machine Learning, ICML 2014, pp. 1278–1286 (2014)
44. Romaszko, L., Williams, C.K., Moreno, P., Kohli, P.: Vision-as-inverse-graphics: obtaining a rich 3D explanation of a scene from a single image. In: Proceedings of the IEEE Conference on Computer Vision and Pattern Recognition, CVPR 2017, pp. 851–859 (2017)
45. Salimans, T., Goodfellow, I., Zaremba, W., Cheung, V., Radford, A., Chen, X.: Improved techniques for training GANs. In: Advances in Neural Information Processing Systems, vol. 29, pp. 2234–2242 (2016)
46. Schoenberg, I.J.: Metric spaces and positive definite functions. Trans. Am. Mathe. Soc. **44**, 522–536 (1938)
47. Schölkopf, B., Smola, A.J.: Learning with Kernels. MIT Press, Cambridge, MA (2002)
48. Sejdinovic, D., Sriperumbudur, B., Gretton, A., Fukumizu, K.: Equivalence of distance-based and rkhs-based statistics in hypothesis testing. Ann. Stat. **41**(5), 2263–2291 (2013)
49. Serfling, R.J.: Approximation Theorems of Mathematical Statistics. Wiley, New York; Chichester (1980)
50. Sriperumbudur, B.: On the optimal estimation of probability measures in weak and strong topologies. Bernoulli **22**(3), 1839–1893 (2016)
51. Sriperumbudur, B.K., Fukumizu, K., Gretton, A., Schölkopf, B., Lanckriet, G.R.: On the empirical estimation of integral probability metrics. Electron. J. Stat. **6**, 1550–1599 (2012)

52. Sriperumbudur, B.K., Fukumizu, K., Lanckriet, G.R.: Universality, characteristic kernels and RKHS embedding of measures. J. Mach. Learn. Res. **12**, 2389–2410 (2011)
53. Székely, G.J., Rizzo, M.L.: Energy statistics: a class of statistics based on distances. J. Stat. Plan. Infer. **143**(8), 1249–1272 (2013)
54. Székely, J.G.: E-statistics: The energy of statistical samples. Technical report, 02–16, Bowling Green State University, Department of Mathematics and Statistics (2002)
55. Theis, L., van den Oord, A., Bethge, M.: A note on the evaluation of generative models. In: International Conference on Learning Representations (2016)
56. Villani, C.: Optimal Transport: Old and New. Grundlehren der mathematischen Wissenschaften. Springer, Berlin (2009)
57. Zinger, A.A., Kakosyan, A.V., Klebanov, L.B.: A characterization of distributions by mean values of statistics and certain probabilistic metrics. J. Sov. Mathe. **4**(59), 914–920 (1992). Translated from Problemy Ustoichivosti Stokhasticheskikh Modelei-Trudi seminara, pp. 47–55 (1989)

Deep Learning in the Natural Sciences: Applications to Physics

Peter Sadowski and Pierre Baldi[(✉)]

Department of Computer Science, University of California, Irvine, Irvine, USA
pfbaldi@uci.edu

Abstract. Machine learning is increasingly being used not only in engineering applications such as computer vision and speech recognition, but in data analysis for the natural sciences. Here we describe applications of deep learning to four areas of experimental sub-atomic physics — high-energy physics, antimatter physics, neutrino physics, and dark matter physics.

1 Introduction

There are of course many applications of machine learning to engineering problems, from computer vision to speech recognition. Braverman with his early interest in machine learning might have foreseen some of them. But what perhaps could have surprised him, is the breadth of the applications of machine learning in general, and deep learning in particular, to problems in the natural sciences.

1.1 Deep Learning in the Natural Sciences

For the past three decades we have been engaged in the application of deep learning methods to problems in the natural sciences, in biology, chemistry, and physics. In biology, we introduced convolutional neural networks for biological imaging [20] and recursive neural networks for problems of protein structure prediction [18,19,21,29,69]. These methods, together with great progress in data availability, have helped solve the problem of predicting protein secondary structure, going from a level of about 60% achieved by Chou and Fasman in the 1970s, to about 95% today [62]. This line of work has also led to more general frameworks for the design of recursive neural networks for problems characterized by variable-size, structured, data [17]. A detailed review of applications of deep learning to biomedical data can be found in [16].

In chemistry, deep learning methods have been developed and applied to the problem of predicting the physical, chemical, or biological properties of small molecules [39,60]. Deep learning methods have also been developed and applied to the problem of predicting chemical reactions [51,52].

In this chapter, we focus on some of the applications of deep learning methods to physics. Even within physics, there are many applications of deep learning at

© Springer Nature Switzerland AG 2018
L. Rozonoer et al. (Eds.): Braverman Readings in Machine Learning, LNAI 11100, pp. 269–297, 2018.
https://doi.org/10.1007/978-3-319-99492-5_12

all scales, from the subatomic to the cosmological. Here we focus on four areas of subatomic physics — high-energy physics, antimatter physics, neutrino physics, and dark matter physics — emphasizing the common aspects that cut across these different areas.

1.2   Deep Learning in Physics

A primary focus of modern physics is the study of the subatomic particles that determine the fundamental nature of matter. Measuring the properties of these particles can require highly-sophisticated experiments, some of which recreate extreme conditions and energies that have not existed in nature since the Big Bang. Yet these engineering feats give rise to major data analysis challenges, as the detectors used in these experiments produce a torrent of high-dimensional data with small signal-to-noise ratios. For example, the ATLAS detector at the Large Hadron Collider (LHC) produces a petabyte of raw data per second, which must be carefully analyzed to search for the faint signals of new particles.

The detectors used in these experiments vary in scale and design, but generally consist of layers of sensors that take measurements of traversing particles. The data recorded for a single event — a particle collision for example — is high-dimensional and structured; it roughly corresponds to an image or video, with correlations between adjacent pixels and time steps. From this high-dimensional event data, physicists must reconstruct the underlying physical processes that produced it.

This is typically performed by applying a series of algorithms that reduce and summarize the raw data, each carefully designed by an expert physicist familiar with a particular part of the detector or data-generating process. These include trigger systems that aim to quickly discard uninteresting events in real time, clustering algorithms that group sets of detector hits that were likely caused by the same particle, tracking algorithms that attempt to reconstruct the trajectory of each particle through the detector layers, and classifiers that attempt to discriminate between different types of particles. Each step in this data-processing pipeline transforms the data in some way, distilling the information from the previous step into something more manageable.

Increasingly, machine learning methods are being employed in these data-analysis pipelines. The underlying physical processes that generate the data are often so complex — involving many intermediate unobserved variables — that knowledge of the system alone is insufficient to write a good event-reconstruction algorithm. Instead, machine learning models are trained to perform tasks like particle classification using calibration and/or simulation data for training. Until recently, the machine learning models in common use were not deep learning methods, but rather things like Support Vector Machines or Boosted Decision Trees. In practice, these methods are sensitive to the choice of input features, and usually require the user to engineer a good set of features for optimal performance. This can be a daunting task when the data is high-dimensional and the scientist has little intuition about its shape. Engineering optimal input features to shallow machine learning models is a common research project and is the subject of many papers in the field.

In contrast, the goal of *deep* learning is to *automatically* learn multiple sequential processing steps in an end-to-end manner, typically using deep artificial neural networks. Deeper networks are able to represent complex functions more efficiently than shallow networks, and the learning problem can be constrained by designing the neural network architecture to exploit known structure in the data. While deep models can be more difficult to train, recent advances in computing hardware and software have made training considerably easier and faster. Deep learning now dominates many machine learning tasks in computer vision, speech recognition, and natural language processing — in part by replacing engineered feature representations with many layers of *learned* representations.

Deep learning is a particularly promising approach in experimental physics for multiple reasons. First, training data is abundant in many important applications, either from highly-accurate Monte Carlo simulations or data produced in calibration experiments. Second, the structure of the detector data is often similar to that of images, so that deep learning techniques developed for computer vision can be applied. Third, end-to-end approaches could greatly simplify the development of data-analysis pipelines by aggregating multiple algorithmic steps, each of which usually requires research and testing. And finally, a deep learning approach could alleviate the common problem of early steps losing information that is relevant to the final task. Information is necessarily discarded in each step of the data-processing pipeline, but deep neural networks trained with stochastic gradient descent seem to balance the need to extract useful information about the target while removing irrelevant information [89].

This chapter describes deep learning applications in a diverse set of antimatter, neutrino, dark matter, and collider experiments. These include classification and regression tasks, as well as supervised, unsupervised, and adversarial learning. In general, we compare traditional approaches that rely on "high-level" human-engineered feature representations to an end-to-end deep learning approach that takes in "low-level" features. In each application, the performance of the deep learning approach matches or exceeds that of the traditional approach, with the latter case revealing how much information is lost by the engineered features. These examples demonstrate the potential of deep learning as a powerful tool in scientific data analysis. Ultimately, it could also help us to better understand the limitations of our detectors, and aid in the design of better ones.

2   Antimatter Physics

In the Standard Model of physics, every matter particle has a corresponding antimatter particle. Matter and antimatter particles are hypothesized to be symmetric with respect to charge conjugation, parity transformation, and time reversal, a property known as CPT symmetry. One of the big open questions in modern physics is why the universe is unbalanced, consisting of more matter than antimatter, which would appear to violate this symmetry. At the forefront of this investigation is the study of the antihydrogen atom, an atom consisting of a positron-antiproton pair. Measuring the properties of antihydrogen and comparing them to those of its matter counterpart (the hydrogen atom) may reveal asymmetries that would violate the CPT symmetry hypothesis and give rise to new physics beyond the Standard Model. However, experimental studies are difficult because a pair of matter-antimatter particles annihilate whenever they come into contact with each other. Thus, antihydrogen experiments require elaborate electromagnetic traps maintained in almost complete vacuum conditions.

Despite these challenges, multiple experiments in recent years have managed to trap or form a beam of antihydrogen atoms in order to test CPT symmetry [10,11,42,54], or test the effects of gravity on antimatter [3,7,70]. Antihydrogen is produced either by injecting antiproton and positron plasmas into cryogenic multi-ring electrode traps where three-body recombination takes place ($\bar{p} + 2e^+ \rightarrow \bar{H} + e^+$), or by the charge exchange processes between positronium, antiproton, and antihydrogen ($Ps^{(*)} + \bar{p} \rightarrow \bar{H} + e^-$ and $\bar{H} + Ps \rightarrow \bar{H}^+ + e^-$). In most cases, the production of an antihydrogen atom is identified by its annihilation signature whereby several charged pions are emitted from the point of annihilation, or vertex [49], and hit a detector system surrounding the trap. These events are distinguished from cosmic or instrumental background events using the hit multiplicity and the inferred positions of the annihilation vertex.

The materials required to maintain the antimatter vacuum trap — layers of vacuum chamber walls, multi-ring electrodes, and thermal isolation — make annihilation detection difficult. The annihilation products are subject to electromagnetic forces and scattering effects before reaching the course-grained detectors, impeding event reconstruction using the traditional tracking and vertex-finding algorithms that have been adopted from other types of physics experiments [8,9,32,33,73,83,92]. Thus, a data analysis pipeline based on traditional tracking and vertex-reconstruction algorithms is likely to be sub-optimal for these experiments. Recently, the authors proposed deep learning as a novel approach to this problem [76], and in the following we discuss its application to the ASACUSA experiment.

2.1   ASACUSA

The ASACUSA antihydrogen experiment [54] aims to directly compare the ground-state hyperfine transition frequency of antihydrogen with that of hydrogen. Measurements of the antihydrogen hyperfine transition frequency are performed by measuring the rate of antihydrogen atom production in a Penning-Malmberg trap while controlling the antiproton injection conditions, the overlap time while mixing antimatter plasmas, and other key parameters which constrain the three-body recombination yield and the level population evolution time [72]. Thus, the precision of the measurement depends directly on the ability to detect antihydrogen.

Antihydrogen is detected through the indirect observation of annihilations occurring on the wall of the trap, which must be distinguished from background events where a lone antiproton annihilates with a residual gas particle in the trap. The ASACUSA experiment traps charged antimatter particles (antiprotons and positrons) on the central axis using electro-magnetic fields. When a neutral antihydrogen atom is produced (predominately via three-body recombination of antiprotons and positrons), it may escape these trapping fields and annihilate on the inner wall of the multi-ring electrodes, emitting charged pions that can be detected. However, continuous annihilation of trapped antiprotons on residual gas atoms produces background annihilation events that emit the same pions. The latter tend to be emitted from the central axis, rather than the inner wall, so we can distinguish between antihydrogen and antiproton events if the position of the annihilation can be inferred from the detector data.

The pions are detected by the Asacusa Micromegas Tracker [73], which has a full-cylinder trigger layer of plastic scintillator bars and two detector layers made of gaseous micro-strip Micromegas technology [28,44]. The two detector layers consist of co-centric half-cylinders, both located on the same side of the detector so that pions can be tracked when they travel through both layers. Each layer is made up of micro-strips, oriented along either the azimuth or axis of the cylinder. When a trigger event occurs, a snapshot of detector activity is recorded. The raw detector data consists of 1,430 binary values per event, one from each micro-strip: 246 inner azimuth strips, 290 outer azimuthal strips, 447 inner axial strips, and 447 outer axial strips.

In the typical vertex-finding approach, each trigger event is processed with a track and vertex reconstruction algorithm, proceeding as follows. First, the single detector channels are searched for entries above threshold, and neighboring channels are iteratively clustered to form hits. The detected hits are used to form and fit tracks, using a Kalman-filtering algorithm [84]. After fitting all hit pair combinations in an event, the track candidates are filtered by their compatibility with the known antiproton cloud position. The filtered track candidates are then paired, and their three-dimensional point of closest approach position is assigned as the vertex position. Finally, event classification is performed based on the radial distance of the reconstructed vertex closest to the central axis (Fig. 1). However, the reconstruction algorithm only succeeds if two distinct tracks can be identified, and if the reconstructed tracks come within close proximity of each other in three-space (a threshold of 1 cm is used). The failure rate is high because the detector only covers half of the trap, such that roughly half of all pions escape, and the stochastic pion trajectories lead to high uncertainty in the reconstructed tracks.

In the deep learning approach, an artificial neural network was trained to classify annihilations directly from the raw detector data. This automated, end-to-end learning strategy can potentially identify discriminative information in the raw data that is typically discarded by the vertex-reconstruction algorithm. Indeed, this approach will provide predictions for events where vertex-reconstruction completely fails, such as those where only a single track can be reconstructed.

Two neural network architecture design choices were made in order to take advantage of the geometry of the detector and constrain the learning problem.

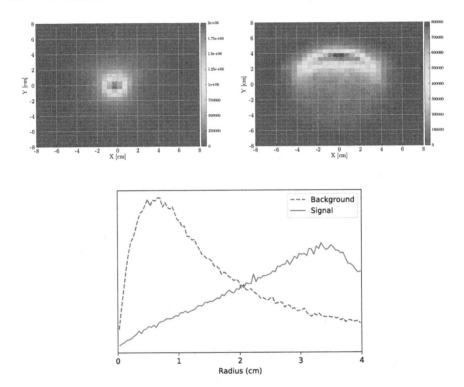

Fig. 1. Empirical distributions of reconstructed vertices for background events (annihilations on the axis at $R = 0$ cm) and signal events (annihilations on the wall of the trap at $R = 4$ cm). Top left: Heat map of background events in XY-space. Top right: Heat map of signal events in XY-space. Bottom: Empirical histogram of vertex radii. The smearing of these distributions reflects the reconstruction error due to pion scattering effects, low detector precision, and limitations in the track and vertex reconstruction algorithm.

First, data from micro-strips situated along the azimuthal and axial dimensions are processed along separate pathways that are then merged by concatenating the hidden representations higher in the architecture. Second, both pathways employ 1-D convolution with locally-connected feature detectors and max-pooling layers. Together, these design choices achieve some amount of translational invariance along each dimension, constraining the learning problem and reducing the total number of network parameters. The highly-correlated inner and outer detector layers are processed together as two associated 'channels' in the first layers of the network, with the outer azimuthal layer downsampled with linear interpolation to match the dimensionality of the inner layer. A diagram of the network architecture is shown in Fig. 2.

The two approaches were compared on Monte Carlo simulations that reflect realistic conditions in the ASACUSA experiment. One million annihilation events of each type were simulated and randomly divided into training (60%),

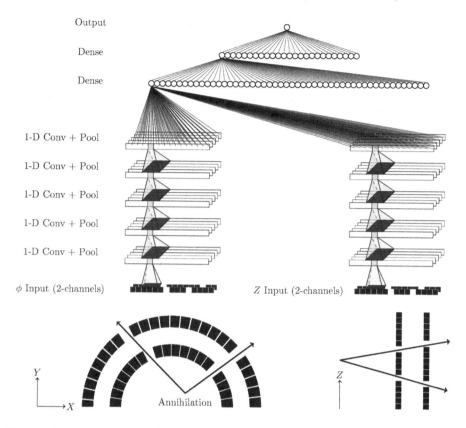

Fig. 2. Deep neural network architecture consisting of two separate pathways that process data from the micro-strips placed along the azimuthal (ϕ) and axial (Z) dimensions. The inner and outer detector layers are treated as two input channels and processed by a sequence of 1-D convolutional layers in each pathway. The two pathways are merged using dense fully-connected layers higher in the network.

validation (20%), and test subsets (20%). The vertex finding (VF) algorithm failed on 75% of these events: 20% did not result in any detector hits; 7% had a hit in only one detector; and 48% had hits in both detectors but a vertex could not be reconstructed. Thus, a direct performance comparison was only possible on the 25% of the test events for which the VF algorithm succeeded.

Deep neural networks were trained with a variety of hyperparameter combinations in order to optimize generalization performance on the validation set. The best architecture had five 1-D convolutional layers with kernel sizes 7-3-3-3-3 (the size of the receptive fields for neurons in each layer), channel sizes 8-16-32-64-128 (the number of distinct feature detectors in each layer), and rectified linear activation [63]. In order to account for translational invariance, each convolution layer is followed by a max-pooling layer with pool size 2 and stride length 2 [45]. The flattened representations from the two pathways are then concatenated and followed by two fully-connected layers of 50 and 25 rectified linear units, then a single logistic output unit with a relative entropy loss. During training, 50% dropout was used in the top two fully-connected layers to reduce overfitting [22,82]. Data was augmented both at training and test time by translating the hits along the axial direction in order to enforce translational invariance along this dimension (since the true axial distribution in real experiments is unknown). The model weights were initialized from a scaled normal distribution as suggested by He et al. [47], then trained using the Adam optimizer [53] ($\beta_1 = 0.9, \beta_2 = 0.999, \epsilon = 1e - 08$) with mini-batch updates of size 100 and a learning rate that was initialized to 0.0001 and decayed by 1% at the end of each epoch. Training was stopped when the validation objective did not improve within a window of three epochs. The models were implemented in KERAS [30] and THEANO [88], and trained on a cluster of Nvidia graphics processors.

Performance comparisons were made by plotting the Receiver Operating Characteristic (ROC) curve and summarizing the discrimination by calculating the Area Under the Curve (AUC). On the 25% of test events for which a vertex could be reconstructed, the deep learning approach increases performance from 0.76 to 0.87 AUC (Fig. 3). On the disjoint set of events for which a vertex could *not* be reconstructed, deep learning achieves 0.78 AUC. This effectively triples

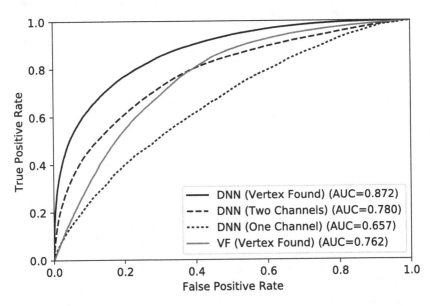

Fig. 3. ROC curves for the vertex finding algorithm (VF) and a deep convolutional neural network (DNN) on the subset of events for which an annihilation vertex can be reconstructed ('Vertex Found,' 25% of all events). Also shown are the curves for DNN predictions on the subset of events in which both the inner and outer detector channels are hit, but no vertex could be reconstructed (48% of all events), and on events in which only one channel is hit (7% of all events).

the event coverage — to 73% of all events — while *simultaneously* improving overall AUC by more than 0.05 (to 0.82 on the union set).

These results provide a clear demonstration of the advantages provided by the deep learning approach. The traditional track and vertex reconstruction algorithm is a multi-step process in which relevant information is inevitably discarded at each step, despite the best efforts of the scientists tuning the algorithm. The deep learning approach is capable of learning this classification in an end-to-end manner, and takes advantage of statistical information that is lost in the traditional approach. In particular, for events in which only a single track can be reconstructed (and thus the vertex finder fails), the deep learning approach learns to predict whether the annihilation occurred on the wall based on whether the observed hits suggest a track that travels through the axis. But even for events in which a vertex *can* be reconstructed, the deep learning approach performs better based on the statistical information learned from the raw data.

Preliminary results indicate that deep learning also performs as expected on real ASACUSA calibration data, and we are currently working to improve the experiment sensitivity with this approach. Furthermore, while this study was performed for the ASACUSA experiment, the deep learning approach can easily be deployed to other ongoing antimatter experiments with different instruments. This even offers a potential new direction in the design of future detectors.

3   High Energy Physics

The Large Hadron Collider is the most complex experimental instrument ever constructed — designed to produce particle collisions at extremely high energies. Through these collisions, physicists aim to test the fundamental properties of matter and discover new, exotic particles that are not part of the Standard Model. Such discoveries will be necessary to answer the biggest mysteries of modern physics, including how the theory of gravity can be unified with the Standard Model, and what particle(s) constitute the "dark matter" that accounts for 85% of all matter in the universe [41, 67].

The high-energy collisions at the LHC interconvert energy and matter, creating unstable particles that quickly decay into lighter, more stable particles. Thus, exotic particles of interest such as the Higgs boson are not observed directly, but rather indirectly via subtle signatures in the trajectories of their decay products. Furthermore, the vast majority of collision events do not produce particles of interest. The LHC produces approximately 10^{11} collisions per hour [13], but for example, only 300 of these collisions will result in a Higgs boson. Therefore, a massive engineering effort is required to record, process, store, and analyze this data.

A single collision event will produce many decay products that travel outward from the point of collision and through the many-layered detector [12]. In ATLAS, over 100 million detector elements capture a "snapshot" of the event — similar to a 3D image. In order to identify collisions that produce particles of interest, each event snapshot must be carefully analyzed using a multitude of

algorithms to identify the different decay products, their trajectories [27], and
the particle interactions that produced them.

In the following, we describe three different applications of deep learning
to this data-processing pipeline. First, we describe the use of deep learning in
exotic particle searches, in which the momentum of the decay products is used to
classify collisions of interest from background processes. Second, we describe the
analysis of calorimeter data, the outer layer of the detector, which lends itself
to deep learning strategies used for computer vision. Third, we describe the use
of adversarial neural networks to train classifiers that are robust to systematic
uncertainties in the experiment.

3.1   Exotic Particle Searches

The goal of exotic particle search studies is to find evidence for new particles or
particle interactions buried in mountains of collision data. This evidence typically
manifests as a faint "bump" in a high-dimensional distribution due to a rare
event, such as the production of a Higgs boson, in an ocean of background
events that produce the same stable decay products but with slightly different
signatures. While a classifier might not be able to distinguish between signal
vs. background events with certainty, the probabilistic output can be used for
dimensionality reduction to aid in the statistical analysis when measuring the
evidence for a new discovery. Here we demonstrate deep learning as an alternative
to the combination of feature-engineering and shallow machine learning that are
traditionally employed [23, 24, 75].

We evaluated the deep learning approach on three benchmark tasks: detecting
Higgs Boson production, detecting the production of supersymmetrical particles,
and detecting Higgs-to-$\tau^+\tau^-$ decay (Figs. 4, 5 and 6). Each task was formulated
as a binary classification problem to distinguish between two physical processes
with identical decay products. The input features could be separated into two
distinct categories: "low-level" features that comprised the 3D momenta the

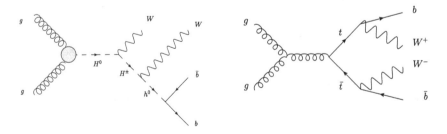

Fig. 4. Feynman diagrams describing the processes in the HIGGS classification task.
The signal involves new exotic Higgs bosons H^0 and $H^\pm$ (left) and the background
process involves top-quarks t (right). In both cases, the resulting particles are two W
bosons and two b-quarks.

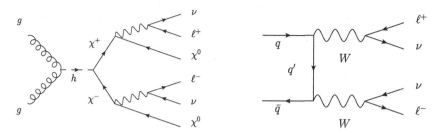

Fig. 5. Feynman diagrams describing the processes in the SUSY classification task. The signal involves hypothetical supersymmetric particles $\chi^{\pm}$ and χ^0 along with charged leptons $\ell^{\pm}$ and neutrinos ν (left) and the background process involves W bosons (right). In both cases, the resulting observed particles are two charged leptons, as neutrinos and χ^0 escape undetected.

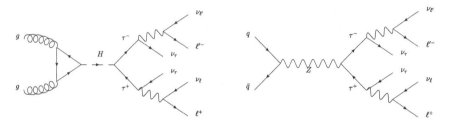

Fig. 6. Feynman diagrams describing the processes in the Higgs-to-$\tau^+\tau^-$ classification task. The signal process (left) involves a Higgs H decaying to leptons $\tau^+\tau^-$, while the dominant background process (right) produces the same decay products.

observed decay products and other measurements that contain additional information, and "high-level," *derivative* features — functions of the low-level features that were engineered to aid in classification.

Deep and shallow neural network architectures — networks with a single hidden layer — were trained and tested on each task with millions of simulated training samples. The hyperparameters were optimized separately for each task and architecture, using the Spearmint Bayesian optimization algorithm [81], on the complete set of 18–28 input features then tested on each feature subset. The results demonstrate that the deep architectures significantly outperform the shallow architectures in terms of AUC (Table 1), and similarly in terms of discovery significance (Table 2), a measure of statistical power. Furthermore, the deep learning models do almost as well on the low-level feature sets as they do with the complete feature set, indicating that the deep learning approach is *automatically* learning useful intermediate data representations, without the aid of high-level engineered features (Fig. 7). In fact, deep architectures trained

Table 1. Performance of shallow and deep neural network architectures on low-level vs. high-level features. The same neural network model was trained on three sets of input features: low-level features, high-level features and the complete set of features. Neural network was trained five times with different random initializations; the table displays the mean AUC, with standard deviations in parentheses.

Data set	AUC			
	Model	Low-level	High-level	Complete
HIGGS	Shallow NN	0.790 (<0.001)	0.794 (<0.001)	0.844 (<0.001)
	Deep NN	0.891	0.801	0.896
SUSY	Shallow NN	0.875 (<0.001)	0.868 (<0.001)	0.879 (< 0.001)
	Deep NN	0.876 (<0.001)	0.870 (<0.001)	0.879 (<0.001)
Higgs-to-$\tau^+\tau^-$	Shallow NN	0.789 (0.001)	0.792 (<0.001)	0.797 (<0.001)
	Deep DNN	0.798 (<0.001)	0.798 (<0.001)	0.802 (<0.001)

Table 2. Discovery significance. Performance in terms of expected significance of a discovery, estimated using 100 signal events and 5000 background events.

Data set	Discovery significance			
	Model	Low-level	High-level	Complete
HIGGS	Shallow NN	2.16σ (0.03)	2.64σ (0.01)	3.28σ (0.02)
	Deep NN	4.57σ	2.87σ	4.82σ
SUSY	Shallow NN	7.86σ (0.06)	7.22σ (0.02)	7.81σ (0.05)
	Deep NN	7.73σ (0.07)	7.58σ (0.08)	7.94σ (0.08)
Higgs-to-$\tau^+\tau^-$	Shallow NN	2.57σ (<0.01)	2.92σ (<0.01)	3.02σ (<0.01)
	Deep DNN	3.16σ (<0.01)	3.24σ (<0.01)	3.37σ (<0.01)

on low-level features for benchmarks 1 and 3 actually outperform the shallow models trained on the complete feature set.

3.2   Jet Substructure Classification

The ATLAS detector's cylindrical outer layer consists of a giant calorimeter, which captures showers of energy released by fragmenting quark and gluon particles called "jets." Sometimes, massive particles can be produced at such high velocities that their hadronic decays are collimated and multiple jets overlap. Classifying jets that are due to single low-mass particles or due to the decay of a massive particle is an important problem in the analysis of collider data [1,2,4,5].

There does not exist a complete analytical model for classifying jets directly from theoretical principles, so traditional approaches to this problem have relied on engineered features that were designed to detect patterns of energy deposition in the calorimeter [35–37,50,55,57,61,68,74,86]. However, the complexity

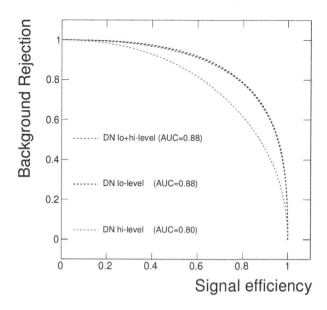

Fig. 7. ROC plot comparing the same deep neural network trained on the HIGGS task using the low-level feature subset, high-level feature subset, and the complete feature set. The high-level engineered features no longer aid in classification, because the model is able to automatically learn useful intermediate representations. This is not the case with the shallow machine learning models trained on the same problem.

of the data makes this task an excellent application for machine learning. Indeed, state-of-the-art approaches use shallow machine learning models such as shallow neural networks and boosted decision trees (BDTs) to aggregate information from multiple expert features. Training data can be produced using tools that simulate the microphysics of jet formation and how these jets deposit energy in the calorimeter [15, 80].

Recently, we and others proposed deep learning with the low-level calorimeter data itself, rather than with high-level engineered features [25, 65]. Calorimeter data can be treated as a two-dimensional image, lending itself to the natural application of the deep-learning strategies developed for computer vision. The following experiments demonstrate this approach on the problem of classifying single jets produced in quark or gluon fragmentation from two overlapping jets produced when a high-velocity W boson decays to a collimated pair of quarks.

A simulated data set was produced with standard physics packages [6, 66, 80]. Dijets from boosted $W \rightarrow q q'$ were generated with a center of mass energy $\sqrt{s} = 14$ TeV using the diboson production, decay process $pp \rightarrow W^+ W^- \rightarrow qqqq$ leading to two pairs of quarks, and with each pair of quarks collimated and leading to a single jet. Jets originating from single quarks and gluons were generated using the $pp \rightarrow qq, qg, gg$ process. In addition, a separate data set was produced that included the effects of pile-up events, a source of noise in which additional pp interactions overlap in time.

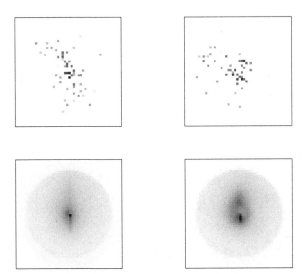

Fig. 8. Top: Typical jet images from class 1 (single QCD jet from q or g) on the left, and class 2 (two overlapping jets from $W \to qq'$) on the right. Bottom: Averages of 100,000 jet images from each class.

To compare our approach to the current state-of-the-art, we calculated six high-level jet-substructure features commonly used in the literature: the invariant mass of the trimmed jet, N-subjettiness [86,87] $\tau_{21}^{\beta=1}$, and the energy correlation functions [56,57] $C_2^{\beta=1}$, $C_2^{\beta=2}$, $D_2^{\beta=1}$, and $D_2^{\beta=2}$. Each of these individual quantities were outperformed by a BDT trained with all six as input.

In the deep learning approach, each jet was represented as a 32×32 pixel image, approximating the resolution of the calorimeter cells. Jets were translated and rotated into a canonical position based on the center of mass and the principal axis (Fig. 8). Because this canonicalization removes the translational and rotational invariance of the original data, the deep neural network structure used locally-connected layers (without the parameter-sharing of convolutional layers). The neural network architecture and learning hyperparameters, as well as the BDT hyperparameters, were optimized on the no-pileup data set using the Spearmint Bayesian Optimization algorithm [81].

Our experiments demonstrate that even without the aid of expert features, deep neural networks match or modestly outperform the current state-of-the-art approach (Fig. 9, Table 3). Furthermore, these performance gains persist when our simulations include pileup effects. Similar results have been reported for other jet classification tasks, and with recurrent neural networks trained on tracking data instead of images [46,58].

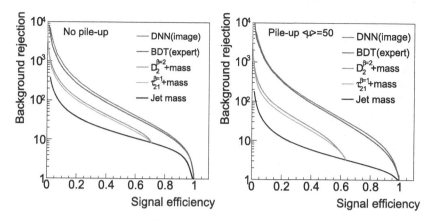

Fig. 9. ROC plots showing performance of deep neural networks (DNN) trained on the images, BDTs trained on the expert features, both with (left) and without pile-up (right). Typical choices of signal efficiency in real applications are in the 0.5–0.7 range. Also shown are the performance of jet mass feature, as well as two other expert variables in conjunction with a mass window.

Table 3. Performance results for BDT and deep networks. Shown for each method are both the signal efficiency at background rejection of 10, as well as the Area Under the Curve (AUC), the integral of the background efficiency versus signal efficiency. For the neural networks, we report the mean and standard deviation of three networks trained with different random initializations.

Technique	Performance	
	Signal efficiency	AUC
No pileup		
BDT on derived features	86.5%	95.0%
Deep NN on images	87.8%(0.04%)	95.3%(0.02%)
With pileup		
BDT on derived features	81.5%	93.2%
Deep NN on images	84.3%(0.02%)	94.0%(0.01%)

3.3 Decorrelated Jet Substructure Tagging with Adversarial Neural Networks

An additional challenge in jet-substructure classification is the systematic uncertainty in the jet invariant mass distribution of the background process. The simulations used to train classifiers necessarily have a fixed distribution, so in order for a classifier to be robust against this uncertainty, it should not use information about the invariant mass in its prediction. However, most of the features used to classify jets are correlated with this quantity, so models fit to these features will learn to take advantage of these correlations when classifying the jets. One proposed solution is to perform classification based on a single engineered feature that can be "decorrelated," but this sacrifices the boost in classification performance that comes from combining information from multiple features [31,38].

Another strategy is to incorporate the decorrelation requirement directly into the machine learning model by penalizing the jet classifier for using information about the nuisance variable. This can be achieved with adversarial training [40, 43,77], and has been proposed to solve the problem of nuisance parameters in physics problems by Ref. [59]. This approach involves training two models together — the original classifier, and a separate, adversarial model — such that the adversary adaptively shapes the loss function of the first model.

For the jet substructure classification problem, we proposed an adversarial neural network classifier that attempts to predict the jet-invariant mass from the output of the jet classifier (Fig. 10) [78]. The loss of the adversary is included as a linear term with a negative coefficient in the loss of the jet classifier, such that it acts as a regularizer. That is, the jet tagger must find a balance between minimizing the classification loss, and maximizing the loss of the adversary,

$$L_{\text{tagger}} = L_{\text{classification}} - \lambda L_{\text{adversary}}.$$

Results on the classification task from Ref. [79] show that this approach yields a jet tagger that is only slightly less powerful in terms of classification performance, but much more robust to changes in the distribution of the background invariant mass (Fig. 11).

This approach can be generalized to include the case where both the classifier and its adversary are parameterized by some external quantity, such as a theoretical hypothesis for the mass of a new particle or a field coupling strength.

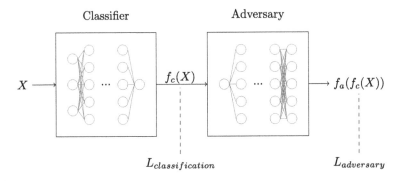

Fig. 10. Adversarial training approach with a jet-substructure classifier network and an adversary network that attempts to predict the invariant mass from the classifier output. The entire system is trained simultaneously with stochastic gradient descent, with a gradient reversal layer in between the classifier and adversary segments. In this experiment, the classifier is a deep neural network with eleven high-level input features and a sigmoid output; the adversarial network has a single input and a softmax output, with the invariant jet mass quantized into ten bins. See Ref. [78] for details.

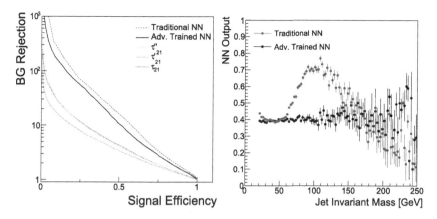

Fig. 11. Left: Signal efficiency and background rejection (1/efficiency) curves for several jet-tagging discriminants: neural networks trained with and without the adversarial strategy to optimize classification while minimizing impact on jet mass, the unmodified τ_{21}, and two "decorrelated" variables τ'_{21} and τ''_{21} [38]. In these experiments, the signal samples have mass $m_{Z'} = 100$ GeV. **Right:** Box plot showing the distribution of tagger predictions for different values of the jet mass. The adversarial training encourages the classifier output to be flat with respect to this quantity.

This is motivated by the fact that resonance searches, such as the one described here, are often performed as a scan over a range of potential particle masses. Generally the optimal classifier for each hypothesized mass will differ; but the simulations used for training can usually only be sampled from a small number of hypothesized values due to computational expense. By including the hypothesized particle mass as an input, we enable the model to interpolate between the discrete values observed in the training data [26,34]. In this case, we condition both the classifier and the adversary networks on the hypothesized particle mass (Fig. 12). Again, the adversarially-training keeps the classifier decorrelated with the jet invariant mass over the range of values upon which it is trained, leading to discovery significance that is robust to uncertainty in the background (Fig. 13).

In these experiments, we have demonstrated an ability to exert *control* over the representations that are learned in deep neural networks. By including an adversarial objective in the loss function, we are able to constrain what information is used to perform the classification task and remove a source of systematic uncertainty from the classifier. This is a powerful technique, both for physics applications and more generally.

4   Neutrino Physics

Neutrinos are uncharged elementary particles that only interact with other matter through the weak subatomic force and gravity. In nature, they are produced by nuclear power plants, cosmic rays, and stars. Neutrinos have recently become

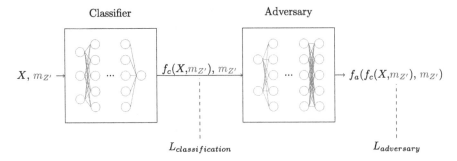

Fig. 12. Architecture for adversarial training approach, parameterized on the mass of the simulated signal particle $m_{Z'}$. Parameterizing the system in this way allows us to train on a discrete set of simulated particle masses and interpolate to intermediate values.

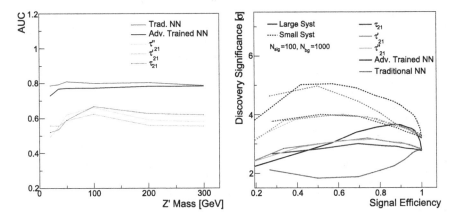

Fig. 13. Left: AUC vs. particle mass for the jet-tagging discriminants. As in Fig. 11, the neural network classification performance is only slightly decreased by adversarial training. **Right:** Statistical significance of a hypothetical signal for varying thresholds on the neural network output. Two scenarios are shown, in which the uncertainty on the background level is either negligible or large, both with $N_{\text{sig}} = 100, N_{\text{bg}} = 1000$. In each case, the adversarial training increases discovery significance over the other approaches.

a topic of intense interest due to the discoveries that they have non-zero mass and that they oscillate between three states, or *flavors*.

Because neutrinos rarely interact with other particles, detecting them in experiments is a challenge. Detectors commonly consist of large volumes of dense materials, such that neutrinos traveling through the volume occasionally hit the nucleus of an atom and initiate a cascade of events that produce photons or other radiation that can be detected by surrounding instruments. The data produced by these detectors require sophisticated processing and analysis. Deep learning is being explored as a tool for this analysis in at least three areas: (1) unsupervised learning for anomaly detection and clustering, (2) event classification, and (3) measuring the energy of decay products. We discuss examples of each application.

Experiments sometimes produce unexpected results, and there is a need for tools that automatically flag anomalies. For example, in the Daya Bay Reactor Neutrino Experiment, post-data-collection analysis revealed an unexpected source of light flashes that turned out to be a malfunction in the detector's photo-multiplier tubes. This required alert physicists to pour through the data and notice a pattern — there is a need for *automated* tools for detecting such anomalies. Deep learning can help in this task by representing complex data distributions in a new way that makes such anomalies more apparent. In one study, unsupervised deep neural networks were used to learn new representations of high-dimensional image-like data from the Daya Bay experiment [71]; a convolutional autoencoder was trained and the low-dimensional embedding was used to visualize and cluster events, revealing structure that corresponded to the different types of physical processes in the detector (Fig. 14).

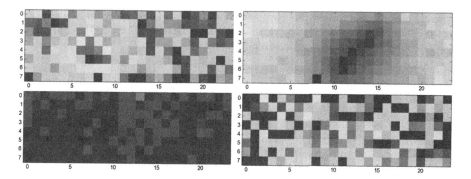

Fig. 14. Example detector data from the Daya Bay Reactor Neutrino experiment, in which a neutrino hits a liquid scintillator and produces a flash of light that is recorded by the surrounding detector elements, producing an image. Deep learning can be used in both a supervised or unsupervised manner to distinguish events of interest (top left) from different types of background events including muons (top right), mechanical malfunctions (lower left), and random fluctuations (lower right).

Deep learning is also being explored as a way to perform the central measurements of neutrino experiments. For example, the NOνA experiment aims to measure the probability of $\nu_\mu \to \nu_e$ oscillations by sending a beam of muon neutrinos from Fermilab in Illinois to a far detector in Ash River, Minnesota. Observing electron neutrinos in the far detector provides a measurement of the oscillation rate, which also depends on the energy of the neutrinos. Thus, the precision of the experiment depends directly on the ability to classify and measure the energy of neutrinos from the high-dimensional data collected by the far detector.

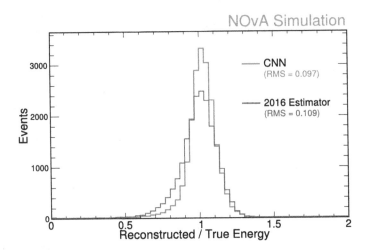

Fig. 15. Empirical distributions of the predicted energy relative to the true energy (reconstructed/ground truth) on a simulated test set. A deep convolutional neural network (CNN) is compared to the standard approach used in a previous analysis (2016 Estimator). Reprinted from "Convolutional Neural Networks for Electron Neutrino and Electron Shower Energy Reconstruction in the NOvA Detectors," by L. H., et al., *NIPS Deep Learning for Physical Sciences Workshop*, 2017.

Recent preliminary work with simulated data from the NOνA experiment has demonstrated that deep learning improves performance over standard methods in both tasks: classification [14] and energy estimation [48]. In the latter, millions of neutrino shower events in the NOνA far detector were simulated, each producing a pair of 896×384 pixel images, where each pixel measures the energy deposited in the corresponding detector cell. These images capture two views of a single inbound particle showering into secondary particles. A simple cropping procedure was used to reduce the dimensionality of each image to 141×151 pixels, and then a Siamese neural network was trained to combine the information from the two detector views and make a prediction. Deep learning improved performance by 11% compared to the specialized algorithms engineered for a previous NOνA analysis (Fig. 15).

Our own preliminary work with the DUNE and LArIAT neutrino experiments show similar results. Like NOνA, the LArIAT detector produces data that can be treated as a pair of high-resolution images (Fig. 16). In classification experiments, we used a Siamese convolutional neural network where each arm is based on the GoogLeNet architecture [85], and demonstrated that the deep learning approach trained on the low-level detector data achieves better performance than a BDT trained on derived features (Fig. 17). For the energy estimation task, the architecture was adapted for a heteroskedastic regression task by replacing the sigmoid output layer with two outputs, $\hat{y}_1$ and $\hat{y}_2$, and a loss function proportional to the negative log likelihood of the target value under a Gaussian distribution with mean $\hat{y}_1$ and standard deviation $e^{\hat{y}_2}$. This allows

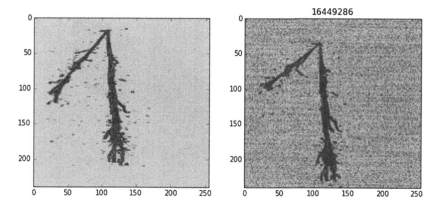

Fig. 16. A simulated LArIAT event. The two panels show the same particle shower from two separate sections of the detector. Each pixel is a single detector element at a particular location (y-axis) and time (x-axis). The shape of the shower can be used to classify the type of incident particle and predict its energy. In the experiments discussed, the two images were fed into separate arms of a Siamese convolutional neural network in order to classify the type of particle and predict its energy.

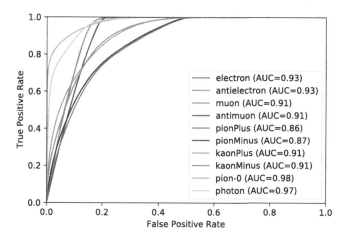

Fig. 17. Performance of the Siamese convolutional neural network on a 10-way particle classification task, trained on 4 million simulated LArIAT events of the type in Fig. 16. One-vs-all ROC curves are shown for a test set of 400 thousand events.

the network to model the uncertainty in its predictions [64]. Preliminary results show that this approach works well (Fig. 18).

While the results discussed in this section are preliminary and proof-of-concept, they suggest that deep learning is a promising approach to these problems. We expect deep learning to play a growing role in the analysis of data from neutrino experiments.

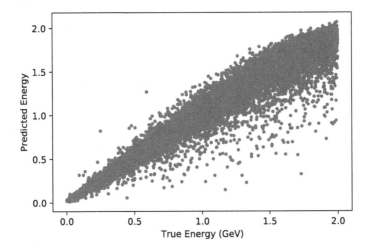

Fig. 18. Predicted energy vs. true energy for a Siamese convolutional neural network trained to predict the energy of simulated electron showers from raw detector data. The mean squared error of these predictions is 0.02 GeV.

5 Dark Matter Physics

It is believed that dark matter is most likely composed of weakly interacting massive particles, or *WIMPs*, that interact with other matter only through gravity and the weak nuclear force. The XENON1T experiment at the Italian Gran Sasso laboratory is attempting to detect these WIMPs directly. Located in an underground facility shielded from cosmic rays by 1,400 meters of rock, the XENON1T consists of a cylindrical tank containing 3.5 tons of ultra-radio-pure liquid Xenon scintillator, achieving record low levels of background radiation [90,91]. Any radiation above the expected background level would indicate the presence of a hypothesized WIMP particle.

When an inbound particle collides with a Xenon atom in the XENON1T detector, recoiling against either the nucleus or an electron, the interaction releases photons that are observed by the photo multiplier tubes (PMTs) located at the top and bottom of the tank (Fig. 19). From the pattern of observed photons, the data is analyzed to determine the type and location of the collision. The standard approach relies on algorithms hand-tuned by physicists on a number of high-level descriptive features, but this approach discards potentially-useful information. Thus, a deep learning approach might be able to improve performance by extracting additional information from the raw data.

We have explored deep learning for performing both event localization and classification in the XENON1T. The raw data from the detector consists of real-valued measurements from the 248 PMTs sampled at 10 ns intervals (Fig. 20). The S2 scintillation response can last up to 15 μs, so the dimensionality of the low-level data is approximately 248 × 1500. From this input, we trained one deep neural network to predict the X,Y location of the collision, and another network

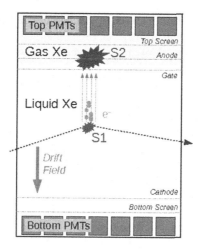

Fig. 19. Diagram of the XENON1T detector cross-section, showing the PMT arrays located at the top and bottom of the cylindrical tank of Xenon. An incoming particle (dotted black line) recoils against a Xe atom, causing an initial scintillation (S1), as well as releasing electrons that rise to the top of the tank and cause a second scintillation (S2). Reprinted from "The XENON1T Dark Matter Experiment," XENON Collaboration, *The European Physical Journal C*, 77(1434-6052), 2017.

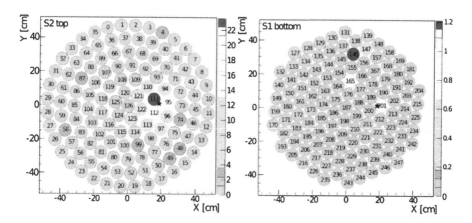

Fig. 20. Example data from a low-energy single-scatter nuclear recoil event recorded by XENON1T during a calibration run. Shown here is the S2 response at the top of the detector (Left), and the S1 response on the bottom of the detector (Right), summed over time. The inferred XY position of the event is indicated by the star marker. In the deep learning approach, the readout from each PMT (circle) at each timestep is treated as an input to a neural network. Adapted from "The XENON1T Dark Matter Experiment," XENON Collaboration, *The European Physical Journal C*, 77(1434-6052), 2017.

to classify examples as containing a single collision or multiple overlapping collisions. The networks were each trained on 180,000 simulated events and tested on a held out set of another 20,000 simulated events. On the event location prediction, the deep learning approach achieves a root mean squared error of 0.68 cm (the cylindrical volume has a radius of 48 cm), and on the classification task the deep learning approach achieves 90% accuracy on a balanced test set.

Our deep learning approach takes advantage of two important invariances in the data. First, the hexagonal arrangement of the PMTs result in a symmetry isomorphic to the dihedral group of order 6 (under the assumption that the PMTs operate at roughly the same efficiency). To account for this, we augmented our training data set by randomly applying rotation and mirror permutations to the input data in each training batch. Second, we account for a limited amount of translational invariance in the time dimension with 1D convolutions and pooling in the network architecture. Note that while the detector data is essentially a video, the PMTs are not arranged in a grid, so standard 2D or 3D convolutions cannot be applied. However, the arrangement of the PMTs does induce local structure, which could be accounted for with graph convolution architectures in future work.

These results are preliminary and have yet to be applied to real data, but they indicate that deep learning is a promising approach to improving the sensitivity at XENON1T as well as future dark matter experiments.

6   Conclusion

This chapter has discussed how deep learning is being applied to diverse data analysis pipelines in antimatter, collider, neutrino, and dark matter experiments. The common theme is that sophisticated data-processing pipelines can be automatically *learned* rather than *engineered*. In many of the studies presented, deep learning is able to extract information from the low-level data that is lost in the traditional approach, leading to improved performance. This could speed up scientific discoveries while simultaneously reducing the labor required to analyze the data.

Deep learning could also potentially influence the design of future experiments. Detectors are designed to satisfy the constraints of the algorithms used to process the data, such as tracking algorithms. These designs are likely suboptimal from an information theory perspective, and deep learning may enable us to design more efficient ones.

In general, we expect deep learning to be especially useful in scenarios where lots of quality training data is available and where other approaches do not capture all the relevant information from the raw data. This makes deep learning a promising approach in a wide range of physics experiments, and in the coming years we expect it to play a growing role in physics data-analysis.

Finally, many of the data sets discussed in this work can be downloaded from the UCI Machine Learning Physics Portal http://mlphysics.ics.uci.edu/.

References

1. Abdesselam, A.: Boosted objects: a probe of beyond the standard model physics. Eur. Phys. J. C **71**, 1661 (2011)
2. Adams, D., Arce, A., Asquith, L., Backovic, M., Barillari, T., et al.: Towards an understanding of the correlations in jet substructure. Eur. Phys. J. C **75**, 409 (2015)
3. Aghion, S.: A moiré deflectometer for antimatter. Nat. Commun. **5**, 4538 (2014)
4. Altheimer, A.: Jet substructure at the Tevatron and LHC: new results, new tools, new benchmarks. J. Phys. **G39**, 063001 (2012)
5. Altheimer, A.: Boosted objects and jet substructure at the LHC. Report of BOOST2012, held at IFIC Valencia, 23rd–27th of July 2012. Eur. Phys. J. **C74**(3), 2792 (2014)
6. Alwall, J.: MadGraph 5: going beyond. JHEP **1106**, 128 (2011)
7. Amole, C.: Description and first application of a new technique to measure the gravitational mass of antihydrogen. Nat. Commun. **4**, 1785 (2013)
8. Amole, C., et al.: The alpha antihydrogen trapping apparatus. Nucl. Instr. Meth. A **735**, 319–340 (2014)
9. Amoretti, M., et al.: The athena antihydrogen apparatus. Nucl. Instr. Meth. A **518**, 679–711 (2004)
10. Andresen, G.B., et al.: Confinement of antihydrogen for 1,000 seconds. Nat. Phys. **7**, 558–564 (2011)
11. Andresen, G., et al.: Trapped antihydrogen. Nature **468**(7324), 673–676 (2010)
12. ATLAS Collaboration: ATLAS experiment at the CERN Large Hadron Collider. JINST 3, S08003 (2008)
13. ATLAS Collaboration: Luminosity determination in PP collisions at $\sqrt{s} = 7$ TeV using the ATLAS detector at the LHC. Eur. Phys.J **C73**, 2518 (2013)
14. Aurisano, A., et al.: A convolutional neural network neutrino event classifier. J. Instrum. **11**(09), P09001 (2016). http://stacks.iop.org/1748-0221/11/i=09/a=P09001
15. Bahr, M.: Herwig++ physics and manual. Eur. Phys. J. C **58**, 639–707 (2008)
16. Baldi, P.: Deep learning in biomedical data science. Ann. Rev. Biomed. Data Sci. **1**, 181 (2018)
17. Baldi, P.: The inner and outer approaches for the design of recursive neural networks architectures. Data Mining Knowl. Disc. **32**, 218–230 (2017). https://doi.org/10.1007/s10618-017-0531-0
18. Baldi, P., Brunak, S.: Bioinformatics: The Machine Learning Approach, second edition edn. MIT Press, Cambridge (2001)
19. Baldi, P., Brunak, S., Frasconi, P., Pollastri, G., Soda, G.: Exploiting the past and the future in protein secondary structure prediction. Bioinformatics **15**, 937–946 (1999)
20. Baldi, P., Chauvin, Y.: Neural networks for fingerprint recognition. Neural Comput. **5**(3), 402–418 (1993)
21. Baldi, P., Pollastri, G.: The principled design of large-scale recursive neural network architectures-DAG-RNNs and the protein structure prediction problem. J. Mach. Learn. Res. **4**, 575–602 (2003)
22. Baldi, P., Sadowski, P.: The dropout learning algorithm. Artif. Intell. **210C**, 78–122 (2014)
23. Baldi, P., Sadowski, P., Whiteson, D.: Searching for exotic particles in high-energy physics with deep learning. Nat. Commun. **5**, Article no. 4308 (2014)

24. Baldi, P., Sadowski, P., Whiteson, D.: Enhanced higgs boson to τ τ search with deep learning. Phys. Rev. Lett. **114**, 111801 (2015)

25. Baldi, P., Bauer, K., Eng, C., Sadowski, P., Whiteson, D.: Jet substructure classification in high-energy physics with deep neural networks. Phys. Rev. D **93**, 094034 (2016). https://doi.org/10.1103/PhysRevD.93.094034

26. Baldi, P., Cranmer, K., Faucett, T., Sadowski, P., Whiteson, D.: Parameterized neural networks for high-energy physics. Eur. Phys. J. C **76**(5), 235 (2016). https://doi.org/10.1140/epjc/s10052-016-4099-4

27. Beringer, J.: Review of particle physics. Phys. Rev. D **86**, 010001 (2012)

28. Cazaux, S., Lerch, T., Aune, S.: Detecteur courbe de particules gazeux, patent App. EP20,130,188,550, April. https://www.google.ch/patents/EP2720252A3?cl=fr

29. Cheng, J., Randall, A.Z., Sweredoski, M., Baldi, P.: Scratch: a protein structure and structural feature prediction server. Nucleic Acids Res. **33**, W72–W76 (2005)

30. Chollet, F.: Keras. GitHub (2015). https://github.com/fchollet/keras

31. CMS Collaboration: Search for light vector resonances decaying to quarks at 13 TeV. CMS-PAS-EXO-16-030 (2016)

32. Corradini, M., et al.: Experimental apparatus for annihilation cross-section measurements of low energy antiprotons. Nucl. Instr. Meth. A **711**, 12–20 (2013)

33. Corradini, M.: Scintillating bar detector for antiproton annihilations measurements. Hyperfine Interact. **233**, 53–58 (2015)

34. Cranmer, K., Pavez, J., Louppe, G.: Approximating likelihood ratios with calibrated discriminative classifiers (2015)

35. Krohn, D., Thaler, J., Wang, L.T.: Jet trimming. JHEP **1002**, 084 (2010)

36. Dasgupta, M., Fregoso, A., Marzani, S., Powling, A.: Jet substructure with analytical methods. Eur. Phys. J. C **73**(11), 2623 (2013)

37. Dasgupta, M., Powling, A., Siodmok, A.: On jet substructure methods for signal jets. JHEP **08**, 079 (2015)

38. Dolen, J., Harris, P., Marzani, S., Rappoccio, S., Tran, N.: Thinking outside the ROCs: designing decorrelated taggers (DDT) for jet substructure. JHEP **05**, 156 (2016)

39. Duvenaud, D., et al.: Convolutional networks on graphs for learning molecular fingerprints. In: Neural Information Processing Systems (2015)

40. Edwards, H., Storkey, A.J.: Censoring representations with an adversary (2016). http://arxiv.org/abs/1511.05897

41. Feng, J.L.: Dark matter candidates from particle physics and methods of detection. Ann. Rev. Astron. Astrophys. **48**, 495–545 (2010)

42. Gabrielse, G., et al.: Trapped antihydrogen in its ground state. Phys. Rev. Lett. **108**, 113002 (2012). https://doi.org/10.1103/PhysRevLett.108.113002

43. Ganin, Y., et al.: Domain-adversarial training of neural networks. J. Mach. Learn. Res. **17**(1), 2096–2030 (2016). http://dl.acm.org/citation.cfm?id=2946645.2946704

44. Giomataris, Y., Rebourgeard, P., Robert, J.P., Charpak, G.: Micromegas: a high-granularity position-sensitive gaseous detector for high particle-flux environments. Nucl. Instr. Meth. A **376**, 29 (1996)

45. Goodfellow, I., Bengio, Y., Courville, A.: Deep Learning. MIT Press, Cambridge (2016). http://www.deeplearningbook.org

46. Guest, D., Collado, J., Baldi, P., Hsu, S.C., Urban, G., Whiteson, D.: Jet flavor classification in high-energy physics with deep neural networks. Phys. Rev. D **94**, 112002 (2016). https://doi.org/10.1103/PhysRevD.94.112002

47. He, K., Zhang, X., Ren, S., Sun, J.: Delving deep into rectifiers: surpassing human-level performance on imagenet classification. In: The IEEE International Conference on Computer Vision (ICCV), December 2015

48. Hertel, L., Li, L., Baldi, P., Bian, J.: Convolutional neural networks for electron neutrino and electron shower energy reconstruction in the nova detectors. In: Deep Learning for Physical Sciences Workshop at Neural Information Processing Systems (2017)

49. Hori, M., Yamashita, K., Hayano, R., Yamazaki, T.: Analog cherenkov detectors used in laser spectroscopy experiments on antiprotonic helium. Nucl. Instr. Meth. A **496**, 102–122 (2003)

50. Kaplan, D.E., Rehermann, K., Schwartz, M.D., Tweedie, B.: Top tagging: a method for identifying boosted hadronically decaying top quarks. Phys. Rev. Lett. **101**, 142001 (2008)

51. Kayala, M., Azencott, C., Chen, J., Baldi, P.: Learning to predict chemical reactions. J. Chem. Inf. Model. **51**(9), 2209–2222 (2011)

52. Kayala, M., Baldi, P.: Reactionpredictor: prediction of complex chemical reactions at the mechanistic level using machine learning. J. Chem. Inf. Model. **52**(10), 2526–2540 (2012)

53. Kingma, D.P., Ba, J.: Adam: a method for stochastic optimization. In: Proceedings of the 3rd International Conference on Learning Representations (ICLR) (2014)

54. Kuroda, N.: A source of antihydrogen for in-flight hyperfine spectroscopy. Nat. Commun. **5**, 3089 (2014)

55. Larkoski, A.J., Marzani, S., Soyez, G., Thaler, J.: Soft Drop. JHEP **1405**, 146 (2014)

56. Larkoski, A.J., Moult, I., Neill, D.: Power counting to better jet observables. JHEP **12**, 009 (2014)

57. Larkoski, A.J., Salam, G.P., Thaler, J.: Energy correlation functions for jet substructure. JHEP **1306**, 108 (2013)

58. Louppe, G., Cho, K., Becot, C., Cranmer, K.: QCD-aware recursive neural networks for jet physics (2017)

59. Louppe, G., Kagan, M., Cranmer, K.: Learning to pivot with adversarial networks (2016)

60. Lusci, A., Pollastri, G., Baldi, P.: Deep architectures and deep learning in chemoinformatics: the prediction of aqueous solubility for drug-like molecules. J. Chem. Inf. Model. **53**(7), 1563–1575 (2013)

61. Dasgupta, M., Fregoso, A., Marzani, S., Salam, G.P.: Towards an understanding of jet substructure. JHEP **9**, 029 (2013)

62. Magnan, C.N., Baldi, P.: SSpro/ACCpro 5: almost perfect prediction of protein secondary structure and relative solvent accessibility using profiles, machine learning, and structural similarity. Bioinformatics **30**(18), 2592–2597 (2014)

63. Nair, V., Hinton, G.E.: Rectified linear units improve restricted Boltzmann machines. In: Furnkranz, J., Joachims, T. (eds.) Proceedings of the 27th International Conference on Machine Learning (ICML 2010), pp. 807–814. Omnipress (2010). http://www.icml2010.org/papers/432.pdf

64. Nix, D.A., Weigend, A.S.: Estimating the mean and variance of the target probability distribution. In: 1994 IEEE International Conference on Neural Networks, IEEE World Congress on Computational Intelligence, vol. 1, pp. 55–60, June 1994

65. de Oliveira, L., Kagan, M., Mackey, L., Nachman, B., Schwartzman, A.: Jet-images – deep learning edition. J. High Energy Phys. **2016**(7), 69 (2016). https://doi.org/10.1007/JHEP07(2016)069

66. Ovyn, S., Rouby, X., Lemaitre, V.: DELPHES, a framework for fast simulation of a generic collider experiment (2009)
67. Planck Collaboration: Planck 2013 results. XVI, Cosmological parameters (2013)
68. Plehn, T., Spannowsky, M., Takeuchi, M., Zerwas, D.: Stop reconstruction with tagged tops. JHEP **1010**, 078 (2010)
69. Pollastri, G., Przybylski, D., Rost, B., Baldi, P.: Improving the prediction of protein secondary strucure in three and eight classes using recurrent neural networks and profiles. Proteins **47**, 228–235 (2001)
70. Pérez, P.: The GBAR antimatter gravity experiment. Hyperfine Interact. **233**, 21–27 (2015)
71. Racah, E., et al.: Revealing fundamental physics from the daya bay neutrino experiment using deep neural networks. In: 2016 15th IEEE International Conference on Machine Learning and Applications (ICMLA), pp. 892–897, December 2016
72. Radics, B., Murtagh, D.J., Yamazaki, Y., Robicheaux, F.: Scaling behavior of the ground-state antihydrogen yield as a function of positron density and temperature from classical-trajectory Monte Carlo simulations. Phys. Rev. A **90**(3), 032704 (2014). https://doi.org/10.1103/PhysRevA.90.032704
73. Radics, B., et al.: The ASACUSA micromegas tracker: a cylindrical, bulk micromegas detector for antimatter research. Rev. Sci. Instrum. **86**, 083304 (2015)
74. Ellis, S.D., Vermilion, C.K., Walsh, J.R.: Recombination algorithms and jet substructure: pruning as a tool for heavy particle searches. Phys. Rev. **D81**, 094023 (2010)
75. Sadowski, P., Collado, J., Whiteson, D., Baldi, P.: Deep learning, dark knowledge, and dark matter. J. Mach. Learn. Res. **42**, 81–97 (2015). Workshop and Conference Proceedings
76. Sadowski, P., Radics, B., Ananya, Yamazaki, Y., Baldi, P.: Efficient antihydrogen detection in antimatter physics by deep learning. J. Phys. Commun. **1**(2), 025001 (2017). http://stacks.iop.org/2399-6528/1/i=2/a=025001
77. Schmidhuber, J.: Learning factorial codes by predictability minimization. Neural Comput. **4**, 863–879 (1991)
78. Shimmin, C., et al.: Decorrelated jet substructure tagging using adversarial neural networks. Phys. Rev. D **96**, 074034 (2017). arXiv: 1703.03507
79. Shimmin, C., Whiteson, D.: Boosting low-mass hadronic resonances. Phys. Rev. D **94**, 055001 (2016). https://doi.org/10.1103/PhysRevD.94.055001
80. Sjostrand, T., et al.: PYTHIA 6.4 physics and manual. JHEP **05**, 026 (2006)
81. Snoek, J., Larochelle, H., Adams, R.P.: Practical Bayesian optimization of machine learning algorithms. In: Pereira, F., Burges, C.J.C., Bottou, L., Weinberger, K.Q. (eds.) Advances in Neural Information Processing Systems, vol. 25, pp. 2951–2959. Curran Associates, Inc. (2012)
82. Srivastava, N., Hinton, G., Krizhevsky, A., Sutskever, I., Salakhutdinov, R.: Dropout: a simple way to prevent neural networks from overfitting. J. Mach. Learn. Res. **15**, 1929–1958 (2014). http://jmlr.org/papers/v15/srivastava14a.html
83. Storey, J.: Particle tracking at 4k: the fast annihilation cryogenic tracking (fact) detector for the aegis antimatter gravity experiment. Nucl. Instr. Meth. A **732**, 437–441 (2013)
84. Strandlie, A., Frühwirth, R.: Track and vertex reconstruction: from classical to adaptive methods. Rev. Mod. Phys. **82**, 1419 (2010)
85. Szegedy, C., et al.: Going deeper with convolutions. In: 2015 IEEE Conference on Computer Vision and Pattern Recognition (CVPR), pp. 1–9, June 2015
86. Thaler, J., Van Tilburg, K.: Identifying boosted objects with N-subjettiness. JHEP **1103**, 015 (2011)

87. Thaler, J., Van Tilburg, K.: Maximizing boosted top identification by minimizing n-subjettiness. JHEP **02**, 093 (2012)
88. Theano Development Team: Theano: a Python framework for fast computation of mathematical expressions. arXiv e-prints, May 2016. http://arxiv.org/abs/1605.02688
89. Tishby, N., Zaslavsky, N.: Deep learning and the information bottleneck principle. In: 2015 IEEE Information Theory Workshop (ITW), pp. 1–5, April 2015
90. XENON Collaboration: First dark matter search results from the XENON1T experiment. Phys. Rev. Lett. **119**(18), 181301 (2017)
91. XENON Collaboration: The XENON1T dark matter experiment. Eur. Phys. J. C **77**(12), 881, December 2017. https://doi.org/10.1140/epjc/s10052-017-5326-3
92. Zhang, Z., Oelert, W., Grzonka, D., Sefzick, T.: The antiproton annihilation detector system of the atrap experiment. Chin. Sci. Bull. **54**, 189–195 (2009)

From Reinforcement Learning to Deep Reinforcement Learning: An Overview

Forest Agostinelli, Guillaume Hocquet, Sameer Singh, and Pierre Baldi[✉]

University of California - Irvine, Irvine, CA 92697, USA
{fagostin,sameer,pfbaldi}@uci.edu

Abstract. This article provides a brief overview of reinforcement learning, from its origins to current research trends, including deep reinforcement learning, with an emphasis on first principles.

Keywords: Machine learning · Reinforcement learning
Deep learning · Deep reinforcement learning

1 Introduction

This article provides a concise overview of reinforcement learning, from its origins to deep reinforcement learning. Thousands of articles have been written on reinforcement learning and we could not cite, let alone survey, all of them. Rather we have tried to focus here on first principles and algorithmic aspects, trying to organize a body of known algorithms in a logical way. A fairly comprehensive introduction to reinforcement learning is provided by [113]. Earlier surveys of the literature can be found in [33, 46, 51].

1.1 Brief History

The concept of reinforcement learning has emerged historically from the combination of two currents of research: (1) the study of the behavior of animals in response to stimuli; and (2) the development of efficient approaches to problems of optimal control.

In behavioral psychology, the term *reinforcement* was introduced by Pavlov in the early 1900s, while investigating the psychology and psychopathology of animals in the context of conditioning stimuli and conditioned responses [47]. One of his experiments consisted in ringing a bell just before giving food to a dog; after a few repetitions, Pavlov noticed that the sound of the bell alone made the dog salivate. In classical conditioning terminology, the bell is the previously neutral stimulus, which becomes a *conditioned stimulus* after becoming associated with the *unconditioned stimulus* (the food). The conditioned stimulus eventually comes to trigger a conditioned response (salivation). Conditioning

---

G. Hocquet—Work performed while visiting the University of California, Irvine.

© Springer Nature Switzerland AG 2018
L. Rozonoer et al. (Eds.): Braverman Readings in Machine Learning, LNAI 11100, pp. 298–328, 2018.
https://doi.org/10.1007/978-3-319-99492-5_13

experiments led to Thorndike's Law of Effect [118] in 1911, which states that: "Of several responses made to the same situation, those which are accompanied or closely followed by satisfaction to the animal will, other things being equal, be more firmly connected with the situation, so that, when it recurs, they will be more likely to recur".

This formed the basis of *operant conditioning* (or instrumental conditioning) in which: (1) the strength of a behavior is modified by the behavior's consequences, such as reward or punishment; and (2) the behavior is controlled by antecedents called "discriminative stimuli" which come to emit those responses. Operant conditioning was studied in the 1930s by Skinner, with his experiments on the behavior of rats exposed to different types of reinforcers (stimuli).

A few years later, in the Organization of Behavior [39] (1949), Hebb proposed one of the first theories about the neural basis of learning using the notions of cell assemblies and "Hebbian" learning, encapsulated in the sentence "When an axon of cell A is near enough to excite cell B and repeatedly or persistently takes part in firing it, some growth process or metabolic change takes place in one or both cells such that A's efficiency, as one of the cells firing B, is increased." These are some of the biological underpinnings and sources of inspiration for many subsequent developments in reinforcement learning and other forms of learning, such as supervised learning.

In 1954, in the context of optimal control theory, Bellman introduced dynamic programming [9], and the concept of value functions. These functions are computed using a recursive relationship, now called the Bellman equation. Bellman's work was within the framework of Markov Decision Process (MDPs), which were studied in detail by [44]. One of Howard's students, Drake, proposed an extension with partial observability: the POMDP models [27].

In 1961, [70] discussed several issues in the nascent field of reinforcement learning, in particular the problem of *credit assignment*, which is one of the core problems in the field. Around the same period, reinforcement learning ideas began to be applied to games. For instance, Samuel developed his checkers player [93] using Temporal Differences method. Other experiments were carried by Michie, including the development of the MENACE system to learn how to play Noughts and Crosses [67,68], and the BOXES controller [69] which has been applied to pole balancing problems.

In the 1970s, Tsetlin made several contributions within the area of Automata, in particular in relation to the n-armed bandit problem, i.e. how to select which levers to pull in order to maximize the gain in a game comprising n slot machines without initial knowledge. This problem can be viewed as a special case of a reinforcement learning problem with a single state. In 1975, Holland developed genetic algorithms [42], paving the way for reinforcement learning based on evolutionary algorithms.

In 1988, [126] presented the REINFORCE algorithms, which led to a variety of policy gradient methods. The same year, Sutton introduced TD(λ) [111]. In 1989, Watkins proposed the Q-Learning algorithm [123].

1.2   Applications

Reinforcement learning methods have been effective in a variety of areas, in particular in games. Success stories include the application of reinforcement learning to stochastic games (Backgammon [117]), learning by self-play (Chess [56]), learning from games played by experts (Go [100]), and learning without using any hand-crafted features (Atari games [72]).

When the objective is defined by a control task, reinforcement learning has been used to perform low-speed sustained inverted hovering with an helicopter [77], balance a pendulum without a priori knowledge of its dynamics [3], or balance and ride a bicycle [88]. Reinforcement learning has also found plenty of applications in robotics [52], including recent success in manipulation [59] and locomotion [97]. Other notable successes include solutions to the problems of elevator dispatching [19], dynamic communication allocation for cellular radio channels [104], job-shop scheduling [129], and traveling salesman optimization [26]. Other potential industrial applications have included packet routing [12], financial trading [73], and dialog systems [58].

1.3   General Idea Behind Reinforcement Learning

Reinforcement learning is used to compute a behavior strategy, a *policy*, that maximizes a satisfaction criteria, a long term sum of *rewards*, by interacting through *trials and errors* with a given environment (Fig. 1).

Fig. 1. The agent-environment interaction protocol

A reinforcement learning problem consists of a decision-maker, called the *agent*, operating in an *environment* modeled by *states* $s_t \in S$. The agent is capable of taking certain *actions* $a_t \in \mathcal{A}(s_t)$, as a function of the current state s_t. After choosing an action at time t, the agent receives a scalar *reward* $r_{t+1} \in \mathbb{R}$ and finds itself in a new state s_{t+1} that depends on the current state and the chosen action.

At each time step, the agent follows a strategy, called the *policy* π_t, which is a mapping from states to the probability of selecting each possible action: $\pi(s, a)$ denotes the probability that $a = a_t$ if $s = s_t$.

The objective of reinforcement learning is to use the interactions of the agent with its environment to derive (or approximate) an optimal policy to maximize the total amount of reward received by the agent over the long run.

Remark 1. *This definition is quite general: time can be continuous or discrete, with finite or infinite horizon; the state transitions can be stochastic or deterministic, the rewards can be stationary or not, and deterministic or sampled from a given distribution. In some cases (with an unknown model), the agent may start with partial or no knowledge about its environment.*

1.4   Definitions

Return. To maximize the long-term cumulative reward after the current time t, in the case of a finite time horizon that ends at time T, the *return* R_t is equal to:

$$R_t = r_{t+1} + r_{t+2} + r_{t+3} + \ldots + r_T = \sum_{k=t+1}^{T} r_k$$

In the case of an infinite time horizon, it is customary instead to use a *discounted* return:

$$R_t = r_{t+1} + \gamma r_{t+2} + \gamma^2 r_{t+3} + \ldots = \sum_{k=0}^{\infty} \gamma^k r_{t+k+1},$$

which will converge if we assume the rewards are bounded and $\gamma < 1$. Here $\gamma \in [0, 1]$ is a constant, called the *discount factor*. In what follows, in general we will use this discounted definition for the return.

Value Functions. In order to find an optimal policy, some algorithms are based on *value functions*, $V(s)$, that represent how beneficial it is for the agent to reach a given state s. Such a function provides, for each state, a numerical estimate of the potential future reward obtainable from this state, and thus depends on the actual policy π followed by the agent:

$$V^\pi(s) = \mathbb{E}_\pi \left[R_t \mid s_t = s \right] = \mathbb{E}_\pi \left[\sum_{k=0}^{\infty} \gamma^k r_{t+k+1} \,\middle|\, s_t = s \right]$$

where $\mathbb{E}_\pi [.]$ denotes the expected value given that the agent follows policy π, and t is any time step.

Remark 2. *The existence and uniqueness of V^π are guaranteed if $\gamma < 1$ or if T is guaranteed to be finite from all states under the policy π [113].*

Action-Value Functions. Similarly, we define the value of taking action a in state s under a policy π as the *action-value function* Q:

$$Q^\pi(s, a) = \mathbb{E}_\pi \left[R_t \mid s_t = s, a_t = a \right]$$
$$= \mathbb{E}_\pi \left[\sum_{k=0}^{\infty} \gamma^k r_{t+k+1} \,\middle|\, s_t = s, a_t = a \right]$$

Optimal Policy. An *optimal policy* π^* is a policy that achieves the greatest expected reward over the long run. Formally, a policy π is defined to be better than or equal to a policy π' if its expected return is greater than or equal to that of π' for all states. Thus:

$$\pi^* = \operatorname*{argmax}_{\pi} V^{\pi}(s) \quad \forall s \in \mathcal{S}$$

Remark 3. *There is always at least one policy that is better than or equal to all other policies. There may be more than one, but we denote all of them by π^* because they share the same value function and action-value function, noted:*

$$V^*(s) = \max_{\pi} V^{\pi}(s) \quad \forall s \in \mathcal{S}$$
$$Q^*(s,a) = \max_{\pi} Q^{\pi}(s,a) \quad \forall s \in \mathcal{S}, \quad \forall a \in \mathcal{A}(s)$$

1.5   Markov Decision Processes (MDPs)

A *Markov Decision Process* is a particular instance of reinforcement learning where the set of states is finite, the sets of actions of each state are finite, and the environment satisfies the following Markov property:

$$Pr(s_{t+1} = s'|s_0, a_0, ...s_t, a_t) = Pr(s_{t+1} = s'|s_t, a_t)$$

In other words, the probability of reaching state s' from state s by action a is independent of the other actions or states in the past (before time t). Hence, we can represent a sequence of actions, states, rewards sampled from an MDP by a decision network (see Fig. 2).

Most reinforcement learning research is based on the formalism of MDPs. MDPs provide a simple framework in which to study basic algorithms and their properties. We will continue to use this formalism in Sect. 2. Then, we will emphasize its drawbacks in Sect. 3 and present potential improvements in Sect. 4.

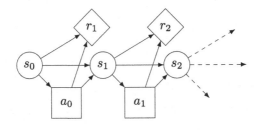

Fig. 2. Decision network representing an episode sampled from an MDP

1.6   A Visualization of Reinforcement Learning Algorithms

An overview of the algorithms that will be presented in this chapter can be found in Fig. 3. While this does not cover all reinforcement learning algorithms, we present it as a tool for the reader to get an overview of the reinforcement learning landscape. Each algorithm is color-coded according to whether it is *model based* or *model free*. Model based methods, such as those presented in Sects. 2.2 and 2.5, require a model of the environment while model free methods, such as those presented in Sects. 2.3 and 2.4, do not require a model of the environment. The functions (value function, action-value function, and/or policy function) that each algorithm uses are displayed beneath the algorithm. As shown in Sect. 5, these functions can take the form of deep neural networks.

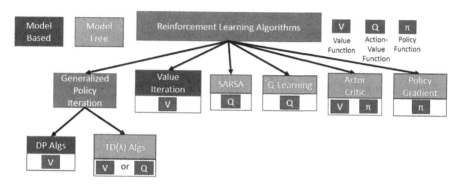

Fig. 3. An overview of the reinforcement learning algorithms that will be presented in this paper. The functions associated with each reinforcement learning algorithm can take the form of a deep neural network.

2   Main Algorithmic Approaches

Given a reinforcement learning problem, we now are going to present different approaches to computing the optimal policy. There are two main approaches: one based on searching in the space of value functions, and one based on searching in the space of policies. Value function space search methods attempt to compute the optimal value function V^* and deduce at the end the optimal policy π^* from V^*. These methods include linear programming, dynamic programming, Monte-Carlo methods, and temporal difference methods. Policy space search methods, on the other hand, maintain explicit representations of policies and update them over the time in order to compute the optimal policy π^*. Such methods typically include evolutionary and policy gradient algorithms. We provide a brief overview of these methods in the following sections.

2.1   Linear Programming

In order to cast the goal of finding the optimal value function as a linear programming problem [89], we treat the value function V as a cost function and then try to minimize the cost from each starting state s. In order to minimize a cost, we need to invert the sign of the rewards. We will note the cost function $g_\pi(s_t) = -r_{t+1}$. Thus here we want to minimize:

$$J^\pi(s) = \mathbb{E}_\pi \left[\sum_{k=0}^\infty \gamma^k g_\pi(s_k) \middle| s_0 = s \right]$$

In order to perform this minimization, we define the optimal Bellman operator T:

$$(TJ)(s) = \min_\pi (g_\pi(s) + \gamma P_\pi(s)J)$$

where J is a vector of states, P_π is the transition matrix with the (s, s') entry representing the probability of reaching s' from s under policy π, and the minimization is carried out component-wise.

The solution that minimizes the cost should verify the Bellman equation:

$$J(s) = (TJ)(s)$$

It can be found by solving the linear programming optimization (using, for example, the simplex algorithm):

$$\min_J \quad \mu^T J$$
$$s.t. \, TJ \geq J$$

where μ is a vector of positive weights, known as the *state-relevance weights*.

From a theoretical perspective, linear programming provides the only known algorithm that can solve MDPs in polynomial time, although in general linear programming approaches to reinforcement learning problems do not fare well in practice. In particular, the main problem for linear programming approaches is that the time and space complexity can be extremely high.

2.2   Dynamic Programming

Dynamic programming algorithms are the simplest way to tackle a reinforcement learning problem, however, this method requires perfect knowledge of the model and is limited by its computational cost. The idea behind the dynamic programming formulation of reinforcement learning is to choose a policy π, estimate its value function V^π (Algorithm 1), deduce a new policy π' from V^π (Algorithm 2), and iterate this process until a satisfying policy is found (Algorithm 3). This process is known as *policy iteration*. Since each step strictly improves the policy, the algorithm is guaranteed to converge to the optimal policy. For computational convenience, one can decide to stop the policy evaluation step when the change in the value function is small between two iterations, as implemented below with the threshold θ:

Algorithm 1. Policy Evaluation

Data: π, the policy to be evaluated
Result: $V \approx V^{\pi}$, an approximation of the value function of π
repeat
   | $\Delta \leftarrow 0$
   | **for** $s \in \mathcal{S}$ **do**
   | | $v \leftarrow V(s)$
   | | $V(s) \leftarrow \sum_a \pi(s,a) \sum_{s'} P^a(s,s')(R^a(s,s') + \gamma V(s'))$
   | | $\Delta \leftarrow \max(\Delta, |v - V(s)|)$
until $\Delta < \theta$;

Remark 4. *At each step* k, *the value function* V_{k+1} *can be computed from the previous one* V_k *in two ways [113]:*

- *Full Backup: using two distinct arrays to store the two functions* V_k *and* V_{k+1}.
- *In Place: using only one array, and overwriting* V_k *when computing* V_{k+1} *for each state.*

The second approach is usually faster.

Algorithm 2. Policy Improvement

Data: π, the policy to be updated
         V, the value function
Result: π, the updated policy
for $s \in \mathcal{S}$ **do**
   | $\pi(s) \leftarrow \underset{a}{\mathrm{argmax}} \sum_{s'} P^a(s,s')(R^a(s,s') + \gamma V(s'))$

Algorithm 3. Policy Iteration

Result: π^*, the optimal policy
Initialization: π chosen arbitrarily
repeat
   | $\pi_0 \leftarrow \pi$
   | $V = \mathrm{Policy\_evaluation}(\pi)$
   | $\pi = \mathrm{Policy\_improvement}(\pi, V)$
until $\pi_0 = \pi$;

One drawback of policy iteration is the policy evaluation step; which requires multiple iterations over every state. Another way to proceed is to combine policy evaluation and policy improvement in the same loop (Algorithm 4). This process is called value iteration. Value iteration is not always better than policy iteration, the efficiency depends on the nature of the problem and the parameters chosen. These differences are discussed in [85].

Algorithm 4. Value Iteration

Result: π^*, the optimal policy
Initialization: V chosen arbitrarily
repeat
  $\Delta \leftarrow 0$
  for $s \in \mathcal{S}$ **do**
    $v \leftarrow V(s)$
    $V(s) \leftarrow \max_a \sum_{s'} P^a(s,s')(R^a(s,s') + \gamma V(s'))$
    $\Delta \leftarrow \max(\Delta, |v - V(s)|)$
until $\Delta < \theta$;
for $s \in \mathcal{S}$ **do**
  $\pi(s) \leftarrow \underset{a}{\mathrm{argmax}} \sum_{s'} P^a(s,s')(R^a(s,s') + \gamma V(s'))$

2.3 Monte-Carlo Methods

The following algorithms correspond to online learning methods that do not require any knowledge of the environment. To estimate the value function V^π of a policy π, we must generate a sequence of actions and states with π, called an *episode*, compute the total reward at the end of this sequence, then update the estimate of V^π, V, for each state of the episode according to its contribution to the final reward, and repeat this process. One way to achieve this is to compute the average of the expected return from each state (Algorithm 5).

When one has a model of the environment, state values alone are sufficient to determine a policy. At any state s, the action taken is:

$$\pi(s) \leftarrow \underset{a}{\mathrm{argmax}} \sum_{s'} P^a(s,s')(R^a(s,s') + \gamma V(s'))$$

However, without a model, we will not have access to the state transition probabilities and/or the expected reward; therefore, we will not be able to find action a that maximizes the aforementioned expression. Therefore, action-value

---

Algorithm 5. MC Policy Evaluation

Data: π, the policy to be evaluated
Result: $V \approx V^\pi$, an approximation of the value function of π
Initialization: V chosen arbitrarily
       $Returns(s) \leftarrow [\,], \forall s \in S$
repeat
  $episode = generate\_episode(\pi)$
  for $s \in episode$ **do**
    $R \leftarrow$ Return following first occurrence of s
    $Returns(s).append(R)$
    $V(s) \leftarrow average(Returns(s))$
until;

Algorithm 6. MC Exploring Starts

Result: π^*, the optimal policy
Initialization: Q chosen arbitrarily
$\qquad\qquad\qquad$ π chosen arbitrarily
$\qquad\qquad\qquad$ $Returns(s, a) \leftarrow [\,], \forall s \in S$, $\forall a \in \mathcal{A}(s)$
repeat
$\quad$ episode = generate_episode_exploring_starts(π)
$\quad$ **for** $s, a \in episode$ **do**
$\quad\quad$ $R \leftarrow$ Return following first occurrence of s, a
$\quad\quad$ $Returns(s, a).append(R)$
$\quad\quad$ $Q(s, a) \leftarrow$ average($Returns(s, a)$)
$\quad$ **for** $s \in episode$ **do**
$\quad\quad$ $\pi(s) \leftarrow \underset{a}{\operatorname{argmax}} \, Q(s, a)$
until;

---

functions are necessary to find the optimal policy. If we are following a deterministic policy, many state-action pairs may never be visited. We present two different methods for addressing this problem: *exploring starts* [113] and *stochastic policies*. Similar to value iteration, the methods we present for *exploring starts* and *stochastic policies* do not wait to complete policy evaluation before doing policy improvement. Instead, policy evaluation and policy improvement are done every episode.

Under the *exploring starts* assumption, each episode starts at a state-action pair and every state-action pair has a nonzero chance of being the starting pair. This algorithm is shown in Algorithm 6.

The exploring starts assumption may often be infeasible in practice. To explore as many state-action pairs as possible, one must consider policies that are stochastic. We distinguish between two different types of policies: The policy that is used to generate episodes (the behavior policy) and the policy that is being evaluated and improved (the estimation policy). The behavior policy must be stochastic in order to ensure new state-action pairs are explored. There are two main types of methods that utilize *stochastic policies*: *on-policy* methods and *off-policy* methods. For *on-policy* methods, the behavior policy and the estimation policy are the same; therefore, the policy that is being evaluated and improved must also be stochastic. Algorithm 7 shows an *on-policy* MC algorithm that utilizes an *ϵ-greedy* policy: with probability ϵ it chooses an action at random, otherwise, it chooses the greedy action.

On the other hand, *off-policy* methods can have a behavior policy that is separate from the estimation policy. The behavior policy should still be stochastic and must have a nonzero probability of selecting all actions that the estimation policy might select, however, the estimation policy can be greedy and always select the action a at state s that maximizes $Q(s, a)$. The downside of *off-policy* methods is that policy improvement is slower because it can only learn from states where the behavior policy and the estimation policy take the same

---

Algorithm 7. MC On-Policy Control

Result: π^*, the optimal policy
Initialization: Q chosen arbitrarily
$\qquad\qquad\quad$ π chosen arbitrarily
$\qquad\qquad\quad$ $Returns(s,a) \leftarrow [], \forall s \in S, \ \forall a \in \mathcal{A}(s)$
repeat
$\quad$ episode = generate_episode(π)
$\quad$ **for** $s, a \in episode$ **do**
$\qquad$ $R \leftarrow$ Return following first occurrence of s, a
$\qquad$ $Returns(s,a).append(R)$
$\qquad$ $Q(s,a) \leftarrow$ average($Returns(s,a)$)
$\quad$ **for** $s \in episode$ **do**
$\qquad$ $a^* \leftarrow \underset{a}{\mathrm{argmax}}\, Q(s,a)$
$\qquad$ **for** $a \in \mathcal{A}(s)$ **do**
$\qquad\quad$ $\pi(s,a) \leftarrow \begin{cases} 1 - \epsilon + \epsilon/|\mathcal{A}(s)| & \text{if } a = a^* \\ \epsilon/|\mathcal{A}(s)| & \text{if } a \neq a^* \end{cases}$
until;

---

actions. Differences between *on-policy* and *off-policy* methods are discussed further in [113]. A well-known off-policy algorithm, Q-learning, will be presented in Sect. 2.4.

Remark 5. *The MC methods presented in this paper are first-visit MC methods. The first-visit method averages the return following the first visit to a state s in an episode, in the case of MC policy evaluation, or following the first occurrence of the state-action pair s, a, in the case of MC exploring starts and MC on-policy control. There are also every-visit methods that use the return from every occurrence of s or s, a. However, these methods are less straightforward because of the introduction of bias [106].*

2.4 Temporal Difference Methods

TD(0). Whereas the Monte-Carlo algorithms are constrained to wait for the end of an episode to update the value function, the TD(0) algorithm (Algorithm 8) is able to compute an update after every step:

$$V(s_t) \leftarrow V(s_t) + \alpha \left[r_{t+1} + \gamma V(s_{t+1}) - V(s_t) \right]$$

When working with action-value functions, a well-known *off-policy* algorithm known as Q-learning (Algorithm 9) approximates Q^* regardless of the current policy.

$$Q(s_t, a_t) \leftarrow Q(s_t, a_t) + \alpha \left[r_{t+1} + \gamma \max_{a'} Q(s_{t+1}, a') - Q(s_t, a_t) \right]$$

Algorithm 8. TD(0)

Data: π, the policy to be evaluated
Result: $V \approx V^\pi$, an approximation of the value function of π
Initialization: V chosen arbitrarily
repeat
    $s \leftarrow get\_initial\_state()$
    while s *not terminal* **do**
        $a \leftarrow get\_action(\pi, s)$
        $s', r \leftarrow get\_next\_state(s, a)$
        $V(s) \leftarrow V(s) + \alpha(r + \gamma V(s') - V(s))$
        $s \leftarrow s'$
until;

Algorithm 9. Q-Learning

Result: π^*, the optimal policy
Initialization: Q chosen arbitrarily
repeat
    $s \leftarrow get\_initial\_state()$
    while s *not terminal* **do**
        $a \leftarrow get\_action(Q, s)$
        $s', r \leftarrow get\_next\_state(s, a)$
        $Q(s, a) \leftarrow Q(s, a) + \alpha(r + \gamma \max_{a'} Q(s', a') - Q(s, a))$
        $s \leftarrow s'$
until;

Remark 6. *An on-policy variant of the Q-Learning algorithm, called the SARSA algorithm [90], consists of choosing a' with respect to the current policy for selecting the next action, rather than the max of the value function for the next state.*

TD(λ) [forward view]. The TD(λ) algorithm, with λ chosen between 0 and 1, is a compromise between the full backup method of the Monte-Carlo algorithm and the step-by-step update of the TD(0) algorithm. It relies on backups of episodes that are used to update each state, while assigning a greater importance to the very next step after each state.

We first define a n-step target: $R_t^{(n)} = \sum_{k=1}^{n} \gamma^{k-1} r_{t+k} + \gamma^n V(s_{t+n})$ Then, we can introduce the particular averaging of the TD(λ) algorithm on a state at time t in an episode ending at time T:

$$R_t^\lambda = (1 - \lambda) \sum_{n=1}^{T-t-1} \lambda^{n-1} R_t^{(n)} + \lambda^{T-t-1} R_t$$

This can be expanded as:

$$\begin{aligned} R_t^\lambda = &(1-\lambda)(r_{t+1} + \gamma V(s_{t+1})) \\ &+ (1-\lambda)\lambda(r_{t+1} + \gamma r_{t+2} + \gamma^2 V(s_{t+2})) \\ &+ (1-\lambda)\lambda^2(r_{t+1} + \gamma r_{t+2} + \gamma^2 r_{t+3} + \gamma^3 V(s_{t+2})) \\ &\quad \cdots \\ &+ \lambda^{T-t-1}(r_{t+1} + \gamma r_{t+2} + \gamma^2 r_{t+3} + \dots + \gamma^{T-1} r_T) \end{aligned}$$

Finally, the update method used is:

$$V(s_t) \leftarrow V(s_t) + \alpha \left[R_t^\lambda - V(s_t) \right]$$

Remark 7. *One can notice that the sum of the weights $(1-\lambda)\lambda^{n-1}$ and λ^{T-t-1} is equal to 1. Moreover:*

- *If $\lambda = 0$, the algorithm corresponds to TD(0).*
- *If $\lambda = 1$, the algorithm corresponds to the MC algorithm.*

TD(λ)[backward view]. The previous description of TD(λ) illustrates the mechanism behind this method. However, it is not computationally tractable. Here, we describe an equivalent approach that leads to an efficient implementation.

We have to introduce for each state the *eligibility trace* $e_t(s)$ that represents how much the state will influence the update of a future encountered state in an episode:

$$e_t(s) = \begin{cases} 0 & if\, t = 0 \\ \gamma\lambda e_{t-1}(s) & if\, t > 0\ and\ s \neq s_t \\ \gamma\lambda e_{t-1}(s) + 1 & if\, t > 0\ and\ s = s_t \end{cases}$$

We can now define the update method to be applied at each step t to all states s_i:

$$V(s_i) \leftarrow V(s_i) + \alpha e_t(s_i) \left[r_{t+1} + \gamma V(s_{t+1}) - V(s_t) \right]$$

yielding Algorithm 10.

Actor-Critic Methods. Actor-Critic methods separate the policy and the value function into two distinct structures [54]. The *actor*, or policy structure, is used to select actions; while the *critic*, or the estimated value function V, is used to criticize those actions in the form of a TD error:

$$\delta_t = r_{t+1} + \gamma V(s_{t+1}) - V(s_t)$$

Algorithm 10. TD(λ)

Result: $V \approx V^\pi$, an approximation of the value function of π
Initialization: V chosen arbitrarily
$$e(s) = 0, \quad \forall s \in \mathcal{S}$$
repeat
  $\quad$ $s \leftarrow$ get_initial_state()
  $\quad$ **while** $s \notin Terminal$ **do**
    $\quad\quad$ $a \leftarrow$ get_action(π, s)
    $\quad\quad$ $s', r \leftarrow get\_next\_state(s, a)$
    $\quad\quad$ $\delta \leftarrow r + \gamma V(s') - V(s)$
    $\quad\quad$ $e(s) \leftarrow e(s) + 1$
    $\quad\quad$ **for** $u \in \mathcal{S}$ **do**
      $\quad\quad\quad$ $V(u) \leftarrow V(u) + \alpha \delta e(u)$
      $\quad\quad\quad$ $e(u) \leftarrow \gamma \lambda e(u)$
    $\quad\quad$ $s \leftarrow s'$
until;

A positive δ_t indicates that the policy's decision to take action a_t in state s_t should be strengthened, on the other hand, a negative δ_t indicates that the policy's decision should be weakened. In a simple case, if the policy for s_t and a_t is just a scalar $p(s_t, a_t)$ that is then normalized across all actions (i.e. using a softmax function), we can adjust the parameters of the policy using δ_t:

$$p(s_t, a_t) \leftarrow p(s_t, a_t) + \beta \delta_t (1 - \pi_t(s_t, a_t))$$

where β is a positive scaling factor.

If $\pi_t(s_t, a_t)$ is a more complicated parameterized function, such as a deep neural network, then δ_t is used for computing gradients.

2.5 Planning

The key difference between dynamic programming methods and temporal difference methods is the use of a model. Dynamic programming methods use a model of the world to update the value of each state based on state transition probabilities and expectations of rewards. However, temporal difference methods achieve this through directly interacting with the environment.

A model produces a prediction about the future state and reward given a state-action pair. There are two main types of models: *distribution models* and *sample models*. A distribution model, like the one used in dynamic programming methods, produces all the possible next states with their corresponding probabilities and expected rewards, whereas a sample model only produces a sample next state and reward. Distribution models are more powerful than sample models; however, sample models can be more efficient in practice [113].

The benefit of a model is that one can simulate interactions with the environment, which is usually less costly than interacting directly with the environment itself. The downside is that a perfect model does not always exist. A model may have to be approximated by hand or learned through real-world interaction with the environment. Any sub-optimal behavior in the model can lead to a sub-optimal policy. [112] presented an algorithm that combines reinforcement learning, model learning, and planning (Algorithm 11) [113]. This algorithm requires that the environment be deterministic. The resulting state and reward of each observed state-action pair is stored in the model. The agent can then use the model to improve the action-values associated with each previously seen state-action pair without having to interact with the environment.

Algorithm 11. Dyna-Q

Result: π^*, the optimal policy
Initialization: Q chosen arbitrarily
          $Model(s, a)$ chosen arbitrarily $\forall s \in \mathcal{S}, \forall a \in \mathcal{A}$
          N some positive integer
repeat
 | $s \leftarrow$ current (nonterminal) state
 | $a \leftarrow get\_action(Q, s)$
 | $s', r \leftarrow get\_next\_state(s, a)$
 | $Q(s, a) \leftarrow Q(s, a) + \alpha(r + \gamma \max_{a'} Q(s', a') - Q(s, a))$
 | $Model(s, a) \leftarrow s', r$
 | $n \leftarrow 0$
 | **repeat**
 | | $s \leftarrow$ random previously seen state
 | | $a \leftarrow$ random action previously taken in s
 | | $s', r \leftarrow Model(s, a)$
 | | $Q(s, a) \leftarrow Q(s, a) + \alpha(r + \gamma \max_{a'} Q(s', a') - Q(s, a))$
 | | $n \leftarrow n + 1$
 | **until** $n >= N$;
until;

A model can be used to improve a value function and policy or it can be used to pick better actions given the current value function and policy. Heuristic search does this by using the value function and policy as a "heuristic" to search the state-space in order to select better actions. Monte Carlo tree search (MCTS) [18,53] is a heuristic search algorithm which uses a model to run simulations from the current state. When searching the state-space, the probability of selecting an action a in state s is influenced by the policy as well as the number of times that state-action pair has been selected. In order to encourage exploration, the probability of selecting a state-action pair goes down each time that pair is selected. Backed up values come from either running the simulation until the end of the episode or from the value of the leaf nodes.

2.6   Evolutionary Algorithms

We now turn to algorithms that search the policy space, starting with evolution-ary algorithms. These algorithms mimic the biological evolution of populations under natural selection (see [74] for more details). In reinforcement learning applications, populations of policies are evolved using a *fitness function*. At each generation, the most fit policies have a better chance of surviving and producing offspring policies in the next generation.

The most straightforward way to represent a policy in an evolutionary algo-rithm is to use a single chromosome per policy, with a single gene associated with each observed state. Each allele (the value of a gene) represents the action-value associated with the corresponding state. The algorithm (Algorithm 12) first gen-erates a population of policies $P(0)$, then selects the best ones according to a given criteria (*selection*), then randomly perturbs these policies (for instance by randomly selecting a state and then randomly perturbing the distribution of the actions given that state) (*mutation*). The algorithm may also create new policies by merging two different selected policies (*crossover*). This process is repeated until the selected policies satisfy a given criteria.

The fitness of a policy in the population is defined as the expected accumu-lated rewards for an agent that uses that policy. During the selection step, we keep either the policies with the highest fitness, or use a probabilistic choice in order to avoid local optima, such as:

$$Pr(p_i) = \frac{\text{fitness}(p_i)}{\sum_{j=1}^{n} \text{fitness}(p_j)}$$

Algorithm 12. Evolutionary Algorithm

Result: $\pi \approx \pi^*$, an approximation of the optimal policy
Initialization: $t = 0$
                    population $P(0)$ chosen arbitrarily
repeat
  | $t \leftarrow t + 1$
  | select $P(t)$ from $P(t-1)$
  | apply_mutation($P(t)$)
  | apply_crossover($P(t)$)
until;

2.7   Policy Gradient Algorithms

While other approaches tend to struggle with large or continuous state spaces, policy gradient algorithms offer a good alternative for complex environments solvable by relatively simple policies. Starting with an arbitrary policy, the idea behind policy gradient is to modify the policy such that it obtains the largest reward possible. For this purpose, a policy is represented by a parametric probability distribution $\pi_\theta(a|s) = P(a|s, \theta)$ such that in state s action a is selected according to the distribution $P(a|s, \theta)$. Hence, the objective here is to tune the parameter θ to increase the probability of choosing episodes associated with greater rewards. By computing the gradient of the average total return of a batch of episodes sampled from π_θ, we can use this value to update θ step-by-step. This approach is exploited in the REINFORCE algorithm [126].

3   Limitations and Open Problems

3.1   Complexity Considerations

So far, we have presented several ways of tackling the reinforcement learning problem in the framework of MDPs, but we have not described the theoretical tractability of this problem.

Recall that **P** is the class of all problems that can be solved in polynomial time, and **NC** the class of the problems that can be solved in polylogarithmic time on a parallel computer with a polynomial number of processors. As it seems very unlikely that **NC** = **P**, if a problem is proved to be **P**-complete, one can hardly expect to be able to find a parallel solution to this problem. In particular, it has been proved that the MDP problem is **P**-complete in the case of probabilistic transitions, and is in **NC** in the case of deterministic transitions, by [82]. Furthermore, in the case of high-dimensional MDPs, there exists a randomized algorithm [50] that is able to compute an arbitrary near-optimal policy in time independent of the number of states.

Remark 8. *Note that* **NC** $\subseteq$ **P**, *simply because parallel computers can be simulated on a sequential machine.*

Other results for the POMDP framework (see Sect. 3.3) are presented in [64]. In particular:

– Computing an infinite (polynomial) horizon undiscounted optimal strategy for a deterministic POMDP is PSPACE-hard (NP-complete).
– Computing an infinite (polynomial) horizon undiscounted optimal strategy for a stochastic POMDP is EXPTIME-hard (PSPACE-complete).

3.2   Limitations of Markov Decision Processes (MDPs)

Despite its great convenience as a theoretical model, the MDP model suffers from major drawbacks when it comes to real-world implementations. Here we list the most important ones to highlight common pitfalls encountered in practical applications.

- **High-dimensional spaces.** For high-dimensional spaces, typical of real-world control tasks, using a simple reinforcement learning framework becomes computationally intractable: this phenomenon is known as the *curse of dimensionality*. We can limit this by reducing the dimensionality of the problem [120], or by replacing the lookup table by a *function approximator* [15]. However, some precautions may need to be taken to ensure convergence [11].
- **Continuous spaces.** A variety of real world problems lead to continuous state spaces or action spaces, yet it is not possible to store an arbitrary continuous function. To address this problem, one has to use *function approximators* [94] to obtain tractable models, value functions, or policies. Two common techniques are *tile coding* [98] and fuzzy representation of the space [62].
- **Convergence.** While we have good guarantees on the convergence of reinforcement learning methods with lookup tables and linear approximators, our knowledge of the conditions for convergence with non-linear approximators is still very limited [119]. This is unfortunate because non-linear approximators are the most convenient and have been very successful on problems like playing backgammon [117].
- **Speed.** One way to speed up the convergence of reinforcement learning algorithms is to modify the reward function during learning to provide guidance toward good policies. This technique, called *shaping*, has been successfully applied to the problem of bike riding, which would not have been tractable without this improvement [88].
- **Stability.** Highly dependent on the parameters, the stability of the process of computing an optimal policy has not been studied sufficiently. However, it is a key element in the success of a learning strategy. Stability and stability guarantees have been studied in the context of kernel-based reinforcement learning methods [81].
- **Exploration vs Exploitation.** To learn efficiently, an agent in general should navigate the tradeoff between exploration and exploitation. Common heuristics such as ϵ-greedy and Boltzmann (softmax) provide means for addressing this trade-off, yet suffer from major drawbacks in terms of convergence speed and implementation (the choice of the parameters is non-trivial). The R-max algorithm [13], relying on the *optimism under uncertainty* bias, and model-based Bayesian exploration [22] offer convenient alternatives for the *exploration-exploitation dilemma*.
- **Initialization.** The choice of the initial policy, or the initial value function, may influence not only whether the algorithm converges, but also the speed of convergence. In some cases, for example, choosing a random initialization leads to drastically long computational times. One way to tackle this issue is to learn first using a simpler but similar task, and then use this knowledge to influence the learning process of the main task. This is the core principle of *transfer learning* which can lead to significant improvements, as shown in [116].

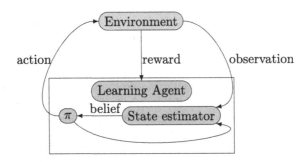

Fig. 4. The POMDP model

3.3   The POMDP Model

The partially observable Markov decision process (POMDP) [130] is a generalization of the MDP model in which the learning agent does not know precisely the current state in which it is operating. Instead, its knowledge relies on *observations* derived from its environment. Formally, a POMDP is an MDP with a finite set of possible observations $\mathcal{Z}$ and an observation model based on the probability $\nu(z|s)$ of observing z when the environment is in state s.

It has been shown in [105], that directly applying the MDP methods to this problem can have arbitrarily poor performance. To address this problem, one has to introduce an internal state distribution for the agent, the *belief state* $b_t(s)$, that represents the probability of being in state s at time t (see Fig. 4). One can then theoretically find an optimal solution to a POMDP problem [16] by defining an equivalent MDP problem, as shown below, and use existing MDP algorithms to solve it.

Assuming that the initial belief state b_0 is known, one can iteratively compute the belief state at any time $t+1$. We denote this operation by $F(b_t, a_t, z_t) = b_{t+1}$ with:

$$b_{t+1}(s') = \frac{\nu(z_t|s') \sum_{s \in \mathcal{S}} b_t(s) P_{a_t}(s, s')}{\sum_{s' \in \mathcal{S}} \nu(z|s') \sum_{s \in \mathcal{S}} b_t(s) P_{a_t}(s, s')}$$

The rewards are then given by:

$$\bar{r}(b) = \sum_{s \in \mathcal{S}} b(s) r(s)$$

In order to compute the transition function, let us first introduce the probability of observing z by applying action a in belief state b:

$$Pr(z|a, b) = \sum_{s' \in \mathcal{S}} \nu(z|s') \sum_{s \in \mathcal{S}} b(s) P_a(s, s')$$

Hence, we can define a transition probability function for the POMDP by:

$$\bar{P}_a(b, b') = \sum_{\substack{z \in \mathcal{Z} \\ F(b_t, a_t, z_t) = b_{t+1}}} Pr(z_t|a_t, b_t)$$

If $\mathcal{B}$ represents the set of belief states, the value function can then be computed as:

$$\bar{V}_{t+1}(b') = \max_a \left[\bar{r}(b') + \gamma \sum_{b \in \mathcal{B}} \bar{P}_a(b, b') \bar{V}_t(b) \right]$$

Remark 9. *This approach is obviously quite limited because of the potentially infinite size of $\mathcal{B}$. Several algorithms have been proposed to improve this, such as region-based pruning [31] and point-based algorithms [108], but they are also unable to deal with very large state spaces. VDCBPI [86] is one of the few efficient heuristics that seems to be able to find reasonable approximate solutions.*

3.4 Multi-agent Paradigm

There are several reasons for studying the case of multiple agents interacting with each other and seeking to maximize their rewards in a reinforcement learning fashion [14]. Many problems in areas as diverse as robotics, control, game theory, and population modeling lend themselves to such a modeling approach. Furthermore, the ability to parallelize learning across multiple agents is also attractive for several reasons, including speed and robustness. In particular, one may expect that if a particular agent fails, the other agents may be able to adapt without leading to a system-wide failure. Lastly, one may be able to improve or speed up learning of similar tasks by sharing experiences between individual learners (*transfer learning*).

However, as can be expected, the multi-agent model comes with significant challenges. By definition the multi-agent model has more variables and thus the curse of dimensionality is heightened. Furthermore, the environment model is more complex and suffers from non-stationarity during learning because of the constantly evolving behavior of each agent, and the problem of coordination between agents in order to achieve the desired results.

The starting model for the multi-agent paradigm corresponds to a *stochastic game*. For a system with n agents, it is composed of a set of states X, the sets of actions U_i for each agent $i = 1, ..., n$ (we let $\mathbf{U} = U_1 \times ... \times U_n$), the state transition function $f : X \times \mathbf{U} \times X \to [0,1]$ and the reward function $\rho_i : X \times \mathbf{U} \times X \to \mathbb{R}$.

There is a large collection of literature with different methods suitable for different multi-agent settings. The two major characteristics of such algorithms are their stability, which is related to their ability to converge to a stationary policy, and their adaptation, which measures how well the agents react to a change in the policy. Usually, it is difficult to guarantee both, and one must favor one over the other. The relationships between the agents can be classified in several classes, including:

– Fully cooperative: all the agents share a common set of objectives that have to be maximized. The *optimal adaptive learning* algorithm [122] has been proven to converge to an optimal Nash equilibrium (a configuration where no agent can improve its expected payoff by deviating to a different strategy) with probability 1. Good experimental results have also been obtained with the *coordinated reinforcement learning* approach [36].

- Fully competitive: the success of each agent directly depends on the failure of the other agents. For such settings, the minimax-Q [63] algorithm has been proposed, combining the minimax strategy (acting optimally while considering that the adversary will also act optimally) with the Q-learning method.
- Mixed: each agent has its own goal. As the objectives of this scenario are not well defined, there exist a significant number of approaches designed to tackle various formulations of this setting. An attempt to organize and clarify this case has been proposed in [87], for instance, along with a comparison of the most popular methods.

4   Other Directions of Research

4.1   Inverse Reinforcement Learning

Inverse reinforcement learning is the task of determining the reward function given an observed behavior. This observed behavior can be an optimal policy or a teacher's demonstration. Thus, the objective here is to estimate the reward attribution such that when reinforcement learning is applied with that reward function, one obtains the original behavior (in the case of behaviors associated with optimal policies), or even a better one (in the case of demonstrations).

This is particularly relevant in a situation where an expert has the ability to execute a given task but is unable, due to the complexity of the task and the domain, to precisely define the reward attribution that would lead to an optimal policy. One of the most significant success stories of inverse reinforcement learning is the apprenticeship of self driving cars [1].

To solve this problem in the case of MDPs, [78] identifies inequalities such that any reward function satisfying them must lead to an optimal policy. In order to avoid trivial answers, such as the all-zero reward function, these authors propose to use linear programming to identify the reward function that would maximize the difference between the value of an optimal action and the value of the next-best action in the same state. It is also possible to add regularization on the reward function to make it simpler (typically with non-zero reward on few actions). Systematic applications of inverse reinforcement learning in the case of POMDPs have not yet been developed.

4.2   Hierarchical Reinforcement Learning

In order to improve the time of convergence of reinforcement learning algorithms, different approaches for reducing the dimensionality of the problem have been proposed. In some cases, these approaches extend the MDP model to semi-Markov Decision Process (SMDP), by relaxing the Markov property, i.e. policies may base their choices on more than just the current state.

The *option* method [114] makes use of local policies that focus on simpler tasks. Hence, along with actions, a policy π can choose an option O. When the option O is chosen, a special policy μ associated with O is followed until a stochastic stop condition over the states and depending on O is reached. After the stop condition is reached, the policy π is resumed. The reward associated with O is the sum of the rewards of the actions performed under μ discounted by γ^τ were τ is the number of steps needed to terminate the option O. These option policies can be defined by an expert, or learned. There has been some work to try to automate this process of creating relevant options, or deleting useless ones [66].

State abstraction [4], used in the MAXQ algorithm [24] and in *hierarchical abstract machines* [83], is a mapping of the problem representation to a new representation that preserves some of its properties, in particular those needed for learning an optimal policy.

4.3   Approximate Linear Programming

As noted before, the linear programming approach to reinforcement learning typically suffers from the curse of dimensionality: the large number of states leads to an intractable number of variables for applying exact linear programming. A common way to overcome this issue is to approximate the cost-to-go function [30] by carefully designing some basis functions $\phi_1, ..., \phi_K$ that map the state space to rewards, and then constructing a linearly parameterized cost-to-go function:

$$\tilde{J}(\cdot, r) = \sum_{k=1}^{K} r_k \phi_k$$

where r is a parameter vector to be approximated by linear programming. In this way, the number of variables of the problem is drastically reduced, from the original number of states to K. The work in [45] proposes automated methods for generating a suitable basis functions ϕ for a given problem.

Using a *dynamic Bayesian network* to represent the transition model leads to the concept of *factored MDP* that can lead to reduced computational times on problems with a large number of states [35].

4.4   Relational Reinforcement Learning

Relational reinforcement learning [28] combines reinforcement learning with a relational representation of the state space, for instance by using inductive logic programming [75]. The goal is to propose a formalism that is able to perform well on problems requiring a large number of states, but can be represented compactly using a relational representation. In particular, experiments highlight the ability of this approach to take advantage of learning on simple tasks to accelerate the learning on more complex ones. This representation allows the learning of more "abstract" concepts, which leads to a reduced number of states that can significantly benefit generalization.

4.5   Quantum Reinforcement Learning

By taking advantage of the properties of quantum superposition, there is a possibility for considering novel quantum algorithms for reinforcement learning. The study in [25] presents potentially promising results, through simulated experiments, in regards to the speed of convergence and the trade-off between exploration and exploitation. Much work remains to be done in relation to modeling the environment, implementing function approximations, and deriving theoretical guarantees for quantum reinforcement learning (Fig. 5).

5   Deep Reinforcement Learning

Neural networks and deep learning approaches have well known universal approximation properties [21,43]. In recent years, and although they are far from new [96], neural networks and deep learning approaches have been used to successfully tackle a variety of problems in engineering, ranging from computer vision [5,20,38,55,109,115] to speech recognition [34], to natural language processing [32,107,110]. Likewise, deep learning is playing an essential role in the natural sciences, in areas ranging from high energy physics [7,92], to chemistry [48,49,65], and to biology [2,6,23,29,131]. Most of these applications use supervised, or semi-supervised learning, with stochastic gradient descent as the main learning algorithm and have benefited from significant increases in the amounts of available training data and computing power, including GPUs, as well as the development of good neural network software libraries. [71] also showed that, in certain cases, it is more efficient to train deep reinforcement learning algorithms using many CPUs instead of just one GPU.

It is therefore natural to try to combine deep learning methods with reinforcement learning methods, possibly in combination with frameworks for massively distributed reinforcement learning, such as Gorila [76]. This has been done, for instance, for the game of Go. The early work in [127,128] used deep learning methods, in the form of recursive grid neural networks, to evaluate the board or decide the next move. One characteristic of this approach is the ability to transfer learning between different board sizes (e.g. learn from games played on 9×9 or 11×11 boards and transfer the knowledge to larger boards). More recently, reinforcement learning combined with massive convolutional neural networks has been used to achieve the AI milestone of building an automated Go player [100] that can outperform human experts. Thus, deep reinforcement learning is a very active current area of research.

5.1   Value-Based Deep Reinforcement Learning

For *value-based deep reinforcement learning*, the value function is approximated by a deep neural network. [72] used Deep Q-networks that combine Q-learning with such a neural representation in order to teach an agent to play Atari video games, without any game-specific feature engineering. In this case, the *state* is represented by the stack of four previous frames, with the deep network consisting of multiple convolutional and fully-connected layers, and the action consisting

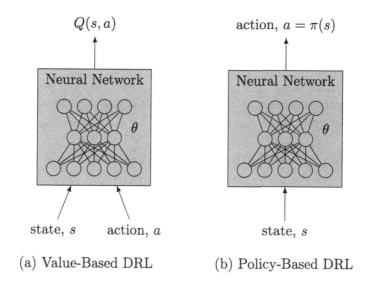

Fig. 5. Two types of deep reinforcement learning

of the 18 joystick positions. Since directly using neural networks as the function approximator leads to instability or divergence, the authors used additional heuristics such as replaying histories to reduce correlations, or using updates that change the parameters only periodically. Agents using this approach learned to play the majority of the games at a level equal or higher than the level of professional human players for the same game. There has been subsequent work to improve this approach, such as addressing stability [8], and applying Double Q-Learning [37] to avoid overestimation of the action-value functions in Deep Q-Networks [121]. Other extensions include multi-task learning [91,95], and rapid learning [10], among others.

5.2   Policy-Based Deep Reinforcement Learning

The second class of approaches, *policy-based deep reinforcement learning*, approximates the policy with deep neural networks. Policy-based approaches, by avoiding the search over the possible actions, converge and train much faster for many problems, especially with high-dimensional or continuous action spaces. The deterministic policy gradient, proposed by [102] and subsequently extended to deep representations by [61], were shown to be more efficient than their stochastic variants, thus extending deep reinforcement learning to continuous action spaces. [71] introduced the asynchronous advantage actor-critic (A3C) algorithm, that lets agents efficiently learn tasks with continuous action spaces, and works both on 2D and 3D games, with both feed-forward and recurrent neural approximators. As an application to robotic grasping, [60] uses a policy-gradient approach with a single deep convolutional network that combines the visual input and the gripper motor control to predict the grasp success probability. For multi-agent reinforcement learning, deep reinforcement learning has been used to learn agents by combining the fictitious self-play (FSP) approach [40] with neural representations [41], and applied to games such as poker.

Of course, both value- and policy-based deep reinforcement learning can be combined together with search algorithms. This is precisely the approach used in [100] for the game of Go.

5.3   Planning with Deep Reinforcement Learning

In Algorithm 11, a look up table served as the model of the environment. However, it is intractable to represent high-dimensional environments, such as images, with a simple lookup table. To address this issue, deep neural networks have been trained to predict the next state and the reward given a state-action pair and thus, perform the task of the model. When the environment takes the form of an image, deep neural networks have been shown to be able to produce realistic images that the agent can use to plan [17,57,79,84,124,125]. However, the predicted images are sometimes noisy and are sometimes missing key elements of the state. An alternative approach is to use a deep neural network to encode the current state into an abstract state and then, given an action, learn to predict the next abstract state along with its value and reward [80,99].

In addition to improving action selection, heuristic search algorithms have been combined with value and policy networks to improve the value and policy networks themselves. When applying deep reinforcement learning to Go, [100] mainly used the MCTS algorithm for action selection while the value and policy networks relied heavily on gameplay from human experts. However, [103] used MCTS to train a value and policy network from scratch by using the heuristic search algorithm for self-play, which resulted in an agent that outperformed all previous Go agents. This approach was also used when learning to play chess and shogi [101].

Acknowledgment. This research was in part supported by National Science Foundation grant IIS-1550705 and a Google Faculty Research Award to PB.

References

1. Abbeel, P., Ng, A.Y.: Apprenticeship learning via inverse reinforcement learning. In: Proceedings of the Twenty-First International Conference on Machine Learning, p. 1. ACM (2004)
2. Agostinelli, F., Ceglia, N., Shahbaba, B., Sassone-Corsi, P., Baldi, P.: What time is it? deep learning approaches for circadian rhythms. Bioinformatics **32**(12), i8–i17 (2016)
3. Anderson, C.W.: Learning to control an inverted pendulum using neural networks. Control Syst. Mag. IEEE **9**(3), 31–37 (1989)
4. Andre, D., Russell, S.J.: State abstraction for programmable reinforcement learning agents. In: AAAI/IAAI, pp. 119–125 (2002)

5. Baldi, P., Chauvin, Y.: Neural networks for fingerprint recognition. Neural Comput. **5**(3), 402–418 (1993)
6. Baldi, P., Pollastri, G.: The principled design of large-scale recursive neural network architectures-DAG-RNNs and the protein structure prediction problem. J. Mach. Learn. Res. **4**, 575–602 (2003)
7. Baldi, P., Sadowski, P., Whiteson, D.: Searching for exotic particles in high-energy physics with deep learning. Nat. Commun. **5**, 4308 (2014)
8. Bellemare, M.G., Ostrovski, G., Guez, A., Thomas, P.S., Munos, R.: Increasing the action gap: new operators for reinforcement learning. In: AAAI, pp. 1476–1483 (2016)
9. Bellman, R.: The theory of dynamic programming. Technical report, DTIC Document (1954)
10. Blundell, C., et al.: Model-free episodic control. arXiv preprint arXiv:1606.04460 (2016)
11. Boyan, J., Moore, A.W.: Generalization in reinforcement learning: safely approximating the value function. In: Advances in Neural Information Processing Systems, pp. 369–376 (1995)
12. Boyan, J.A., Littman, M.L., et al.: Packet routing in dynamically changing networks: a reinforcement learning approach. In: Advances in Neural Information Processing Systems, pp. 671–671 (1994)
13. Brafman, R.I., Tennenholtz, M.: R-max-a general polynomial time algorithm for near-optimal reinforcement learning. J. Mach. Learn. Res. **3**, 213–231 (2003)
14. Busoniu, L., Babuska, R., De Schutter, B.: A comprehensive survey of multiagent reinforcement learning. IEEE Trans. Syst. Man Cybern. Part C Appl. Rev. **38**(2), 156–172 (2008)
15. Busoniu, L., Babuska, R., De Schutter, B., Ernst, D.: Reinforcement Learning and Dynamic Programming Using Function Approximators, vol. 39. CRC Press, Boca Raton (2010)
16. Cassandra, A.R., Kaelbling, L.P., Littman, M.L.: Acting optimally in partially observable stochastic domains. In: AAAI, vol. 94, p. 1023–1028 (1994)
17. Chiappa, S., Racaniere, S., Wierstra, D., Mohamed, S.: Recurrent environment simulators. arXiv preprint arXiv:1704.02254 (2017)
18. Coulom, R.: Efficient selectivity and backup operators in Monte-Carlo tree search. In: van den Herik, H.J., Ciancarini, P., Donkers, H.H.L.M.J. (eds.) CG 2006. LNCS, vol. 4630, pp. 72–83. Springer, Heidelberg (2007). https://doi.org/10.1007/978-3-540-75538-8_7
19. Crites, R., Barto, A.: Improving elevator performance using reinforcement learning. In: Advances in Neural Information Processing Systems, vol. 8. Citeseer (1996)
20. Cun, Y.L., et al.: Handwritten digit recognition with a back-propagation network. In: Touretzky, D. (ed.) Advances in Neural Information Processing Systems, pp. 396–404. Morgan Kaufmann, San Mateo (1990)
21. Cybenko, G.: Approximation by superpositions of a sigmoidal function. Math. Control Signals Syst. (MCSS) **2**(4), 303–314 (1989)
22. Dearden, R., Friedman, N., Andre, D.: Model based Bayesian exploration. In: Proceedings of the Fifteenth Conference on Uncertainty in Artificial Intelligence, pp. 150–159. Morgan Kaufmann Publishers Inc. (1999)
23. Di Lena, P., Nagata, K., Baldi, P.: Deep architectures for protein contact map prediction. Bioinformatics **28**, 2449–2457 (2012). https://doi.org/10.1093/bioinformatics/bts475. First published online: July 30, 2012

24. Dietterich, T.G.: An overview of MAXQ hierarchical reinforcement learning. In: Choueiry, B.Y., Walsh, T. (eds.) SARA 2000. LNCS (LNAI), vol. 1864, pp. 26–44. Springer, Heidelberg (2000). https://doi.org/10.1007/3-540-44914-0_2

25. Dong, D., Chen, C., Li, H., Tarn, T.J.: Quantum reinforcement learning. IEEE Trans. Syst. Man Cybern. Part B Cybern. **38**(5), 1207–1220 (2008)

26. Dorigo, M., Gambardella, L.: Ant-Q: a reinforcement learning approach to the traveling salesman problem. In: Proceedings of ML-95, Twelfth International Conference on Machine Learning, pp. 252–260 (2014)

27. Drake, A.W.: Observation of a Markov process through a noisy channel. Ph.D. thesis, Massachusetts Institute of Technology (1962)

28. Džeroski, S., De Raedt, L., Driessens, K.: Relational reinforcement learning. Mach. Learn. **43**(1–2), 7–52 (2001)

29. Esteva, A., et al.: Dermatologist-level classification of skin cancer with deep neural networks. Nature **542**(7639), 115–118 (2017)

30. de Farias, D.P., Van Roy, B.: The linear programming approach to approximate dynamic programming. Oper. Res. **51**(6), 850–865 (2003)

31. Feng, Z., Zilberstein, S.: Region-based incremental pruning for POMDPs. In: Proceedings of the 20th Conference on Uncertainty in Artificial Intelligence, pp. 146–153. AUAI Press (2004)

32. Goldberg, Y.: A primer on neural network models for natural language processing. J. Artif. Intell. Res. **57**, 345–420 (2016)

33. Gosavi, A.: Reinforcement learning: a tutorial survey and recent advances. INFORMS J. Comput. **21**(2), 178–192 (2009)

34. Graves, A., Mohamed, A., Hinton, G.: Speech recognition with deep recurrent neural networks. In: 2013 IEEE International Conference on Acoustics, Speech and Signal Processing (ICASSP), pp. 6645–6649. IEEE (2013)

35. Guestrin, C., Koller, D., Parr, R., Venkataraman, S.: Efficient solution algorithms for factored MDPs. J. Artif. Intell. Res. **19**, 399–468 (2003)

36. Guestrin, C., Lagoudakis, M., Parr, R.: Coordinated reinforcement learning. In: ICML, vol. 2, pp. 227–234 (2002)

37. Hasselt, H.V.: Double q-learning. In: Advances in Neural Information Processing Systems, pp. 2613–2621 (2010)

38. He, K., Zhang, X., Ren, S., Sun, J.: Deep residual learning for image recognition. arXiv preprint arXiv:1512.03385 (2015)

39. Hebb, D.O.: The Organization of Behavior: A Neuropsychological Approach. Wiley, New York (1949)

40. Heinrich, J., Lanctot, M., Silver, D.: Fictitious self-play in extensive-form games. In: International Conference on Machine Learning (ICML), pp. 805–813 (2015)

41. Heinrich, J., Silver, D.: Deep reinforcement learning from self-play in imperfect-information games. arXiv preprint arXiv:1603.01121 (2016)

42. Holland, J.H.: Genetic algorithms and the optimal allocation of trials. SIAM J. Comput. **2**(2), 88–105 (1973)

43. Hornik, K., Stinchcombe, M., White, H.: Multilayer feedforward networks are universal approximators. Neural Netw. **2**(5), 359–366 (1989)

44. Howard, R.A.: Dynamic programming and Markov processes (1960)

45. Hutter, M.: Feature reinforcement learning: Part I. Unstructured MDPs. J. Artif. Gen. Intell. **1**(1), 3–24 (2009)

46. Kaelbling, L.P., Littman, M.L., Moore, A.W.: Reinforcement learning: a survey. J. Artif. Intell. Res. **4**, 237–285 (1996)

47. Kandel, E.R., Schwartz, J.H., Jessell, T.M.: Principles of Neural Science, vol. 4. McGraw-hill, New York (2000)

48. Kayala, M., Azencott, C., Chen, J., Baldi, P.: Learning to predict chemical reactions. J. Chem. Inf. Model. **51**(9), 2209–2222 (2011)
49. Kayala, M., Baldi, P.: Reactionpredictor: prediction of complex chemical reactions at the mechanistic level using machine learning. J. Chem. Inf. Model. **52**(10), 2526–2540 (2012)
50. Kearns, M., Mansour, Y., Ng, A.Y.: A sparse sampling algorithm for near-optimal planning in large Markov decision processes. Mach. Learn. **49**(2–3), 193–208 (2002)
51. Keerthi, S.S., Ravindran, B.: A tutorial survey of reinforcement learning. Sadhana **19**(6), 851–889 (1994)
52. Kober, J., Bagnell, J.A., Peters, J.: Reinforcement learning in robotics: a survey. Int. J. Robot. Res. **32**, 1238–1274 (2013). p. 0278364913495721
53. Kocsis, L., Szepesvári, C.: Bandit based Monte-Carlo planning. In: Fürnkranz, J., Scheffer, T., Spiliopoulou, M. (eds.) ECML 2006. LNCS (LNAI), vol. 4212, pp. 282–293. Springer, Heidelberg (2006). https://doi.org/10.1007/11871842_29
54. Konda, V.R., Tsitsiklis, J.N.: Actor-critic algorithms. In: NIPS. **13**, 1008–1014 (1999)
55. Krizhevsky, A., Sutskever, I., Hinton, G.E.: Imagenet classification with deep convolutional neural networks. In: Advances in Neural Information Processing Systems, pp. 1097–1105 (2012)
56. Lai, M.: Giraffe: Using deep reinforcement learning to play chess. arXiv preprint arXiv:1509.01549 (2015)
57. Leibfried, F., Kushman, N., Hofmann, K.: A deep learning approach for joint video frame and reward prediction in atari games. arXiv preprint arXiv:1611.07078 (2016)
58. Levin, E., Pieraccini, R., Eckert, W.: A stochastic model of human-machine interaction for learning dialog strategies. IEEE Trans. Speech Audio Process. **8**(1), 11–23 (2000)
59. Levine, S., Finn, C., Darrell, T., Abbeel, P.: End-to-end training of deep visuomotor policies. J. Mach. Learn. Res. **17**(39), 1–40 (2016)
60. Levine, S., Pastor, P., Krizhevsky, A., Quillen, D.: Learning hand-eye coordination for robotic grasping with deep learning and large-scale data collection. In: International Symposium on Experimental Robotics (2016)
61. Lillicrap, T.P., et al.: Continuous control with deep reinforcement learning (2016)
62. Lin, C.T., Lee, C.G.: Reinforcement structure/parameter learning for neural-network-based fuzzy logic control systems. IEEE Trans. Fuzzy Syst. **2**(1), 46–63 (1994)
63. Littman, M.L.: Markov games as a framework for multi-agent reinforcement learning. In: Proceedings of the Eleventh International Conference on Machine Learning, vol. 157, pp. 157–163 (1994)
64. Littman, M.L.: Algorithms for sequential decision making. Ph.D. thesis, Brown University (1996)
65. Lusci, A., Pollastri, G., Baldi, P.: Deep architectures and deep learning in chemoinformatics: the prediction of aqueous solubility for drug-like molecules. J. Chem. Inf. Model. **53**(7), 1563–1575 (2013)
66. McGovern, A., Barto, A.G.: Automatic discovery of subgoals in reinforcement learning using diverse density. Computer Science Department Faculty Publication Series, p. 8 (2001)
67. Michie, D.: Trial and error. In: Science Survey, Part 2, pp. 129–145 (1961)
68. Michie, D.: Experiments on the mechanization of game-learning part I. Characterization of the model and its parameters. Comput. J. **6**(3), 232–236 (1963)

69. Michie, D., Chambers, R.A.: Boxes: an experiment in adaptive control. Mach. Intell. **2**(2), 137–152 (1968)
70. Minsky, M.: Steps toward artificial intelligence. Proc. IRE **49**(1), 8–30 (1961)
71. Mnih, V., et al.: Asynchronous methods for deep reinforcement learning. In: International Conference on Machine Learning (ICML) (2016)
72. Mnih, V., et al.: Human-level control through deep reinforcement learning. Nature **518**(7540), 529–533 (2015)
73. Moody, J., Saffell, M.: Reinforcement learning for trading. In: Advances in Neural Information Processing Systems, pp. 917–923 (1999)
74. Moriarty, D.E., Schultz, A.C., Grefenstette, J.J.: Evolutionary algorithms for reinforcement learning. J. Artif. Intell. Res. (JAIR) **11**, 241–276 (1999)
75. Muggleton, S., De Raedt, L.: Inductive logic programming: theory and methods. J. Logic Program. **19**, 629–679 (1994)
76. Nair, A., et al.: Massively parallel methods for deep reinforcement learning. arXiv preprint arXiv:1507.04296 (2015)
77. Ng, A.Y., et al.: Autonomous inverted helicopter flight via reinforcement learning. In: Ang, M.H., Khatib, O. (eds.) Experimental Robotics IX. STAR, vol. 21, pp. 363–372. Springer, Heidelberg (2006). https://doi.org/10.1007/11552246_35
78. Ng, A.Y., Russell, S.J., et al.: Algorithms for inverse reinforcement learning. In: ICML, pp. 663–670 (2000)
79. Oh, J., Guo, X., Lee, H., Lewis, R.L., Singh, S.: Action-conditional video prediction using deep networks in atari games. In: Advances in Neural Information Processing Systems, pp. 2863–2871 (2015)
80. Oh, J., Singh, S., Lee, H.: Value prediction network. In: Advances in Neural Information Processing Systems, pp. 6120–6130 (2017)
81. Ormoneit, D., Sen, Ś.: Kernel-based reinforcement learning. Mach. Learn. **49**(2–3), 161–178 (2002)
82. Papadimitriou, C.H., Tsitsiklis, J.N.: The complexity of Markov decision processes. Math. Oper. Res. **12**(3), 441–450 (1987)
83. Parr, R., Russell, S.: Reinforcement learning with hierarchies of machines. In: Advances in Neural Information Processing Systems, pp. 1043–1049 (1998)
84. Pascanu, R., et al.: Learning model-based planning from scratch. arXiv preprint arXiv:1707.06170 (2017)
85. Pashenkova, E., Rish, I., Dechter, R.: Value iteration and policy iteration algorithms for Markov decision problem. In: AAAI 1996, Workshop on Structural Issues in Planning and Temporal Reasoning. Citeseer (1996)
86. Poupart, P., Boutilier, C.: VDCBPI: an approximate scalable algorithm for large POMDPs. In: Advances in Neural Information Processing Systems, pp. 1081–1088 (2004)
87. Powers, R., Shoham, Y.: New criteria and a new algorithm for learning in multi-agent systems. In: Advances in Neural Information Processing Systems, pp. 1089–1096 (2004)
88. Randløv, J., Alstrøm, P.: Learning to drive a bicycle using reinforcement learning and shaping. In: ICML, vol. 98, pp. 463–471. Citeseer (1998)
89. Ross, S.M.: Introduction to Stochastic Dynamic Programming. Academic press, Norwell (2014))
90. Rummery, G.A., Niranjan, M.: On-line Q-learning using connectionist systems. University of Cambridge, Department of Engineering (1994)
91. Rusu, A.A., et al.: Policy distillation. In: International Conference on Learning Representations (ICLR) (2016)

92. Sadowski, P., Collado, J., Whiteson, D., Baldi, P.: Deep learning, dark knowledge, and dark matter. In: Journal of Machine Learning Research, Workshop and Conference Proceedings, vol. 42, pp. 81–97 (2015)

93. Samuel, A.L.: Some studies in machine learning using the game of checkers. II. Recent progress. IBM J. Res. Dev. **11**(6), 601–617 (1967)

94. Santamaría, J.C., Sutton, R.S., Ram, A.: Experiments with reinforcement learning in problems with continuous state and action spaces. Adapt. Behav. **6**(2), 163–217 (1997)

95. Schaul, T., Horgan, D., Gregor, K., Silver, D.: Universal value function approximators. In: International Conference on Machine Learning (ICML), pp. 1312–1320 (2015)

96. Schmidhuber, J.: Deep learning in neural networks: an overview. Neural Netw. **61**, 85–117 (2015)

97. Schulman, J., Moritz, P., Levine, S., Jordan, M., Abbeel, P.: High-dimensional continuous control using generalized advantage estimation. In: Proceedings of the International Conference on Learning Representations (ICLR) (2016)

98. Sherstov, A.A., Stone, P.: On continuous-action Q-learning via tile coding function approximation. Under Review (2004)

99. Silver, D., et al.: The predictron: end-to-end learning and planning. arXiv preprint arXiv:1612.08810 (2016)

100. Silver, D., et al.: Mastering the game of go with deep neural networks and tree search. Nature **529**(7587), 484–489 (2016)

101. Silver, D., et al.: Mastering chess and shogi by self-play with a general reinforcement learning algorithm. arXiv preprint arXiv:1712.01815 (2017)

102. Silver, D., Lever, G., Heess, N., Degris, T., Wierstra, D., Riedmiller, M.: Deterministic policy gradient algorithms. In: International Conference on Machine Learning (ICML) (2014)

103. Silver, D., et al.: Mastering the game of go without human knowledge. Nature **550**(7676), 354 (2017)

104. Singh, S., Bertsekas, D.: Reinforcement learning for dynamic channel allocation in cellular telephone systems. In: Advances in Neural Information Processing Systems, pp. 974–980 (1997)

105. Singh, S.P., Jaakkola, T.S., Jordan, M.I.: Learning without state-estimation in partially observable Markovian decision processes. In: ICML, pp. 284–292 (1994)

106. Singh, S.P., Sutton, R.S.: Reinforcement learning with replacing eligibility traces. Mach. Learn. **22**(1–3), 123–158 (1996)

107. Socher, R., et al.: Recursive deep models for semantic compositionality over a sentiment treebank. In: Proceedings of the Conference on Empirical Methods in Natural Language Processing (EMNLP), vol. 1631, p. 1642. Citeseer (2013)

108. Spaan, M.T., Spaan, M.T.: A point-based POMDP algorithm for robot planning. In: 2004 IEEE International Conference on Robotics and Automation, Proceedings, ICRA 2004, vol. 3, pp. 2399–2404. IEEE (2004)

109. Srivastava, R.K., Greff, K., Schmidhuber, J.: Training very deep networks. In: Advances in Neural Information Processing Systems, pp. 2368–2376 (2015)

110. Sutskever, I., Vinyals, O., Le, Q.V.: Sequence to sequence learning with neural networks. In: Advances in Neural Information Processing Systems, pp. 3104–3112 (2014)

111. Sutton, R.S.: Learning to predict by the methods of temporal differences. Mach. Learn. **3**(1), 9–44 (1988)

112. Sutton, R.S.: Integrated architectures for learning, planning, and reacting based on approximating dynamic programming. In: Machine Learning Proceedings 1990, pp. 216–224. Elsevier (1990)

113. Sutton, R.S., Barto, A.G.: Reinforcement Learning: An Introduction. MIT Press, Cambridge (1998)

114. Sutton, R.S., Precup, D., Singh, S.: Between MDPs and semi-MDPs: a framework for temporal abstraction in reinforcement learning. Artif. Intell. **112**(1), 181–211 (1999)

115. Szegedy, C., et al.: Going deeper with convolutions. In: Proceedings of the IEEE Conference on Computer Vision and Pattern Recognition, pp. 1–9 (2015)

116. Taylor, M.E., Stone, P.: Cross-domain transfer for reinforcement learning. In: Proceedings of the 24th International Conference on Machine Learning, pp. 879–886. ACM (2007)

117. Tesauro, G.: Temporal difference learning and TD-Gammon. Commun. ACM **38**(3), 58–68 (1995)

118. Thorndike, E.L.: Animal Intelligence: Experimental Studies. Transaction Publishers, New York (1965)

119. Tsitsiklis, J.N., Van Roy, B.: An analysis of temporal-difference learning with function approximation. IEEE Trans. Autom. Control **42**(5), 674–690 (1997)

120. Van Der Maaten, L., Postma, E., Van den Herik, J.: Dimensionality reduction: a comparative. J. Mach. Learn. Res. **10**, 66–71 (2009)

121. Van Hasselt, H., Guez, A., Silver, D.: Deep reinforcement learning with double q-learning. In: AAAI, pp. 2094–2100 (2016)

122. Wang, X., Sandholm, T.: Reinforcement learning to play an optimal Nash equilibrium in team Markov games. In: Advances in Neural Information Processing Systems, pp. 1571–1578 (2002)

123. Watkins, C.J., Dayan, P.: Q-learning. Mach. Learn. **8**(3–4), 279–292 (1992)

124. Watter, M., Springenberg, J., Boedecker, J., Riedmiller, M.: Embed to control: a locally linear latent dynamics model for control from raw images. In: Advances in Neural Information Processing Systems, pp. 2746–2754 (2015)

125. Weber, T., et al.: Imagination-augmented agents for deep reinforcement learning. arXiv preprint arXiv:1707.06203 (2017)

126. Williams, R.J.: Simple statistical gradient-following algorithms for connectionist reinforcement learning. Mach. Learn. **8**(3–4), 229–256 (1992)

127. Wu, L., Baldi, P.: A scalable machine learning approach to go. In: Weiss, Y., Scholkopf, B., Editors, J.P. (eds.) NIPS 2006. MIT Press, Cambridge (2007)

128. Wu, L., Baldi, P.: Learning to play go using recursive neural networks. Neural Netw. **21**(9), 1392–1400 (2008)

129. Zhang, W., Dietterich, T.G.: High-performance job-shop scheduling with a time-delay td network. In: Advances in Neural Information Processing Systems, vol. 8, pp. 1024–1030 (1996)

130. Zhang, W.: Algorithms for partially observable Markov decision processes. Ph.D. thesis, Citeseer (2001)

131. Zhou, J., Troyanskaya, O.G.: Predicting effects of noncoding variants with deep learning-based sequence model. Nat. Methods **12**(10), 931–934 (2015)

Personal and Beyond

A Man of Unlimited Capabilities
(in Memory of E. M. Braverman)

Lev I. Rozonoer[✉]

Institute of Control Sciences RAS, Profsoyuznaya, 65,
Moscow, Russian Federation
levroz@gmail.com

In our politically correct era, E. M. Braverman would be called a man of limited capabilities, since he lost his leg while still a child (it was cut off up to the knee by a streetcar by which Musya – his home nickname – got hit) and was wearing an extremely uncomfortable and heavy artificial leg (at the time these were the only kind available). The artificial leg would often rub against the remaining flesh of the limb, and E. M. would feel and bear the pain at every move.

What is "a man of limited capabilities" incapable of?

To dance? – Yet, E. M. danced quite well (excluding, of course, the really fast dances).

To ski or to skate? – Yet, E. M. could both ski and skate fairly well.

To swim? – Yet, E. M. was a fine swimmer, and once he even got to save me, when we were swimming in a pond near Moscow with our classmates from the 7th grade and one of the guys, fooling around, had pulled me under the water, whence, having swallowed too much water, I could not emerge.

We would often go together for water tourism (including rivers with rapids) on flatboats and kayaks.

What concerns his intellect, E. M. was one of the few people I met in my lifetime who possessed such an original and independent intellect. He was not afraid of putting forward and contemplating ideas that at the first glance seemed unnatural and even absurd. To illustrate my point, I could cite this seemingly strange at first sight idea of his about using in algorithms of potential functions' method the sequence of coefficients whose sum diverges but whose square sum converges (at the time we knew nothing about the method of stochastic approximation where these very coefficients are being used). The everyday advice that he would give his friends, were also remarkable for their unexpectedness and originality.

Remembering E. M., I would call him a man of unlimited capabilities.

Braverman and I met in 1944, when I came back to Moscow from the evacuation area and entered the 6th grade of the junior high school on the Yauza-river bank, where Misha (this is how his friends would call him) had already been a student. He enjoyed indisputable authority among his classmates not only by virtue of his abilities and at the same time his ever-readiness to come to help, but also by virtue of the fact that he was a physically strong, well-developed boy. In particular, he would keep the anti-Semites in their place and would hit for an anti-Semitic remark without giving the bully a break, and many other boys would support him. As a result, in our class there was not a trace left of anti-Semitism.

© Springer Nature Switzerland AG 2018
L. Rozonoer et al. (Eds.): Braverman Readings in Machine Learning, LNAI 11100, pp. 331–332, 2018.
https://doi.org/10.1007/978-3-319-99492-5_14

Afterwards, we went to the high school in Syromyatniki neighborhood from where we successfully graduated (with medals) in 1949. Later on we both graduated from college in 1955 (E. M. graduated from MEPhI). A few years later we both began to work at the Institute of Automatics and Telemechanics (in IAS) which was later called the Institute of Advanced Systems (ISP).

E. M. was always an interesting person to talk to. We had many friends in common. It is with great enjoyment that I remember the first-floor apartment in Zaryadye, where Misha lived with his parents and his sister, and where we would often gather together with friends for infinite communication – games and talks.

Almost from the very beginning after my graduation from college, I worked at M. A. Ayzerman's lab, whereas E. M. was at first a graduate student of M. A. Ayzerman and then he stayed to work at the lab. The common work for the three of us on machine learning problems started after E. M.'s doctoral thesis on image recognition, one of the first works on the topic not only in the Soviet Union but worldwide.

Sharing the work for the three us would happen as follows. On the eve of writing a text, upon discussing together the next fragment of the theme, M. A. on his own and E. M. and myself, the two of us, would think through the exact content of the fragment. The necessary formulae would be brought in by each of us. The text would always be written by M. A.. Prior to putting down yet another sentence, one of the three of us (most often, M. A. himself) would say it out-loud and, no objections following, M. A. would write it down. If objections came up, a discussion that could take infinitely long ensued.

We composed the main bulk of the book "The Method of Potential Functions in Machine Learning Theory" at a rest house near Moscow, where nothing would distract us from the work on this book. At the same time we would not forget about recreation: since it was wintertime, we would go country skiing.

In conclusion, I would like to mention several lessons that I learned from my long-standing friendship with E. M..

(1) To not be afraid of strange and at first glance even grotesque ideas, but try to consider them with an open mind.

(2) Not to be afraid of stern criticism and extreme statements; to listen to them if they are just and motivated by the concern for the cause. At the same time, not to be shy to criticize others sharply in case of necessity.

(3) To think about how you could be of help for a person in need of help, even if he is not asking himself for help. At the same time, not to be shy to receive help from others.

Unfortunately, I cannot say that I learned all these lessons well. Yet, E. M. himself would strictly abide by these principles.

Braverman and His Theory of Disequilibrium Economics

Mark Levin[(⊠)]

National Research University Higher School of Economics, Moscow, Russia
mlevin05@gmail.com

The Teacher. I can unreservedly call Emmanuel Braverman my Teacher with the capital T. He shaped not only my academic career, my research interests and teaching but many other important things as well. There are many people in my life to whom I owe the debt of gratitude, people who I looked up to and continue to look up to.

However, my most important teacher-student relationship remains with Emmanuel Braverman who did not reach his 46th birthday, dying on April 26, 1977. By the time we met, rather by chance, in the Spring of 1970, Braverman was 39. He was already a Doctor of Science (a prestigious academic rank in the Soviet Union and current Russia) and was a world-renowned expert in the area of image recognition and artificial intelligence. He worked in one of the best research institutions of his time – The Institute of Control Sciences. However, I knew none of this at the time.

As I was starting my PhD at the Department of Computer Science and Engineering at the National University of Science and Technology (MISiS), Moscow, I had an urgent problem of finding a supervisor for my project.

I didn't know anyone at the Department, but luckily a friend of mine recommended Braverman who he rated very high as a scientist and even called him a Classic!

Stanislav Emelyanov, Department's Head and later member of the Academy of Sciences of the USSR, happened to know Braverman well and arranged for us to meet.

At our first meeting I saw a limping man below average height. Fortunately, I had enough tact not ask what happened to his leg. As I discovered later, Braverman had a leg prosthesis since childhood, which, nevertheless, did not preclude him from enjoying skating, skiing and cycling. He asked me to tell him about myself. I said that I had graduated from Gubkin University of Oil and Gas, had already worked at the Moscow State University's Biology Department and The Radio Engineering Research Institute. I told him that, at the moment, I was studying Engineering at the Faculty of Mechanics and Mathematics of Moscow State University.

As Braverman told me later he would have not taken me on the face of his first impressions. He thought of me as "an odd young man recommended to him by authorities". However, in the end he was persuaded by the fact that I started doing research work early - on my second year of being an undergraduate student, as well as by the prestige of MSU's Mechanics and Mathematics Department.

He also asked me about the title of my thesis proposal. Eager to impress him, I came up with some idiotic topic. Braverman promptly recommended literature I was to get quickly acquainted with and none of which I had heard of before. I quickly understood that I made a total fool of myself. "And what topic would YOU recommend for me?" I asked. Braverman suggested I attempt to write a thesis proposal on "The Model of General Economic Equilibrium" – something that I had never heard of either.

L. Rozonoer et al. (Eds.): Braverman Readings in Machine Learning, LNAI 11100, pp. 333–340, 2018.
https://doi.org/10.1007/978-3-319-99492-5_15

He suggested I read the seminal book by Samuel Carlin, "Mathematical Methods and Theory in Games, Programming, and Economics", as well as another slim volume which I later found with great difficulty. I later understood that this was Braverman's "FIRST LESSON."

I started writing my proposal with great difficulty, understanding very little of what I was reading and struggling with economic vocabulary: 'demand', 'supply', 'equilibrium', 'utility'- all these concepts were novel to me! Still, by September I was able to finish the thesis proposal, which Braverman, on the whole, quite liked. Hurray - I thought, I passed the first test!

Thus in December of 1970 Braverman became my PhD project supervisor, something that I am still proud of and grateful for to this day. Formally, our working relationship lasted only seven years, but for me that still continuous.

Emmanuel Braverman offered me to work in a radically new field - non-equilibrium economic systems. Braverman started developing his 'own' theory of such systems, which put him at the frontier of economic science, something I was still ill aware of at the time.

Now I realize what luck it was: thanks to my Teacher I was able to participate, if only in an assistant role, in the development of a new area of research in Economics! Braverman suggested I tackle one of the problems in this rapidly developing field. This was the time when Braverman taught me "THE SECOND LESSON" and the true beginning of our collaboration. Studying his papers that had not been yet published, I came across an idea that helped me to come up with a solution to the problem that Braverman tasked me to solve. I spent three or four months trying to come up with a theorem, Braverman studied my calculations but found them wanting. However, the formulation as such looked promising and Braverman decided that we should work on the problem further!

We were meeting almost every week. After Braverman's lectures on image recognition I would often come to his apartment where his wife Elena treated us to delicious dinners. (For 47 years by now Elena Braverman has continued to welcome me in her London and now Boston home for which I am forever grateful).

After dinner we would spend two or there hours trying to come up with different solutions to the problem. In truth the use of 'we' is perhaps misleading here, Braverman was coming up with ideas and I wrote them down and then was given a specific task to work on until our next meeting. Gradually, a whole new area of research started to emerge: analysis of bottlenecks in economic systems. It felt like a breakthrough!

Earlier, when I was struggling with the theorem, I once complained to Braverman that it has been a long time since I started my PhD but nothing was working out and it was disheartening. He replied: "It is always like that in research before the first successful result!" Three years later he confessed to me that at the start of our collaboration he allocated eight or nine months to achieve a tangible result. He then decided that if we could not do it within that time period, he would change the topic of my PhD thesis, moving it to the area of Image Recognition as not to jeopardize my chances of earning the PhD.

The lessons of informal communication. I tried to use every opportunity to communicate with Braverman. Looking back, I think such surprising foresight on my part was partly due to losing my father-in-law, a truly extraordinary person, shortly after I met Braverman. This was the first time when I realized that everything could turn into ruin very quickly, that there was very little time indeed. I only had three years to complete my PhD, I thought, I would either do it or squander both time and opportunity.

Opportunities. At the time, once a week Braverman had a full day of giving lectures. He was a superb lecturer: logical, precise, accessible but without sacrificing complexity of the subject. We often had a chat during the breaks. After the lectures we would ride in his new car to his flat to continue working. Braverman recently changed his tiny, rattling 'Zaporozhets' for a roomier and faster 'Zhiguli', a very basic car by modern standards, but then it seemed the height of luxury to me!

We spoke in his car, although I mostly listened, not wanting to waste the precious opportunity of hearing my teacher talk and, at the same time, being paranoid of saying something stupid and as a result be banished forever from these car trips.

Thanks to Braverman I was also able to work (alongside him) in a world-class research division headed by Mark Aizerman of The Institute of Control Studies. Until this day, the plainly called Laboratory 25, stands in my mind as the gold standard of team work. It employed postgraduate students (there I met my future close friend R. Sheinyn) from within the Institute as well as students from other institutions like me. There were also some affiliated young but very promising researchers like B. Mirkin and V. Polterovich.

Some of these researchers were a little younger than me but most were much older. It was a truly all-star team. Let me name just a few of them: Mark Aizerman, Andrey Malishevsky, Ilya Muchnik, Alexander Dorofeyuk, and of course, Lev Rosonoer, now based in Boston, and whose advice I still consider priceless.

Braverman introduced me to a professional circle where the benchmark for the expected quality of work was set very high in both teaching and academic publishing. Those belonging to this circle often chose to forego climbing the academic rank ladder as it required a lot of fuss and administrative meddling. One's real value was known there regardless of academic titles. It was possible to be incredibly esteemed by fellow colleagues, as Braverman or Lev Rosonoer had been, despite holding a relatively modest academic rank. Being cultured was also something that people of this circle shared: they were avid readers of quality fiction, theater goers and art lovers.

Another Lesson – To Work Regardless of Circumstances! After about a year since I started my PhD thesis, Braverman became seriously ill, he was fighting throat cancer and was often hospitalized for two or three weeks at a time. I visited him several times a week. We would sit together at a large hospital dining table, surrounded from all sides by patients and hospital personnel, and work on our academic publications. We would rewrite the text of an article a hundred times. This is in fact how I learnt to write. One day, I realized that a theory was born, which allowed me to complete a PhD thesis and later still write a book on the subject.

In the Spring of 1975 I defended my PhD thesis at The Central Economics-Mathematics Institute (CEMI) where I was very well received despite being somewhat of an outsider. There was a nasty surprise however: it took whole two years for the Higher Attestation Committee of the USSR to consider my dissertation for approval. In the end I was handed a rejection, apparently due to a negative review by an anonymous reviewer. It was a terrible blow - I now had to resubmit my thesis in front of a special panel - The Expert Members of the Committee. The confidential memo by the Committee's Head read: 'The Dissertation in question is of excellent quality, however the thesis will not be validated as the quota for such approvals, alas, has already been filled for this year".

Braverman was in hospital. His vocal cords as well as part of his gullet had already been surgically removed. He was unable to speak and was breathing heavily. He was about to turn 46. I came to see Braverman in utter distress, showed him the negative review and explained that I had to face the Committee on the 26th of April. "Is this it? – I asked – "Shall we have to start everything from scratch?" He wrote on a piece of paper: "When will you finally understand that in serious matters, unlike in love, the result is much more important than the process?!" Then he put together a detailed plan for my further actions. Braverman coached me how to tackle the negative review comments in front of the panel members and wrote a list of questions I absolutely had to be able to answer. A few days later we met again, as it turned out, for the last time. Braverman gave me his final advice on how to behave in front of the Committee as well as his suggestions regarding the book draft that I had already started to write.

After a week of intense preparation, armed with Braverman's instructions, I felt much more confident and ready to tackle any questions the panel could throw at me. Braverman however was already in coma. On the morning of 26th of April Braverman died. In the afternoon of the same day, another his student, Roman Shenyin drove me to face the Committee. It went smoothly, within an hour I was told my PhD had been approved. Afterwards Shenynin and I rode straight to Braverman's home. It was full of mourners: his friends, colleagues and relatives. Despite the tragic occasion many of them approached and congratulated me – it was mostly Braverman's victory after all!

(For 40 years now, every 26th of April, first in Moscow, later in Israel, and now in America there is a remembrance day devoted to Braverman that gathers people who knew and loved him).

A Few Entertaining Lessons. After two years of working with Braverman I once told him of a minor problem – I needed to pump my son's bicycle tires but lacked a pump. "Would you lend me your own, I recall you had one?" I asked apologetically. Braverman said 'of course' and later called and said that he would drop by my place (for the first time!) and pump the tires himself! He probably did not believe (justifiably!) that I was able to do it on my own! Another time we rode to a small obscure book shop as it was the only place where Braverman could locate a book he wanted to buy - "Geometry for Young". "Children MUST know Topology! This is our responsibility to teach them this area of geometry, and not the schools!" Braverman proclaimed. I guess his devotion to education was the reason why both of his sons became world renowned mathematicians! He added: "By the way I found an English tutor for your son, I remember you were looking for one".

I would like to conclude with three more of the many lessons that my Teacher taught me:

- Research work and teaching are a calling and should be approached with utter dedication.
- It's important to help one's students to the best of one's ability yet never indulge in showing superiority.
- To live and to die honorably

Whatever lessons I was not able to learn from my Teacher are for others to judge. I can only add that all the research work I have conducted after 1977 is my attempt to master and develop Braverman's ideas in the area of Economic Systems Analysis.

1   The Disequilibrium Theory of Economic Systems - a Conceptual Description of Emmanuel Braverman's New Approach

Emmanuel Braverman's basic paradigm is based on the abandonment of the idea of flexible prices as a means of aligning supply and demand in economic analysis of partial and general equilibrium. Braverman proposed including in the analysis quantitative restrictions on the consumption of certain goods, which are (potentially) scarce, and the production of others, which are (potentially) in surplus. This would maintain equilibrium and an equality of demand and supply in non-price-based mechanisms. This fundamental proposition was in many ways contrary to the traditional development of mathematical-economic models and demanded a revision of some basic concepts of the traditional economic theory, primarily, the concepts of demand function, supply function, and equilibrium at the level of individual economic agents - producers and consumers, as well as at the level of the whole system. Braverman developed the economic models, in which quota, i.e. natural (quantitative) restrictions on consumption of some types of goods and production of the other ones is used as a primary management tool. The first are so-called goods in "shortage", and the second are goods in "excess". The separation of products into these two groups is not defined a priori, but determined as endogenous parameters in equilibrium. Equilibrium, according to Braverman, is a condition of the economic system, in which supply and demand of each product are equal, but the demand for goods in "shortage" is limited from above by the consumption quota, and the supply of goods in "excess" is limited from above by the production quota. Prices for all goods are assumed to be fixed at non-equilibrium levels in the traditional sense, i.e. they don't necessarily balance supply and demand in the market. Change in the paradigm of the economic mechanism for equilibrium, associated with the implementation of market equilibrium under disequilibrium prices and quotas, created the need to construct appropriate so-called disequilibrium models, and brought Braverman to develop and implement a programme of modification of economic concepts and creation of new ones.

Among them:

1. Demand and Supply Functions in Disequilibrium Economy Depending Not on Prices, but on Quotas

Braverman established a link between the criterial approach based on maximization of the utility function under quotas and the axiomatic approach, where the choice is considered rational, if it satisfies appropriate axioms concerning revealed preferences. He developed a similar concept to describe the choice of producers maximizing their profits, which enabled the creation of a modified theory of supply and demand functions for factors of production.

2. The Role of Profit, Which Guides Producers When Choosing their Production Plan under Technological Constraints and Quotas

Braverman proved, that the profit exhibits the properties of vector efficiency of total net output for the whole system and thereby provides consumers with maximum possibilities. This feature of the economic system is essentially close to the so-called Second Welfare Theorem. As for the First Welfare Theorem, that situation is much more complicated as you will see it later.

3. Balance Through Quotas

Braverman showed that under standard assumptions regarding consumption and production sets, such a separation into goods in shortage and goods in excess exists, as well as quotas on consumption of the first and production of the second, at which (even if prices are fixed arbitrarily) the aggregate demand equals the aggregate supply for each product. It turned out, that there may be an infinite number of such equilibria - let us call them BP (Braverman equilibrium with fixed prices and quotas) in the same economic system and that equilibria arbitrarily small net outputs exist among them. The set of BP-equilibria can be ordered by vectors of the net outputs of each product (vector ordering). Braverman assigned and largely solved the problem of determining conditions allowing "an improvement" in the equilibrium. In other words, he was able to indicate the conditions of "non-improvability", or efficiency, of the equilibrium. A search for the necessary and/or sufficient conditions for the efficiency of equilibrium led to elaboration of the concept of "unsaturated producers" and production networks.

4. Production Networks, Unsaturated Producers and an Improvement in the State of the Economy

Braverman proposed a method of determining the type of equilibrium (improvable and non-improvable) through the analysis of structural features of the existing state of the system. In his first article on the subject, Braverman had already proven that if the system of producers can be represented as a connected tree graph, then the equilibrium is Pareto optimum. But what if the production system is a closed chain? Is any equilibrium in such a system improvable? An answer was obtained only two years later and was not as simple as it first seemed. Braverman and Levin showed that if all elements of the system (all producers) are unsaturated, then the equilibrium is improvable. i.e. there exists a different quota system, in which the net output of all products can be increased. The very concept of unsaturation is an "enhanced profitability", i.e. the possibility of profit growth of a finite value together with the increase in consumption of some products without reducing the costs for all other products. Furthering the concept of the structure of an economic system, they formulated stronger conditions of effectiveness. It turned out, that by changing quotas (increasing quotas for some products and reducing them for others) and by determining in which of subsystems the unsaturated elements are, it is possible to formulate the conditions, providing improvement of equilibrium. They investigated the characteristics of production sets, for which the necessary and sufficient conditions of efficiency coincide. Thereby they generalized the concept of productivity in the input-output model for nonlinear non-equilibrium systems. This approach to the analysis of economic systems is fundamentally important for the theory and practice of Industrial Organization, as it takes into account the structural features of the production systems. It turned out that for verifying the effectiveness, it is enough to have only qualitative (not quantitative) information – so it is not necessary to make detailed quantitative calculations. This last proposition provides new insights for managing complex integrated production systems.

5. Constructive Feasibility of the Collapse of Economy

Using the concept of equilibrium under quotas, Braverman investigated, what might happen in the economic system if monopolists, using the lack of some products could, and would like to, increase prices for their products, used both for final consumption and by other firms as production factors. He determined the structures of the economy, in which the firm's policy leads to the disequilibrium economy, to unlimited price increase by reducing the net production to zero, i.e. to economic collapse. So he managed to extend the proposed disequilibrium approach to the analysis of dynamic economic systems, which lead to fundamental macroeconomic findings.

So, within just a few years - from 1969 to 1977 - Braverman not only created a new direction in economic theory – an original theory of non-equilibrium economic systems, but also developed its fundamental provisions. The main results of his research can be found in the articles [1–7, 9] and the monographs [8]. It should be noted that since the 1970s, other economists have concurrently undertaken studies of non-equilibrium economic systems. At present, the theories of non-equilibrium have become a subject in economics. A sufficiently complete bibliography and review of these works can be found, for example, in [10]. The survey of some of Braverman's results can be found also in [11], Part I.

Unfortunately, the works of Emmanuel Braverman on the modelling of non-equilibrium economic systems remain little known to the community of mathematical economists at the present time.

References

1. Braverman, E.M.: Production model with unbalanced prices. Ekonomika i Matem. Metody **8** (2), 175–190 (1972). (In Russian)
2. Braverman, E.M.: The model of the price change in production network. Ekonomika i Matem. Metody **9**(2), 218–230 (1973). (In Russian)
3. Braverman, E.M.: Efficient states in the network of production elements. Autom. Rem. Contr. **36**(3), 88–94 (1975)
4. Braverman, E.M.: Model of consumer's choice with fixed prices. Autom. Rem. Contr. **37**(5), 729–740 (1976)
5. Braverman, E.M., Levin, M.I.: Identification of effective networks, states of industrial elements, part I. Autom. Rem. Contr. **39**(6), 67–82 (1978)
6. Braverman, E.M., Levin, M.I.: Identification of effective networks, states of industrial elements, part II. Autom. Rem. Contr. **39**(7), 79–86 (1978)
7. Braverman, E.M., Levin, M.I.: Identification of effective networks, states of industrial elements, part III. Autom. Rem. Contr. **39**(9), 90–101 (1978)
8. Braverman, E.M., Levin, M.I.: Disequilibrium Models of Economic Systems. Nauka, Moscow (1981). (In Russian)
9. Levin, M.I.: Non-improvable states in the model of industrial work. Autom. Rem. Contr. **35** (8), 1212–1220 (1974)
10. Herings, P.J.J.: Static and Dynamic Aspects of General Disequilibrium Theory. Springer, New York (1996). https://doi.org/10.1007/978-1-4615-6251-1
11. Makarov, V.L., Levin, M.J., Rubinov, A.M.: Mathematical Economic Theory: Pure and Mixed Types of Economic Mechanisms. Elsevir, Amsterdam (1995)

Misha Braverman: My Mentor and My Model

Boris Mirkin[1,2(✉)]

[1] Department of Data Analysis and Artificial Intelligence, National Research
University Higher School of Economics, Moscow, Russian Federation
bmirkin@hse.ru
[2] Department of Computer Science, Birkbeck University of London,
London, UK

Before I begin my story of Misha Braverman, let me introduce myself. I graduated, in Mathematics, from a provincial University in Russia, and in a couple of years defended there a PhD thesis in Abstract Automata theory, after which I relocated, in 1967, to the Academgorodok near Novosibirsk in Siberia, a newly developed analogue to a big American University Campus, with many Russian quirks which are not that important here. While there, I was mostly engaged in conducting new research projects in what is now data science and decision support. I met Misha Braverman in April 1969, thanks to Ilya Muchnik. The latter had visited our institution in Academgorodok to see Prof. Tatiana Zaslavskaya, the head and leader of our Sociology Department, regarding a joint research project between her group and that of Braverman. I attended an Ilya's presentation explaining mathematical methods underlying the project. In the end, I asked him as a speaker, whether there was any novelty in their methods at all since I could see none in his narrative. Ilya seemed pleasantly surprised that among the audience was somebody who was able to follow his technical explanations through. He made several remarks explaining the novelties involved in the project. Then he had a long conversation with me asking, among many other things, about my approaches and results. Then Ilya said they would invite me to speak at the Seminar in Moscow, led by Emmanuel Braverman, Mark Aiserman and Lev Rozonoer.

This is how Misha's and my meeting happened, among several others that Ilya had arranged for me. I use the diminutive Misha instead of Emmanuel here and further on because this is how he called himself, as well as every member of his circle. This account not only about myself and Misha – it is of Ilya Muchnik as well, which is much important for my narrative, because in many cases my meetings with Misha involved all the three of us. Sometimes, Misha passed his messages to me through Ilya. Moreover, Ilya gave me, for almost 50 years of our interaction, a lot of his own energy, knowledge, wisdom, support and friendship, so that my memories are not of M. Braverman only; they are of Ilya Muchnik as well. However, there is a big difference between my attitudes to Ilya Muchnik and Misha Braverman: Ilya is my colleague, whereas Misha was my mentor.

My first impression of Misha – of course I had known his name before the meeting because of his publications – was quite favorable – a muscular, open face, attentive bright eyes. I did not notice that he was crippled, nor did I see that he was of a relatively short height. I told him of my only, at that time, mathematics result in the field, a characterization of what now is known as interval orders, mentioning measurement errors as a motivation for the subject. He immediately asked me a question. "Can you explain me more precisely what this has to do with errors in measurement?" I told of

L. Rozonoer et al. (Eds.): Braverman Readings in Machine Learning, LNAI 11100, pp. 341–348, 2018.
https://doi.org/10.1007/978-3-319-99492-5_16

the motivation to many people before but nobody asked me of the substance. People just were saying if that was of interest to them or not. At least I had never gave a thought to the issue. Thus I gave an answer to Misha, which was immediately invalidated by him as something rather meaningless. Almost immediately, he proposed a right answer, too.

Feeling rather lonely in my research field back in Novosibirsk, I felt I found a right milieu for myself and I decided to keep to Misha Braverman's judgement, as so that of an adviser. I think he fully embraced me as his disciple, too. For example, he did say, more than once, in my presence, in his paradoxical, slightly self-ironic, manner: "I conduct my research for Mirkin", meaning, in my view, that he did count on my opinion when developing his projects and/or writing his papers. I do not want to give an impression that he liked all my actions or thoughts. Sometimes he could be much critical. First of all, he and Ilya did not trust in my psychological strengths. On my openly expressed desire to eventually develop and defend the Russian second research degree, the Doctorate of Science, which requires much more politics, publicity and results than the habitual French or German "Habilitation" degree needed to get a permission for supervising PhD projects, both Misha and Ilya were quite pessimistic – "You do not have guts enough to overcome the obstacles – too weak. You cannot hold against odds or enemies. Say, Victor Varshavsky has obtained his Science Doctorate only at the third attempt having two previous theses failed – how can you match such a stamina?[1]" Once, at a meeting with a painter and his drawings, he heard me boasting that I could draw similar paintings by myself. "Oh, yes," – he said to me sarcastically, "Like you can play guitar." Indeed, my guitar playing – whatever my technical skills I had boasted of – was hopeless because I am tone-deaf. Once we had a larger argument – about the quality of our papers. In my institution, I had a relative freedom in collecting and editing collections for publication as books with the "Nauka, Siberian Branch" Publishers, in Russian, of course. Misha noted that I had published a rather weak result in one of such collections. He said: "This cannot be published in "Automation and Remote Control" journal, thus, should not be published at all." I kept claiming that my wider publishing possibilities opened ways for me to publish a wider range of papers. In a year or so, I was rewarded: Mr. Joseph Mullat from Tallinn (Estonia) invented a beautiful method in discrete optimization citing that paper of mine as the only substantive reference – an ostensive demonstration of the advantage of my wider publishing practice.

Once Misha got real angry with me. That was related to the thesis by his PhD student in fix-price economics, Mark Levin – of this I am going to give more detail later in the story. According to Russian customs and regulations, at any PhD defense there must be three pre-assigned sides present, among others, each to provide an independent critical assessment of the thesis. These three sides are two "opponent" speakers and a so-called "leading enterprise". My Siberian institution was the assigned leading enterprise in this case, and I was the author of the critical review of the thesis endorsed by the leading enterprise. Moreover, I came to Moscow to be present at the defense in

---

[1] I am pleased to say that indeed it took even more than three – in fact, four attempts and 16 years for me to get an ScD degree, but I did hold on, perhaps because of the warning.

person. When I entered Misha's office, he said that he liked my review very much and was pleased with it; the review proved to him that I was the only person among all involved who had read the thesis quite attentively and understood it quite well. I said: "But I have not read it either" – which was true, on the first glance, since I forgot to mention that I had discussed Mark's results with Mark several times, which did give me a deeper understanding of them than I would have obtained if I read the thesis by myself. Misha got real frustrated and disappointed: "Then you are superficial", he cried at me – to this I could not object and only meekly smiled. Later Misha, Ilya and I went by foot to a dinner party organized by Mark, after the defense, for relatives, friends and colleagues at a distant place. The weather was not good: quite snowy, windy and frosty. While passing a supermarket, Misha proposed to us to have a drink at a juice stand over there. "I understand," – he said – "Boris does not drink "portwein", he is an 'intelligentsia,' so that he consumes only dry vines; but we here prefer portwein to get warm at such a cold weather." (It should be noted that the Russian "portwein" was an in-house rather sweet and strong liquor, an imitation of the Portuguese port, then unavailable in the USSR. I am not sure whether this drink is produced in Russia currently.) Of course, I strongly denied my intelligentsia habits and joined the portwein drinking effort: each received a simple 200 ml glass of the drink, which did warm up us so that we were able to reach the destination, in the end. By that time I forgot of the Mark Levin's thesis incident, and I am not sure whether the Misha's comment about me was a reflection of that or not.

The very first and, probably, most important for me psychologically, support I received from Ilya and Misha was on the issue of validity of data analysis. Let me remind the reader that data analysis seeks patterns in data being analyzed, with the general purpose of knowledge enhancing. At that time, a half-century ago, the only reasonable approach to such a goal was a scientific one: you develop a model of the process and see if that model brings forth the patterns you have found with data analysis. That seemed to me a rather tall order indeed: a valid model can be developed only for relatively simple processes; the issues of my data analysis concerns were by far too complicated and lacking any reasonable explanations to expect any timely progress along the scientific approach. This is why I was much happy to hear from Misha and Ilya a downright denial of any value of that approach in data analysis. Moreover, Misha approved, as well, my approach by then to modeling patterns in similarity data – via approximating them by binary relations of a simplified structure. "Yes," he said, "your approach is much more relevant here than the approaches being under development by my colleagues."

Of course I cannot help but mention how excited I was with the approach to modeling hidden factors and clusters that Misha and his team were developing starting from about 1970. He was especially pleased with the analysis of international developments (using a dataset of about sixty countries with about twenty features related to their industrial or economic developments). He showed me a graph of distribution of countries on the plane of two data-derived "hidden" factors, one being of the level of economic development, and the other, of the level of "openness" of the economy. "See a pattern?" - He asked me. "It appears that a country to take off for raising the level of income, she needs first to accumulate a degree of openness and market developments. Such is a solution of the optimal control problem in simplified linear models of economics: to maximize the income in the end of a period, they need first accumulate the investment and only then switch to consumption. We always thought that such a simplistic model had nothing to do with the reality. But this picture demonstrates just the opposite. I am much pleased."

About that time Misha began his exploration of uncharted terrains of modeling in fix-price economics. That is an interesting phenomenon! Out of a hundred or two first-rate mathematicians employed by the Soviet Academy of Sciences for developing mathematics for economics, none did anything at all in this direction.

I think that the cause was not just the lack of interest by mathematicians but rather the lack of interest, or perhaps of the ability, by economists. Before doing any mathematics about a fix-price system of economics, one needs a set of well-developed economic notions, analogous to those in a balanced system – demand, supply and the like. Somehow this was never undertaken before, I guess because good economists did not consider any fix-price system as natural; they just patiently waited till the Soviet stage of history of Russia, with its imposed fixed prices, would be over. In this respect, Misha's work is a clear-cut example of a quixotic attempt at filling in the lacuna. There was no visible reason for him to tackle the issue. Nobody asked him anything of the sort. In my opinion, that was just an outward spring out of a man who felt that needed to be done and that he was capable of doing so. Another cause, I think was Misha's constant desire to be socially active, which was otherwise denied to him because of the Soviet oppressive regime admitting the only non-dissident way for social participation, via well-established channels of the governmental trade-unions and the Communist party. No dissidents or liberals in the approved movement, please!

The only help with this project was his PhD student of that time, currently Professor of Economics, Mark Levin. Misha spoke to me once of his fix-price demand functions and their properties. "See -", he said. "That is a very different type of mathematics here, than I was doing previously. It is purely logical. The proofs are quite short, but inventing them is rather difficult. I am used to a very different mathematics: very long chains of equations transforming this or that integral, separating their fragments, and proving that some are relatively small and negligible."

I think he did not see much help from economists either. I asked a leading economist sympathizing with mathematical thinking in economics and West-oriented developments, Prof. Aron Katzenelinboigen, of his attitude towards modeling the "socialist economy" of the Soviet Union. Aron said: "I can see no value in this. In my view, the Soviet system is hopeless and should be changed for an open market economy as soon as possible."

I think Misha, indeed like me myself, was not convinced of this at all. Perhaps he saw that the Soviet system suited the Russian mentality quite well. As all of us, his circle and followers, he probably believed that the Soviet regime was going to remain, almost intact, for many decades to come. I think one should take into account the prevailing mood of the "intelligentsia" in early 70es. I think that is perfectly captured in a song by Alexander Galitch, a most influential dissident author in those times performing his verses in the manner of a Russian romance (see https://www.youtube.com/watch?v=OyKIMn22Mww, accessed on 13 December 2017). His "somewhat muffled", as he said of himself, voice finishes a song with a stanza (a word-by-word translation by myself follows):

And a guest would say:
"These jokes send cold shivers down.
The author is wrong to think that
He could take on the devil himself!"
"No worry, Ivan Petrovich,"-
Would answer the lady of the house –
"There is nothing for the author to get afraid of –
He died quite a while ago, a hundred years back, I guess."

I did not understand at that time, but now I think that Misha's rather ironic attitude towards Aron Katzenelinboigen and the like did reflect this difference of opinions. He never expressed the irony openly, but one can see by oneself from this episode, for example. At a lunch break of a friendly research conference, Aron says: "Let us discuss now this open research issue!" Misha interrupts: "Let me better tell you a Jewish joke. A little girl comes home from her primary school. "Mom, the teacher said that we would have a feast in our school soon! Each must attend being dressed in an attire of their origin or ethnicity." Mother cries to her husband: "You hear, Elijah? She already wants a mink fur coat!" Not much disappointed, Aron says to a handsome girl sitting next to him: "Lena, we are quite ignorant here, in the USSR, because of multiple prohibitions and red tapes. For example, I am much interested to learn why is that geishas are so popular in Japan. What can they do that the others cannot?" Braverman immediately enters the conversation: "Yes, and I am much interested too – to have sex with a giant woman." Not much disappointed, Aron continues: "Yes, Lena, indeed. I am interested to learn more of this novel jazz music style which is almost forbidden here." "My other wish,"- says Braverman loudly – "The same with a midget!" And now Aron gets silent.

Misha Braverman was instrumental in bringing my results to the attention of researchers. Let me give two examples: (a) his help in "defending a PhD thesis"; (b) his help in writing and publishing a book.

The very first PhD thesis prepared under my supervision was that of Leonid Cherny (1973) who demonstrated that our approach of "the partition metric space" could be used as an elegant and workable tool in the analysis of nonmetric data including expert judgements and sociology surveys. Leonid's defense was organized in the Institute of Control Problems where both Ilya Muchnik and Misha Braverman were working. In Russia, such an event is usually pre-prepared quite carefully, like an opera theater production. This, however, was organized rather loosely. I did not want to have anything to do with the technical side of the undertaking. It suffices to mention the poor quality of a necessary artifact, the "auto-referat" (synopsis) brochure, about a hundred copies of a 15-page long account of the thesis contents, to be available for reading through and critics in all organizations concerned, according to the rules. Instead of a final copy, a very preliminary draft, with a lot of corrections and text removals made by hand, was printed for circulation. (Please be reminded, that was well before the digital era advancement!). A sarcastic question by a member of the council can give a good illustration to the sloppiness of our preparation: "How about criteria weighting in your thesis? Say, at a thesis defense – may a thesis have good contents and poor illustrative materials?" Just before the defense meeting started, Ilya said to me: "Misha is preparing a serious conversation with you, of the errors and blunders you have made as the supervisor." I expected Misha to have spoken to me on that, but he never called on me. Just before my departure to Novosibirsk I asked Ilya: "What happens? When Misha is going to speak to me?" Ilya said that Misha decided that no conversation was needed. "See, he said, the defense went smoothly, interesting questions were raised, and several members of the Council spoke favorably to support the thesis. Probably, Boris did all right with his undoing, even if not according to our customs. Then there is no point in reproaching him."

Upon hearing from me of my theorem extending the celebrated "dictatorship" paradox[2] by Kenneth Arrow, a would-be Nobel Prize winner, he said. "You must write a monograph on social choice for the Soviet research community. In this book, you should bring together your results and the body of knowledge that have been developed internationally."

"This subject is getting popular, and it would be a good idea to show to our 'cognoscenti' that there is no point in reinventing the wheel," – he said. Misha took me to Vladimir Levantovsky, the Head of the corresponding department in the USSR Main Editorial of Physics and Mathematics Publishers – the department's publications were well respected and translated to other languages worldwide. "See," - he said to me, - "The Mathematics part of the Publisher is taken over by the Academician Lev Pontryagin, who is notoriously anti-Semitic. You, as a Jew, would have no chances to get published[3]. But the Department of Cybernetics we are getting published at is within the Department of Mechanics, not Department of Mathematics. This Department is controlled by Physicists who are not anti-Semitic." V. Levantovsky told us that any book, from the very moment its manuscript is submitted, takes at least two years to get in print. I was much disappointed by the length of the period and the wait expected, but Misha reassured me: "Is the publication period long? Indeed. But it is long for all, not for you only. The earlier you submit, the earlier you get published." Also, he spoke to me at length of some secrets for "good writing". "One has to write so that reading their writing is as interesting as reading a detective story. For example, you may wish to write: "A is B; let us prove that". This is boring. You do not touch the human curiosity with such a style. You should better write this: "One may think that A is C; the other, that A is D. But these both are wrong. Indeed, A cannot be C because of this, and A cannot be D because of that. Some say that there is no way to know if there exists any relation between A and B. However, A is B indeed, although some may claim just the opposite.To prove that A is B, we need to take a look at the following." Misha's advises regarding this book as well as of other publications have made a great impact onto my writing. I wish he could have lived longer to continue advising me, as my life continued giving me lessons for writing, sometimes in a really hard way.

---

[2] A few seemingly innocent properties of a social choice function imply that the social choice, to satisfy them, must be either dictatorial or imposed externally.

[3] Indeed, in a few years V. Levantovsky told me that L. Pontryagin did visit him immediately after my book, Group Choice, was published (1974, English translation 1979). He requested a copy of the book for checking, claiming that he had been informed that the author was not a well-qualified mathematician so that a lot of mistakes should be expected. Probably he found none, as V. Levantovsky did not hear from L. Pontryagin of the book anymore.

One of the main ideas Misha was imposing upon me was the idea of positioning, which seems to me now, after my extensive international travels, self-evident: one should not invent too many concepts. One should always try finding an earlier analogue to a concept and then explaining what did they do about it to deserve to get published. I must admit, at that time the idea was not quite clear to me; I considered then, as still many Russians do currently, that whatever has been done already on the subject of one's current research, it is worthless, without ever trying to take a closer look at the accumulated knowledge.

I would end my account by mentioning that Misha Braverman was not isolated within the sciences – just the opposite. He was open to the world of which he knew much more than I ever could have known. Everybody around him was seeking Misha's advice for this or that life situation. Misha's word was always brief, wise and sprinkled with good humor. Sometimes he just played with us. In 1971, he showed us, his circle of assistants and younger scientists, an issue, number 4, of the "thick" literary journal "New World" (Новый мир) and said: "This issue has been ordered to get withdrawn from circulation. What do you think of the reason?" We took the issue and looked through it, quite attentively. No reason could be seen. "Ok," – Misha said – "I give you a tip. The reason is in the verses of this poet." We looked at the verses – nothing extraordinary; verses by a virtually unknown poet about an old buoy keeper whom other boatmen welcome "Hi, Isaich!" Here is the reason, Misha explains: Isaich is the patronym of Alexander Solzhenitsyn[4] who is, thus, being likened to a leader, while being formally expelled from the USSR's Writers Union and, later, from the USSR too. A novel by E. Hemingway was not published in the USSR, according to Misha, because a character in the novel mentioned a Spanish communist leader, Dolores Ibarruri, who kept her son in the USSR, so that he could escape the participation in the Spanish Civil War (1936–1939) advocated by her[5].

Several times I heard Misha speaking of general issues emerging at specific events. As usual, his comments were wise, brief and paradoxical. Two examples follow.

Regarding a monument to the memory of 26 Commissars from the capital of Azerbaijan, Baku, erected in the beginning of 70es on the street he was living at, Misha said: "The monument looks like an erect penis. What is its relation to the 26 Commissars murdered back in 1918?"

Regarding an impressionistic movie "The Mirror" (1975) by a celebrated Russian film director Andrei Tarkovsky, Misha said: "The Mirror movie is being shown at only one cinema theater, probably to make things more convenient for the authorities. They can solve the dissidents problem at once by just putting machine-guns at the exit gates to slash down all the exiting spectators." (The movie was not appreciated by simpleminded audiences because of its loose structure; it was a hit with non-conformist "intelligentsia" members only.)

---

[4] A.I. Solzhenitsyn is a celebrated Russian dissident writer whose documentary masterpiece "Gulag Archipelago" written in 60-70es under the nose of KGB, has been perhaps the most influential adjudication of the Soviet Union as a social phenomenon.

[5] Ironically, the son did participate, later, in the WWII and died in combat 1942.

I think I should finish with a quotation by Misha Braverman about Ilya Muchnik, which Misha uttered in the beginning of our acquaintanceship. "Take a look at Ilya Muchnik – How does he make it? One only can admire with his abilities. Everybody knows him, everybody accepts him, and everybody likes him. Comes to him an idea to speak to the Director Executive? At that very moment he goes, he enters the office, he has a conversation. If I want to see the Director Executive, I need a week to get an appointment with the Director. Take into account that I am a Science Doctor, whereas Ilya has no even a PhD degree!" I have been lucky to have Ilya as my roommate for a few years in mid-nineties when we shared an apartment rented near Rutgers University. Of course we were helping each other and tried some actions together. Sometimes I could see his eyes and hear his voice telling me – "You did it brilliantly, almost like Misha".

List of Braverman's Papers Published in the "Avtomatika i telemekhanika" Journal, Moscow, Russia, and Translated to English as "Automation and Remote Control" Journal

Ilya Muchnik[✉]

Rutgers University, Piscataway, NJ, USA
ilyamuchnik@ymail.com

1. Braverman E.M. *Certain problems in the design of machines which classify objects according to an identifying feature which is not specified a priori.* Automation and Remote Control 21, 971–978 (1960)
2. Braverman, E.M. *Experiments on machine learning to recognize visual patterns.* Automation and Remote Control 23, 315–327 (1962)
3. Bashkirov, O.A.; Braverman, E.M.; Muchnik, I.B. *Potential function algorithms for pattern recognition learning machines.* Automat. Remote Control 25, 5, 629–631 (1964)
4. Aizerman, M.A; Braverman, E.M; Rozonoer, L.I. *Theoretical foundations of the potential function method in pattern recognition learning.* Automation and Remote Control 25, 6, 821–837 (1964)
5. Aizerman, M.A.; Braverman, E.M. ; Rozonoer, L.I. *The probability problem of pattern recognition learning and the method of potential functions.* Automation and Remote Control 25, 9, 1175–1190 (1964)
6. Aizerman, M.A.; Braverman, E.M.; Rozoner, L.I. *The method of potential functions for the problem of restoring the characteristic of a function converter from randomly observed points.* Automation and Remote Control 25, 12, 1546–1556 (1964)
7. Aizerman, M.A.; Braverman, E.M.; Rozonoer, L.I. *The Robbins-Monroe process and the method of potential functions.* Automation and Remote Control 26, 1882–1885 (1965)
8. Braverman, E.M. *On the method of potential functions.* Automation and Remote Control 26, 12, 2130–2138 (1965)
9. Braverman, E.M.; Pyatnitskii, E.S. *Estimation of the rate of convergence of algorithms based on the potential function method.* Automation and Remote Control 27, 80–100 (1966)
10. Braverman, E.M. *Determination of a plant's differential equation during its normal operation.* Automation and Remote Control 27, 425–431 (1966)
11. Braverman, E.M. *The method of potential functions in the problem of training machines to recognize patterns without a teacher.* Automation and Remote Control 27, 1748–1771 (1966)

© Springer Nature Switzerland AG 2018
L. Rozonoer et al. (Eds.): Braverman Readings in Machine Learning, LNAI 11100, pp. 349–351, 2018.
https://doi.org/10.1007/978-3-319-99492-5_17

12. Braverman, E.M. *Note on the article by E.M. Braverman: "Method of potential functions in the problem of training machines to recognize patterns without a trainer"*. Automation and Remote Control 28, 6, 999 (1967)

13. Aizerman, M.A.; Braverman, E.M.; Rozonoer, L.I. *The choice of potential function in symmetric spaces*. Automation and Remote Control 28, 10, 1520–1550 (1967)

14. Braverman, E.M.; Rozonoer, L.I. *Convergence of random processes in learning machines theory. Part I*. Automation and Remote Control 30, 1 44–64 (1969)

15. Braverman, E.M.; Rozonoer, L.I. *Convergence of random processes in the theory of learning machines. II*. Automation and Remote Control 30, 3, 386–402 (1969)

16. Braverman, E.M. *Methods for the extremal grouping of parameters and the problem of determining essential factors*. Automation and Remote Control 31, 1, 108–116,(1970)

17. Braverman, E.M.; Rozonoer, L.I. *On the article "Convergence of random processes in learning machine theory, I"*. Automation and Remote Control 31, 2, 332–332 (1970)

18. Aizerman, M.A.; Braverman, E.M.; Rozonoer, L.I. *Extrapolative problems in automatic control and the method of potential functions*. Am. Math. Soc., Transl., II. Ser. 87, 281–303 (1970)

19. Braverman, E.M.; Pyatnitskii, E.S. *The passage of a random signal through absolutely stable systems*. Automation and Remote Control 32, 202–206 (1971)

20. Braverman, E. M.; Kiseleva, N.E.; Muchnik,I.B.; Novikov,S.G. *Linguistic approach to analyzing large data sets*. Automation and Remote Control 35, 1768–1788 (1974)

21. Braverman, E.M.; Meerkov, S.M.; Pyatnitskii, E.S. *A small parameter in the problem of justifying the harmonic balance method (in the case of the filter hypothesis). I*. Automation and Remote Control 36, 1–16 (1975)

22. Braverman, E.M.; Meerkov, S.M.; Pyatnitskii, E.S. *A small parameter in the problem of justifying the harmonic balance method (in the case of the filter hypothesis). II*. Automation and Remote Control 36, 189–196 (1975)

23. Braverman, E.M. *Economic states of a network of production units*. Automation and Remote Control 36, 431–436 (1975)

24. Braverman, E.M.; Litvakov, B.M.; Muchnik, I.B.; Novikov, S.G. *Stratified sampling in the organization of empirical data collection*. Automation and Remote Control 36, 1629–1641 (1975)

25. Braverman, E.M. *Model of consumer choice with fixed prices*. Automation and Remote Control 37, 729–740 (1976)

26. Braverman, E.M.; Meerkov, S.M.; Pyatnitskii, E.S. *Conditions for applicability of the method of harmonic balance for systems with hysteresis nonlinearity (in the case of filter hypothesis)*. Automation and Remote Control 37, 1640–1650 (1976)

27. Braverman, E.M.; Levin, M.I. *Identification of effective network states of industrial elements. I.* Automation and Remote Control 39, 833–847 (1978)
28. Braverman, E.M.; Levin, M.I. *Identification of effective network states of industrial elements. II.* Automation and Remote Control 39, 1001–1008 (1978)
29. Braverman, E.M.; Levin, M.I. *Identification of effective network states of industrial elements. III.* Automation and Remote Control 39, 1335–1345 (1979)
30. Braverman, E.M.; Muchnik, I.B.; Chernyavskij, A.L. *Approximate approach to the solution of systems of structural regression equations.* Automation and Remote Control 39, 1677–1685 (1979)

Author Index

Printed in the United States
By Bookmasters